LINEAR CIRCUIT THEORY

Matrices in Computer Applications

LINEAR CIRCUIT THEORY

Matrices in Computer Applications

Jiri Vlach, PhD

Apple Academic Press

TORONTO   NEW JERSEY

Apple Academic Press Inc. | Apple Academic Press Inc.
3333 Mistwell Crescent | 9 Spinnaker Way
Oakville, ON L6L 0A2 | Waretown, NJ 08758
Canada | USA

©2014 by Apple Academic Press, Inc.

First issued in paperback 2021

Exclusive worldwide distribution by CRC Press, a member of Taylor & Francis Group
No claim to original U.S. Government works

ISBN 13: 978-1-77463-290-1 (pbk)
ISBN 13: 978-1-926895-61-1 (hbk)

Typeset by Accent Premedia Services (www.accentpremedia.com).

Library of Congress Control Number: 2013947030

Library and Archives Canada Cataloguing in Publication

Vlach, J. (Jiri), 1922-, author
Linear Circuit Theory: Matrices in Computer Applications/Jiri Vlach, PhD.

Includes bibliographical references and index.
ISBN 978-1-926895-61-1
1. Electric circuits, Linear. I. Title.

TK454.V63 2013   621.319'2   C2013-905492-8

Apple Academic Press also publishes its books in a variety of electronic formats. Some content that appears in print may not be available in electronic format. For information about Apple Academic Press products, visit our website at **www.appleacademicpress.com** and the CRC Press website at **www.crcpress.com**

ABOUT THE AUTHOR

Jiri Vlach, PhD

Dr. Jiri Vlach is a Distinguished Professor Emeritus of Electrical and Computer Engineering at the University of Waterloo in Ontario, Canada. He was a visiting professor at the University of Illinois at Urbana-Champaign, Illinois, USA. A Life Fellow of the IEEE, he has written several books on circuit analysis and design. He received his PhD equivalent degree in Prague, Czech Republic.

CONTENTS

PREFACE

Circuit theory, developed in the twentieth century, supports today a huge industrial section of analog networks. Most designs are done by computers, but the need to thoroughly understand the theory remains.

In about five hundred pages, this book introduces most subjects of the linear circuit theory. Some methods, like the Laplace transformation, sensitivities, active network design and modified nodal formulation, will be taught in later years and the student can keep the book throughout the undergraduate program.

Emphasis of the book is on solving small examples by hand, because only practice leads to understanding and knowledge. Almost all problems in the book lead to only two or three linear equations. A complete solutions manual is available.

The style of explanation is also important and here is what makes this book different. Solutions by matrix algebra are started in the third chapter. First year students, most likely, have not heard about matrices, so we start with an absolute minimum: how to write two or three linear equations in matrix form, how to evaluate their determinants and how to apply the Cramer's rule. This can be taught in a one hour lecture. Additional subjects of the matrix algebra can be taught later or left to mathematics courses. Cramer's rule avoids divisions and fractions of the elimination method and is suitable for up to four linear equations. Matrices have to be used for more complicated problems anyway and the students will be acquainted with their use.

When designing a network for a certain performance, we may not know numerical values of some or all elements and we must keep them as variables (letters). This is called symbolic analysis. To use elimination is almost unthinkable. Cramer's rule makes the process reasonably simple, as demonstrated on many examples. Programs for any kind of analysis, including symbolic, do exist, but are not useful for introductory studies.

Another simplification in the presentation is based on the fact that linear networks can be scaled to unit element values and unit frequency (see the Appendix). To learn circuit analysis, there is no need to use kiloohms, picofarads and miliampers. In all problems we use basic units, ohms, farads and henries, and entire numbers for element values.

Network equations can be based on mesh or nodal analysis. In this book, both are given approximately the same amount of attention, but they are not equally important. The mesh analysis has severe limitations and nodal formulation is preferred. All widely available circuit analysis programs are based on its extension, called Modified Nodal

Formulation, MNF (Chapter 17). These programs automatically convert all linear resistors into conductances, to keep the number of equations small. In this book, when dealing with nodal formulation, we adhere to this practice and use conductances, G, to avoid fractions of the type $1/R$.

It is advantageous to use different letters for circuit variables and for independent sources. The book uses V and I for variables and E and J for independent sources. This is a minor point, but it helps the student to avoid confusion of what should be on the left or right side of the equal sign.

Circuits with capacitors and inductors are described by differential equations, but the classical mathematical theory of differential equations is rarely used. We introduce these elements in Chapter 6 by means of the s variable. It is used in the Laplace transformation. If we need frequency domain responses, all we do is replace s by $j\omega$.

The Laplace transformation chapter has a section on multiple poles. Theoretically, it is important, simply because multiple poles can exist. From a practical point of view, networks with multiple poles are always inferior to networks with simple poles. The instructor may decide to skip the section on multiple poles.

Active networks can be analyzed by nodal formulation, as explained in Chapter 12, while Chapter 16 shows the importance of sensitivity analysis. These subjects are for later years of undergraduate studies.

In Chapter 15, the student is exposed for the first time to modeling of actual devices. Only linear models are given for field-effect and bipolar transistors. Nonlinearities are introduced by means of the simplest nonlinear device, the diode. We explain the standard Newton-Raphson iteration and use it to solve a few primitive nonlinear networks. We also added a section on numerical time domain integration, to show that integration with the backward Euler formula leads to the substitution of s by $1/h$, where h is the integration step size.

Modified Nodal Formulation, MNF, a method used in computers, is in Chapter 17. It leads to larger matrices and is not suitable for hand calculations. Most computer programs use it. Some calculators may perform time domain and frequency domain evaluations. The students are, of course, welcome to use them, but we do not require them.

In Chapter 18, we briefly introduce the Fourier series and Fourier transformation. Fourier series is important because it shows that any periodic signal can be decomposed into a sum of harmonic frequencies and in linear networks each frequency can be handled separately.

Fourier integral is presented in a limited version, for signals starting at $t = 0$. This is the case of all practical signals and, with this simplification, Fourier integral does not differ from the Laplace transformation. Everything we have learned in Chapter 9 can be applied; all we have to do is substitute $j\omega$ for s to obtain spectra of signals.

To summarize, the book prepares the student for later studies by using matrix algebra for all solutions. It explains most subjects of the linear circuit theory and prepares the student for further studies of nonlinear systems.

The author of this book, with a colleague, Kishore Singhal, wrote an advanced book, "Computer Methods for Circuit Analysis and Design," (Van Nostrand Reinhold, Second Edition, 1994). Its title indicates the content. It can be considered as a continuation of this introductory text.

1   BASIC CONCEPTS

INTRODUCTION

This chapter starts with the concept of electric current and voltage, introduces basic terminology and shows how networks are drawn. It gives a brief overview of all linear elements
which are used in the network theory, plus a few of the most important practical devices.
After this informal explanation we give definitions of the independent current and voltage
sources, of resistors, of power delivered or consumed. Three fundamental laws are stated:
the Ohm's law and the Kirchhoff's current and voltage laws. Based on this knowledge we
study the voltage and current dividers and also explain how to obtain equivalent resistances
for combinations of resistors.

1.1.   VOLTAGES AND CURRENTS

Network theory deals with voltages, currents, network elements and signals; in this section we will consider in detail the first two, and also introduce some basic concepts and
typical signals.

Electric *current* is a flow of electrons and is measured in *amperes*, denoted by the letter *A*. Voltage is a force which causes electrons to flow and is measured in *volts*, denoted
by *V*. The concept of current is usually somewhat easier to grasp because we can visualize
it as a current of water.

When we talk about the flow of water, we associate it with a certain direction and we
know, from daily experience, that the direction is from a higher point to a lower point. We
also accept without difficulties that there must be some force which determines the direction of the flow. A similar situation exists in electrical networks and the force that pushes
the current through the elements is the voltage. The word *potential* can be used as well.

In the early days of electricity, only chemical batteries were available and the concept of electrons was unknown. One of the connecting points of the battery was arbitrarily
marked with a + sign, the other with a − sign and it was assumed that the current flowed
into the external network from the + terminal to the − terminal. The current direction is
usually marked by an arrow, similarly as in Fig. 1.1.1 where the box denotes some element
not yet specified. It is a standard practice to use letters *i* or *I* for the current and letters *v*
or *V* for the voltage.

Since voltage is a force and force must always be referred to some point with respect
to which it acts, we must indicate a reference point when talking about voltage. Such point

is usually *ground*. In electrical instruments ground is understood to be the metal base of the equipment and not earth, to which the instrument may or may not be connected. We will use the term ground even if the equipment is in a flying aircraft.

The concept of a positive direction of current is fundamental to network theory. If the arrow denoting the current is directed from a point with higher voltage to a point with lower voltage, e.g. from + to −, then we say that the current has *positive* direction. If the arrow points from − to +, a negative sign is associated with the current. Positive direction of current is indicated in Fig. 1.1.1. We will say more about the current direction and its implications later in this chapter.

Each idealized element has its own symbol but for a start we will represent the simplest one by a small box with leads as shown in Fig. 1.1.1. The lines coming out of the box have no special electrical properties; their only purpose is to connect the element to other elements. The small circles at the ends of the lines are called *terminals*. The element has two terminals and is thus called a *two-terminal* element. We will see later that we may have three- and four-terminal elements as well.

Several elements may be connected together to form a network, for instance as shown in Fig. 1.1.2a. The drawing, although showing correctly what we intended to do, is not very pleasing to the eye and is also not simple to draw. In order to avoid difficulties in drawing schematics of networks and also in understanding them, we have a convention that a line which connects elements has no special meaning; it just indicates a connection. Using this rule, we can redraw the same network as shown in Fig. 1.1.2b. This figure is clearly much easier to produce and is also easier to interpret. The full dots indicate that there is a true connection of the lines (we could think that they are soldered together). A point connecting two or more elements is called the *node*. Connection of two lines is indicated in Fig. 1.1.3a.

In more complicated situations we may not be able to draw the lines without some intersections, although there is no electrical connection at the point where they intersect.

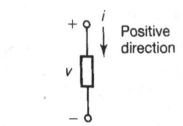

FIGURE 1.1.1   A two-terminal element.

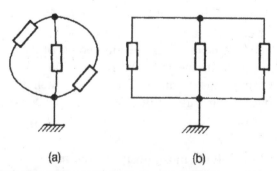

FIGURE 1.1.2   Drawing Networks: (a) Possible way. (b) Accepted method.

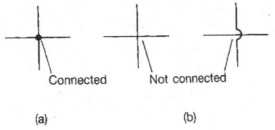

FIGURE 1.1.3   Conventions for connections: (a) Electrically connected.
(b) Electrically not connected.

In order to clearly indicate that the lines are *not* connected, there are two conventions: Intersection *without* a full dot means that the lines are not connected and can be thought of as running one above the other. This is shown in Fig. 1.1.3b. Another convention draws a small semicircle, shown in the same figure. True connection of *three or more* wires will *always* be marked by a full dot. We may use any of the above two conventions to indicate that the lines are not connected.

Another common expression says that a certain point is *grounded*. *Ground* is marked by the symbol at the bottom of Fig. 1.1.2. It is a somewhat misleading concept because we do not ground the network in the way the lightning rod is grounded. The symbol only indicates a reference point. Every network must always have one such reference point. In many network drawings, the bottom long line is automatically assumed to represent the reference point.

We still need a few more fundamental concepts. A voltage at a node, measured with respect to ground, is called *nodal voltage*. Two nodes are sketched in Fig. 1.1.4a. It is a common practice to give the node a name, usually a positive integer number, and the voltage may be indicated at the node, as shown in the Fig. 1.1.4a. The two nodes are not connected and we use the expression *open circuit* for such a situation. Rather obviously, no current can flow through an open circuit and arbitrary voltages may exist at the two nodes and thus across the open circuit. The opposite situation would be a *short circuit*, shown in Fig. 1.1.4b. It is represented by a connecting line, the same line as used in Fig. 1.1.2. Two nodes connected by a short circuit have the same voltage (zero voltage difference between the two nodes) and an arbitrary current can flow through a short circuit in any direction. The node numbers may be in circles, as in Fig. 1.1.4b.

The current (or voltage) can have a value which does not change with time, or it can vary with time. We can draw a diagram in which time is measured on the horizontal axis and current is marked on the vertical axis. If the current value does not change with time,

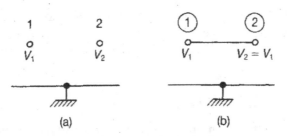

FIGURE 1.1.4   Two nodes: (a) Open circuit, (b) Short circuit.

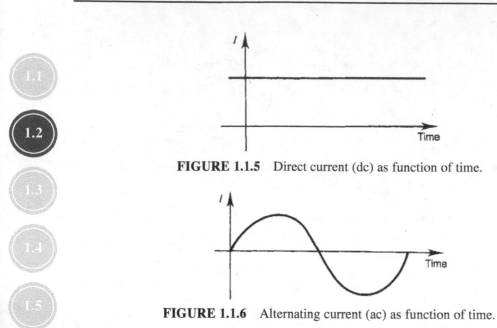

FIGURE 1.1.5   Direct current (dc) as function of time.

FIGURE 1.1.6   Alternating current (ac) as function of time.

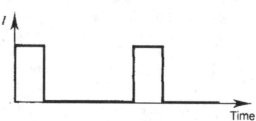

FIGURE 1.1.7   Graph of current in pulses.

such diagram will be a horizontal line, shown in Fig. 1.1.5. We speak about *direct current* and denote it by *dc*. A dc current is supplied, for instance, by a car battery. In other situations we may have currents whose instant values change. Fig. 1.1.6 shows another typical case called *alternating current*, denoted as *ac*. This type of current is supplied by the outlet at your home. It is a periodic function which repeatedly changes its value. An example of still other type of current is in Fig. 1.1.7 where the current flows in short pulses with periods of time where the current does not flow at all. The three examples belong to more typical cases. There are many ways in which current can change with time. Also, whatever we said about currents, will also apply to the voltages.

1.2.   NETWORK ELEMENTS

Our first section introduced some of the most important conventions and explained the concepts of currents and voltages. In this section we will give a brief, informal overview of elements which are used in network theory. All the elements will be discussed in detail later.

1.2.1.   Independent Voltage and Current Sources

An electric network can function only if a source of a voltage or current is attached to it. Network theory uses two sources: the *independent voltage source*, and the *independent*

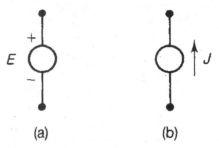

(a)   (b)

FIGURE 1.2.1   Symbols for independent voltage (a) and current (b) sources.

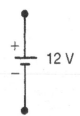

FIGURE 1.2.2   Symbol for battery.

current source. Their symbols are in Fig. 1.2.1. The same symbols are usually used for a *dc*, *ac* or, in fact, any type of source. Sometimes, in order to clearly indicate that there is a special source, another symbol may be used. For instance, it is common to use the symbol in Fig. 1.2.2 for a battery.

1.2.2.   Resistor

A resistor is an element produced from some material which "resists" the flow of electrons. The property is called *resistance*. At normal temperatures, all materials exhibit some resistance. Silver and copper have small resistances and copper is commonly used in wires for connections. Some alloys or carbon mixed with other materials have larger resistances. The letter R is used for the resistance; it is measured in Ohms and the symbol is Ω. As an example, if a resistor has the resistance of ten Ohms, we would write next to its symbol $R = 10\ \Omega$, $R = 10$, or just $10\ \Omega$. The symbol for a resistor is shown in Fig. 1.2.3. The voltage across the resistor and the current flowing through the resistor are coupled by the Ohm's law:

$$V = RI,$$

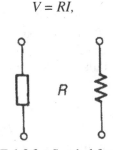

FIGURE 1.2.3   Symbol for a resistor.

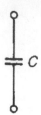

FIGURE 1.2.4   Symbol for a capacitor.

Sometimes we use the inverted value of the resistance,

$$G = \frac{1}{R}.$$

It is called *conductance*, is measured in Siemens and the symbol for the unit is S. We will learn more about resistors later in this chapter.

1.2.3.   Capacitor

The symbol for a capacitor is in Fig. 1.2.4 and in schematics is denoted by the letter C. It is an element formed by two layers of conducting material, separated by a material which does not conduct, an insulator. Many materials can serve as insulators and even waxed paper is sometimes used. Electrons can be stored on the capacitor; in such case we say that there is a charge on the capacitor. Charge is usually denoted by the letter Q or q, and is measured in Coulombs, C (same letter as for a capacitor). If the charge is changing with time, a current flows through the capacitor according to

$$i(t) = \frac{dq(t)}{dt}$$

The equation indicates that no current can flow through the capacitor if the charge does not change with time (derivative of a constant is zero). If the capacitor is independent of any external influences, the equation simplifies to

$$i(t) = C\frac{dv(t)}{dt}$$

where C is the symbol for the *capacitance*, the property of the capacitor to store electrons. *Capacitance* is measured in Farads, F. If the capacitor has a capacitance of 2 Farads, we would write next to the symbol either $C = 2F$, or $C = 2$ or just $2F$. The element will be introduced later in this book, with all details postponed until then.

1.2.4.   Inductor

The symbol for an inductor is shown in Fig. 1.2.5. The classical form of an inductor is a coil of insulated wire, possibly placed into a core of ferromagnetic material, such as iron.

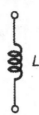

FIGURE 1.2.5 Symbol for an inductor.

A current flowing through the inductor creates a magnetic field, or flux. Flux is usually denoted by the letter Φ and is measured in Webers, *Wb*. There is a relationship between the flux and the voltage across the inductor,

$$v(t) = \frac{d\Phi(t)}{dt}$$

This equation indicates that if the flux is constant (created by a dc current), then there is no voltage across the inductor, because the derivative of a constant is zero. If the inductor is independent of external influences, then the above equation simplifies to

$$v(t) = L\frac{di(t)}{dt}$$

where L is the property of the inductor, called the *inductance*. Inductance is measured in Henries, H. In schematics, the element is denoted by the letter L. For instance, if an inductor has an inductance of 3 Henries we would write next to its symbol $L = 3H$ or $L = 3$ or just $3H$. This element will also be discussed in greater detail later in the book.

1.2.5. Dependent Sources

Unlike the independent sources introduced above, dependent sources deliver voltages or currents whose values depend on a voltage or current somewhere else in the network. This means that the elements have two controlling terminals, where they do the sensing, and two controlled terminals, where they deliver the voltage or current. There are four possible types of dependent sources:

1. *Voltage controlled voltage source, VV*, Fig. 1.2.6a. The voltage difference at the terminals on the left, V_1, is multiplied by a coefficient, μ, and delivered by the source on the right:

$$V_2 = \mu V_1$$

The coefficient μ is often called the *gain* of the VV.

2. *Voltage controlled current source, VC*, Fig. 1.2.6b. The voltage difference at the terminals on the left, V_1, is multiplied by a conversion constant, g, and is delivered as a current at the terminals on the right:

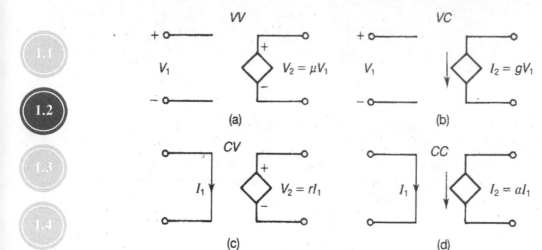

FIGURE 1.2.6 Symbols for dependent sources: (a) Voltage controlled voltage source, VV. (b) Voltage-controlled current source, VC. (c) Current-controlled voltage source, CV. (d) Current-controlled current source, CC.

$$I_2 = gV_1$$

3. *Current controlled voltage source*, CV, Fig. 1.2.6c. The current through the short circuit on the left, I_1, is multiplied by a conversion constant, r, and applied as a voltage to the terminals on the right:

$$V_2 = rI_1$$

4. *Current controlled current source*, CC, Fig. 1.2.6d. The current flowing through the short circuit on the left, I_1, is multiplied by a constant, α, and applied as a current on the terminals on the right:

$$I_2 = \alpha I_1$$

The elements will be discussed in detail in Chapter 4. They are very important because they are used to model properties of useful devices in real-life situations.

1.2.6.   Transformer

If two (or more) coils are near each other, or are placed on one ferromagnetic core, then they form a transformer. There are many forms of transformers, but the most common one has two coils and its symbol is in Fig. 1.2.7. The element will be discussed in detail later.

1.2.7.   Operational Amplifier

Operational amplifier, OPAMP, is a complicated electronic device whose simplified properties can be summarized as follows: it acts as a voltage controlled voltage source, with one fundamental difference. In the ideal case, the amplification factor, μ, approaches infinity. The most common symbol for the operational amplifier is in Fig. 1.2.8 and we will use it in our studies of active networks.

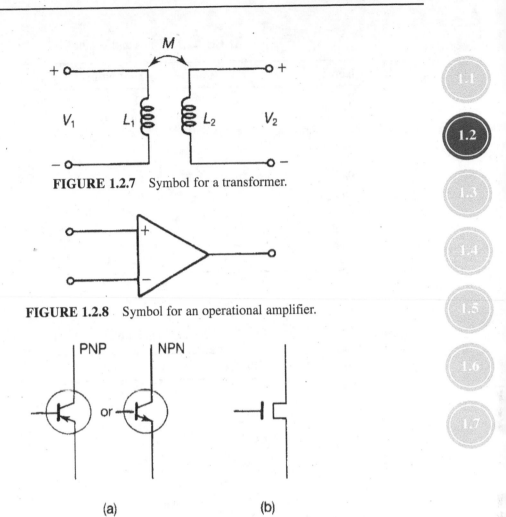

FIGURE 1.2.7 Symbol for a transformer.

FIGURE 1.2.8 Symbol for an operational amplifier.

FIGURE 1.2.9 Symbols for transistors: (a) Bipolar transistor, (b) Field-effect transistor, FET.

1.2.8. Other Elements

The above elements are all the elements used in network theory. In various combinations, they are used to describe the properties of actual, physically existing elements. We will not go into details now, but we will mention two more elements, because they are widely used. They are the field effect and bipolar transistors. In Fig. 1.2.9a is the symbol for a bipolar transistor, and in Fig. 1.2.9b is the symbol for a field effect transistor, FET. Their properties are modeled by combinations of dependent sources, resistors, capacitors, and possibly inductors. Transistors are also used in the design of operational amplifiers.

1.2.9. Units and Their Prefixes

To complete this section we summarize the terminology introduced so far. It is in Table 1.2.1. Theoretical units like Ohms for a resistor, or Henries for the inductor, may in some cases be either too large or too small in practical applications. For instance, a unit to measure capacitance is one Farad, $1F$. This happens to be an impractically large unit. To express that a practical value is, for instance, 10^{-12} times smaller, we write the letter p before the

TABLE 1.2.1   Units and symbols.

Name	Notation	Unit Name	Unit Symbol
Resistance	R	Ohm	Ω
Conductance	G	Siemens	S
Capacitance	C	Farad	F
Inductance	L	Henry	H
Current	I,i	Ampere	A
Voltage	V,v	Volt	V
Energy	W,w	Joule	J
Power	P,p	Watt	W
Charge	Q	Coulomb	C
Flux	Φ	Weber	Wb
Time	t	Second	s

TABLE 1.2.2   Prefixes for electrical units.

Prefix	Name	Value
f	femto	10^{-15}
p	Pico	10^{-12}
n	nano	10^{-9}
μ	micro	10^{-6}
m	milli	10^{-3}
k	kilo	10^{3}
M	mega	10^{6}
G	giga	10^{9}
T	tera	10^{12}

symbol F and have one picofarad, $1\ pF = 10^{-12}F$. The prefixes for all units are summarized in Table 1.2.2.

1.2.10.   Examples of Networks

Since we already know the elements which we will be using in this book, we now give some examples. They will show how we draw the networks and what information we usually write into the schematics.

Figure 1.2.10 shows a network with one independent voltage source, $E = 5V$, two resistors, two capacitors, one inductor and a current controlled current source. If there are several elements of the same type, we either give them different names or use subscripts, like in the figure. Subscripts are not needed here for the inductor, L, and the current controlled current source, CC, but can be given. The controlling current is denoted by I and its direction is indicated by the arrow. It is the current flowing through the resistor R_2. The source part of the CC, connected between ground and node four, has a current gain $\alpha = 3$. We have marked the nodes of the network by numbers in circles; this is not a standard way, but is quite convenient. We have also written nodal voltages to these nodes. The ground node, or reference node, is at the bottom of the picture. Note especially the short circuit

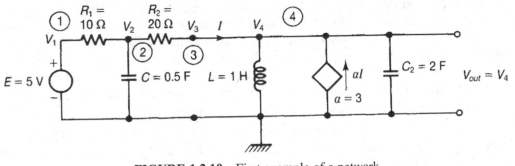

FIGURE 1.2.10 First example of a network.

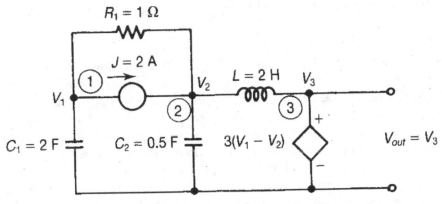

FIGURE 1.2.11 Second example of a network.

between nodes 3 and 4. It is the sensing part of the current controlled current source and is indicated separately. Nodes 3 and 4 are connected by a short circuit.

The voltage source is the *input* of the network. Also marked, on the right side, is the *output*, in this case a voltage, V_{out}. We will use this notation quite often. By output we mean that the variable (voltage or current) will be measured (or calculated) because we need to know it.

Another network is in Fig. 1.2.11. It has a floating current source. The word *floating* refers to elements which are not connected to ground. In this case the source current is $J = 2$ A. The network has two capacitors, an inductor, a resistor and a voltage controlled voltage source. The controlling voltage is taken between nodes 1 and 2 and the gain of the dependent source is $\mu = 3$. The output voltage is $V_{out} = V_3$. The nodes as well as the nodal voltages are written into the figure. In some schematics the symbol for ground is omitted and the bottom line is understood to be the reference node. We have done so in this example.

In general, the output can be any nodal voltage, a difference of two nodal voltages, for instance $V_{out} = V_2 - V_3$, or the current through any of the elements. A network may have several outputs.

1.3. INDEPENDENT VOLTAGE AND CURRENT SOURCES

The first set of idealized elements which we discuss in detail are the independent voltage and current sources. An ideal *independent voltage source* is an element for which we will use the symbol shown in Fig. 1.3.1a. The word *independent* means that the properties of the source will not change due to any external influence. A voltage source always has the symbol + at one of its terminals and the symbol − at the other terminal. Because of the rule we introduced in section 1.1, positive direction of the current flowing through the voltage source will be from the + terminal to the − terminal. It has to be understood that our definition does not mean that the current will actually flow that way; more about this will be said when we have covered the concept of resistors and power.

We will *always* denote an independent voltage source by the letter E. The use of this letter may differ from other books, but there is a strong advantage to give the independent source a special letter. When we get later to the point that we will write the equations, we will see that the independent source always appears on the other side of the equation than the network variables, voltages and currents. By assigning special letters we eliminate many possible mistakes.

The voltage of an independent voltage source is given, but we can say nothing about the current flowing through it. An ideal voltage source can supply *any* amount of current and the current can flow in *any* direction, depending on the network to which it is connected. Since the voltage delivered by the independent voltage source is fixed and any amount of current can flow through it in any direction, we can draw a graph (sometimes also called the characteristic) describing these properties. If we measure current on the horizontal axis and the voltage on the vertical axis, then the characteristic is a horizontal line, shown in Fig. 1.3.1b. If the voltage available from the source changes, for instance from 2 to 4 volts, the horizontal line will shift upwards. Should we apply a voltage − 2 V without changing the positions of the + and − signs, the horizontal line would shift below the horizontal axis. One very important case is what happens if we have a voltage source which is delivering zero voltage. In such case the horizontal line in Fig. 1.3.1 drops down on the horizontal axis. Such element will have zero voltage across it but a current of arbitrary magnitude can flow through it in any direction. If we recall the definition of the short circuit from our previous discussion, then we see that *a voltage source with zero voltage is equivalent to a short circuit*.

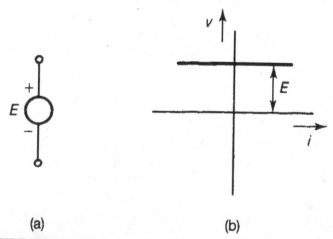

(a)   (b)

FIGURE 1.3.1   Independent voltage source: (a) Symbol. (b) Its characteristic.

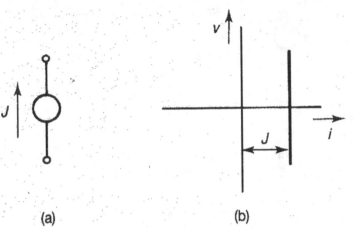

FIGURE 1.3.2   Independent current source: (a) Symbol. (b) Its characteristic.

An ideal *independent current source* is an element for which we will use the symbol shown in Fig. 1.3.2a. Similarly as in the above case, there exists no possibility of changing the amount of the current delivered by some external influence. The direction of the current is indicated by an arrow. We will *always* denote an independent current source by the letter J, for similar reasons as discussed above for the independent voltage source. The current of an independent current source is given, but we can say nothing about the voltages at its terminals, because they depend on the network to which the source is connected. Since the current delivered by the independent current source is fixed, but any voltage difference can be measured between its terminals, we can draw a graph (characteristic) describing these properties. It is shown in Fig. 1.3.2b. If the amount of current delivered by the source changes from, say, $2A$ to $4A$, the straight line will shift to the right. Should we apply, say, $-2A$ without changing the direction of the arrow in the figure, the vertical line will shift to the left of the vertical axis. If we consider a current source which delivers zero current, then the vertical line will coincide with the vertical axis. In such case no current can flow through the element but arbitrary voltage can appear across its terminals; we conclude that *a current source with zero current is equivalent to an open circuit.*

A battery is an example of a nonideal *dc* source, because the voltage at the terminals depends on the amount of current we draw from it. Nevertheless, in many cases we take it as an ideal source and denote it by the letter E. The outlet at your home is another source of voltage, again non-ideal, but it delivers an alternating current, *ac*. This voltage changes with time but we will still use the symbol E.

Independent sources cannot be connected together arbitrarily and we now discuss the acceptable and the prohibited cases. If we connect two voltage sources *in series*, as shown in Fig. 1.3.3a, then the voltage between the external terminals will be $E_1 + E_2$. Should one of them have reversed direction, as shown in Fig. 1.3.3b, then the voltage between the external terminals will be $E_1 - E_2$. We can *never* connect two *different* voltage sources *in parallel*, something like in Fig. 1.3.4, because it is not possible to have simultaneously two different voltages between two nodes. There is, of course, the possibility of connecting in parallel two voltage sources with *exactly the same voltage* and this is sometimes used in theoretical considerations, as we will learn later. From a practical point of view, this is a meaningless situation, since an ideal voltage source can deliver any current and can thus do the same job as two or more such sources in parallel.

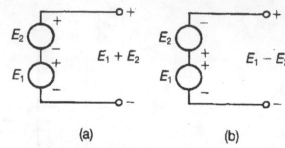

(a)   (b)

FIGURE 1.3.3   Permitted connections of voltage sources: (a) The voltages add.
(b) The voltages subtract.

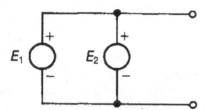

FIGURE 1.3.4   Prohibited connections for $E_1 \neq E_2$.

external terminals will be $E_1 - E_2$. We can *never* connect two *different* voltage sources *in parallel*, something like in Fig. 1.3.4, because it is not possible to have simultaneously two different voltages between two nodes. There is, of course, the possibility of connecting in parallel two voltage sources with *exactly the same voltage* and this is sometimes used in theoretical considerations, as we will learn later. From a practical point of view, this is a meaningless situation, since an ideal voltage source can deliver any current and can thus do the same job as two or more such sources in parallel.

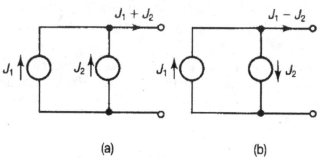

(a)   (b)

FIGURE 1.3.5   Permitted connections of current sources: (a) The currents add;
(b) The currents subtract.

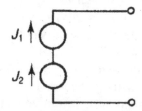

FIGURE 1.3.6   Prohibited connection for $J_1 \neq J_2$.

this is sometimes used in theoretical considerations, but is meaningless practically, because one ideal source can do exactly the same job as two equal sources in series.

Connecting one voltage and one current source either in series or in parallel is not prohibited theoretically, but is not useful practically. Finally, we should mention that in a network schematic a voltage source should never be shortcircuited. By our definition, it should still maintain the same voltage across its terminals, but the current flowing into a short circuit would be infinitely large. Similarly, a current source should never be left without a connection to some circuit which can return the current back to the other terminal of the source.

Example 1.

Consider the combination of voltage sources in Fig. 1.3.7a. What is the voltage we obtain at the terminals of the network? The voltage sources E_1 and E_3 have one orientation, the other sources act in opposite direction. The resulting voltage is

$$E = E_1 - E_2 + E_3 - E_4 - E_5 = -7$$

as shown in Fig. 1.3.7b. The same is achieved by reversing the + and − signs, as shown on the right of Fig. 1.3.7c, and making the voltage positive 7V.

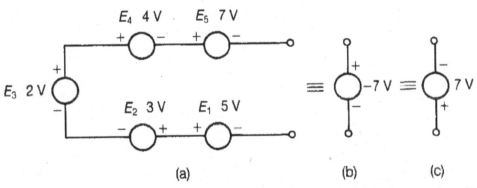

FIGURE 1.3.7   Equivalent voltage of five voltage sources: (a) Original connection. (b) and (c) Two equivalent results.

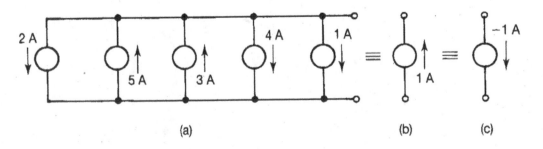

FIGURE 1.3.8   Equivalent current of five current sources: (a) Original connections. (b) and (c) Two equivalent results.

Example 2.

Consider the parallel connection of 5 current sources, Fig. 1.3.8a. The sum of currents flowing upwards is 8, the sum flowing downwards is 7. Thus the equivalent current source will be 1 *A* upwards, as shown n Fig. 1.3.8b. Equivalently, as shown in Fig. 1.3.8c, this could also be drawn as a negative current of 1 *A* pointing downwards.

1.4.   RESISTORS AND OHM'S LAW

In this section we consider the simplest of all elements, the resistor, and the way it relates the voltage across it to the current flowing through it.

The electrical property of the resistor, called the *resistance*, can be best understood by considering a small experiment, at least in our mind, if not in reality. Take several materials (silver, copper, iron, wood, china, glass etc.), make thin rods with exactly equal dimensions, connect them to the same source of voltage and measure the current flowing through them. We will discover that a large current will flow through the rod made of silver, less current through copper, still less through iron, very little through wood and next to nothing through china or glass. The properties of the materials are such that they offer different resistances to the flow of current. More detailed measurements would lead us to the conclusion that the voltage, the current, and the properties of the material are related by Ohm's law which states

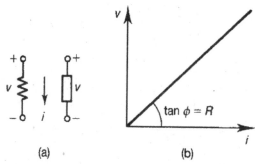

(a)   (b)

FIGURE 1.4.1   Linear resistor: (a) symbol. (b) Its characteristic.

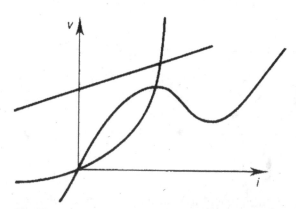

FIGURE 1.4.2   Characteristics of nonlinear resistors.

$$V = IR \tag{1.4.1}$$

Here V is the voltage across the element, measured in *Volts*, V, I is the current flowing through the element, measured in *Amperes*, A, and R is the resistance, measured in *Ohms*, Ω. A resistor will have a resistance of 1 Ω if the voltage across it is 1 V and the current flowing through it is 1 A. For the resistor we will use the symbol shown in Fig. 1.4.1a.

If we draw a graph with perpendicular axes, measure the current on the horizontal axis and the voltage on the vertical axis, then Eq. (1.4.1) represents a straight line passing through the origin. We call it the *characteristic* of the resistor, see the graph in Fig. 1.4.1b. The tangent of the angle, ϕ, is the value of the resistance. The equation describes a *linear* resistor, a resistor whose value is independent of any external influences. In true life, nothing is as simple as that. The resistor will change its value with time or temperature or with the current which flows through it. Fig. 1.4.2 shows several cases of nonlinear resistors. Note that the straight line which *does not* pass through the origin would also represent a nonlinear resistor.

Since the resistance of a linear resistor is a constant and Eq. (1.4.1) represents a straight line, we can invert the equation and write

$$I = \frac{1}{R}V = GV \tag{1.4.2}$$

The inverted value

$$G = \frac{1}{R} \tag{1.4.3}$$

is called the *conductance* and is measured in *Siemens*, S.

Should the resistor change with time, t, we could denote it by the symbol $R(t)$. Its resistance can also depend on the voltage across it. In such case the resistor depends on v and we could express the dependence by writing $R(v)$. Whatever the situation may be, the Ohm's law is always valid. In practical life we talk mostly about resistors and their values in ohms (Ω), kiloohms ($1k\,\Omega = 10^3\Omega$) or megaohms ($1M\Omega = 10^6\Omega$). The notation is standardized and is summarized in Tables 1.2.1 and 1.2.2.

Example 1.

A voltage source delivering 100 V is connected to the resistor having the value $R = 10\ k\Omega$. What is the current through the resistor? The answer is $I = \dfrac{10^2}{10^4} = 10^{-2}\ A = 10 \times 10^{-3}\ A = 10\ mA$.

Example 2.

Let the resistor be described by the relationship $V = I^3$. If the current through through it is 0.5 A, what is the voltage across it? Answer: is $V = 0.5^3 = 0.125\ V$.

1.5. POWER AND ENERGY

Power and energy are the next two fundamental concepts to be covered in this chapter. Energy, usually denoted by $w(t)$ or W, is measured in joules, J; it is the amount of work which a source of energy can deliver. In network theory, energy is delivered by the voltage or current sources and is consumed by resistors. The notation $w(t)$ indicates that the energy is changing with time while W indicates that it is constant.

Closely related to energy is power, denoted by $p(t)$ or P and measured in watts, W. The two definitions are related by

$$p(t) = \frac{dw(t)}{dt}$$

$$w(t) = \int_{t_1}^{t_2} p(t) \tag{1.5.1}$$

The relation of their units is $1W = 1J/s$. In electrical terms, power is expressed by the product of current and voltage,

$$p(t) = v(t)i(t) \tag{1.5.2}$$

and the expression is valid at any instant of time. Substituting from the Ohm's law we can also write

$$p(t) = \frac{v^2(t)}{R} = i^2(t)R \tag{1.5.3}$$

When dealing with direct current, then

$$P = VI = \frac{V^2}{R} = I^2 R \tag{1.5.4}$$

or replacing $R = 1/G$,

$$P = VI = V^2 G = \frac{I^2}{G} \tag{1.5.5}$$

In applications, we must distinguish between power which is delivered by the source and power consumed by the network or element. A resistor *always* consumes power and the current flows through it from + to −. If the network has only one voltage source, then the source delivers power into the network and the current will flow from the + terminal into the network and return through the network to the − terminal. If the network has a single current source, then the terminal at the arrow will be the positive one and the other will be negative.

Example 1.

Consider the network in Fig. 1.5.1 and calculate the currents and powers delivered and consumed in the network.

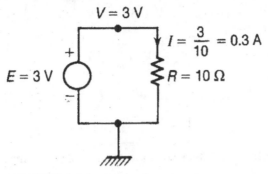

FIGURE 1.5.1 Power from a voltage source.

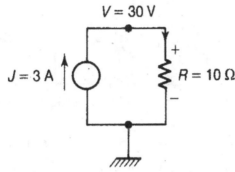

FIGURE 1.5.2 Power from a current source.

Applying the Ohm's law we get the current $I = \dfrac{3}{10} = 0.3$ A. It will flow in the indicated direction. The power *consumed* by the resistor will be $P = VI = 0.3 \times 3$ W and the same amount of power will be *delivered* by the source.

Example 2.

In the network Fig. 1.5.2 the current source delivers 3 A. Calculate the voltages and powers delivered and consumed.

The voltage across the resistor is $V = 3 \times 10 = 30$ V. The power consumed by the resistor (and converted into heat) is $P = V \times I = 3 \times 30 = 90$ W. The same power is delivered by the current source.

The situation becomes more complicated if there are more than one independent source. In a general case, the question how the powers are distributed can be answered only by setting up the appropriate equations and by solving them. We will learn how to do that later. Here we will take a simple network where the answer can be found by the theory covered until now.

Example 3.

Consider the network in Fig. 1.5.3 and calculate the current and powers delivered or consumed by the elements.

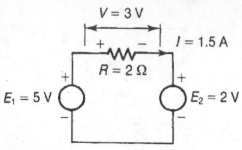

FIGURE 1.5.3   Network with two voltage sources.

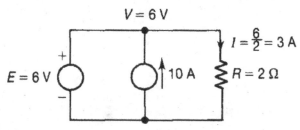

FIGURE 1.5.4   Network with current and voltage source.

The voltage across the resistor is $V = 5 - 2 = 3$ V and the current in the network is $I = \dfrac{3}{2} = 1.5$ A. The power consumed by the resistor is $P_R = I\,V = 3 \times 1.5 = 4.5$ W. The source on the left is delivering the power $P = 5 \times 1.5 = 7.5$ W. On the right, the current flows through the source in the direction from $+$ to $-$ and thus this source *consumes* power. The amount is $P = 2 \times 1.5 = 3$ W. The power delivered by the left source is exactly equal to the power consumed by the resistor and the other source.

Example 4.

A parallel connection of a voltage and current source is in Fig. 1.5.4. Calculate the powers delivered and consumed by the respective elements.

Due to the voltage source, the voltage across the resistor *must* be 6 V and the current through the resistor must be, due to the Ohm's law, $I_R = \dfrac{6}{2} = 3$ A. The current source is supplying 10 A. Three amperes will be forced through the resistor, the rest must flow through the voltage source, from $+$ to $-$. The powers are distributed as follows: The current source delivers $P = 6 \times 10 = 60$ W. Of this the resistor consumes $P = 3 \times 6 = 18$ W while the voltage source consumes $P = 6 \times 7 = 42$ W. As always, the power delivered is equal to the power consumed.

1.6. KIRCHHOFF'S LAWS

The laws discovered by Kirchhoff are fundamental to the theory of electricity. They are valid in *any* situation, even if the elements are nonlinear or time varying.

1.6.1. Kirchhoff Current Law

The Kirchhoff current law (*KCL*) states that *the sum of currents flowing away from a node is zero*. The definition remains valid if we replace the words *away from* by the word *to*. A *node* is a point where we connect several elements and one such case is shown in Fig. 1.6.1a. Since the node is only a connection of wires, it is easily understood that it cannot store electrons and thus any amount of electrons flowing into the node through some of the connecting lines must also leave through some other lines. In the statement of the law we used the word *away* because in this book we will consider a current flowing *away* from a node as having *positive* direction.

In Fig. 1.6.1a we have five elements connected to the central node. *At this node*, the currents i_1, i_2 and i_3 have positive directions while currents i_4 and i_5 have negative directions. The Kirchhoff current law can be written as

$$i_1 + i_2 + i_3 - i_4 - i_5 = 0 \tag{1.6.1}$$

We drew the directions of the currents in the figure arbitrarily. There is, however, one situation where we do not have this freedom of choice, namely when there are independent current sources connected to the node. Fig. 1.6.1b shows a node with two independent current sources and three elements which are not sources. We arbitrarily select directions of currents for the resistors. Following the *KCL* it must be true that

$$i_1 - i_2 + i_3 + J_1 - J_2 = 0 \tag{1.6.2}$$

In mathematics, it is common to write unknown variables on the left side and known quantities on the right side of the equation. Because the currents and directions of independent current sources are known, we can follow the usual mathematical strategy and rewrite the above equation in the form

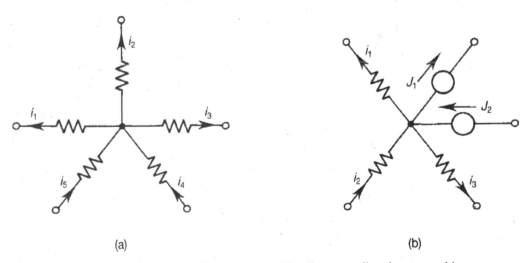

(a)   (b)

FIGURE 1.6.1   A node with five elements: (a) All current directions are arbitrary.
(b) Directions of current sources given, rest arbitrary.

$$i_1 - i_2 + i_3 = -J_1 + J_2 \qquad (1.6.3)$$

We now see the reason why we used different letters to describe properties of independent sources: without much thinking what is dependent and what is independent, we know immediately that the letter J must go to the right side of the equation *with a change in sign*.

1.6.2.   Kirchhoff Voltage Law

The Kirchhoff voltage law (*KVL*) states that **the sum of voltages around a loop is zero**. A loop is such a connection of elements that we can walk around it and reach the same point from which we started. An example of a loop with five elements is shown in Fig. 1.6.2a. We know nothing about the voltages and currents in this loop, because they will be influenced by the connections to other parts of the network (indicated in the figure by the openended lines out of the loop). We can assign, completely arbitrarily, the signs + and − to the elements, as was done in the figure. However, once we have done that, there is nothing arbitrary any more, except for the direction of our walk around the loop. Let us choose the one indicated in the figure by the arrow. If this direction goes through the element from + to −, then we add the voltage of this element. If the direction of our walk is such that we go through the element from − to +, we subtract the voltage. For the network in Fig. 1.6.2a this results in the equation

$$-v_1 + v_2 - v_3 + v_4 - v_5 = 0 \qquad (1.6.4)$$

We have the freedom to assign the directions of voltages to all elements except the independent voltage sources where the + and − signs are given.

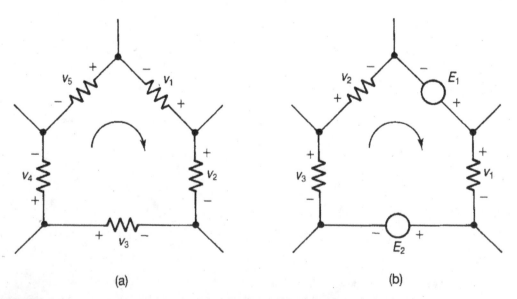

(a)   (b)

FIGURE 1.6.2   A loop with five elements: (a) Individual voltages arbitrary. (b) Voltages of voltage sources given, rest arbitrary.

The network in Fig. 1.6.2b has two voltage sources with their + and − specified. For the remaining elements we choose arbitrary positions of + and −, for instance as shown. Once we have done this, the only arbitrary thing is the direction of the walk around the loop. If we decide to walk as shown, the equation must be written as

$$-E_1 + v_1 + E_2 - v_3 + v_2 = 0 \tag{1.6.5}$$

Since the independent sources are known, we follow the usual mathematical practice and write their voltages on the right side of the equation. The sum of voltages around the loop can be rewritten as

$$v_1 + v_2 - v_3 = +E_1 - E_2 \tag{1.6.6}$$

Here we take advantage of our rule that independent voltage sources are marked by the letter E. Since we use different letters for the unknowns, v, and for the knowns, E, we do not need much thinking to determine what should go to the right side of the equation *with opposite sign* and what should stay on the left side.

Example 1.

The sketch in Fig. 1.6.3 represents part of some larger network. The nodal voltage at the center is V, the other have subscripts 1 through 4. Current directions are also marked. Applying the KCL we can write for the node in the center

$$-I_1 + I_2 - I_3 + I_4 = 0 \tag{1.6.7}$$

From these directions of the currents we conclude that $V_1 > V$, $V_2 < V$, $V_3 > V$ and $V_4 < V$. Express the currents in terms of the voltages across the conductances:

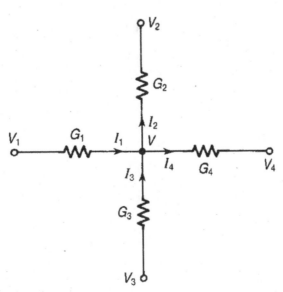

FIGURE 1.6.3   A node with specified current directions.

$$I_1 = (V_1 - V)G_1$$

$$I_2 = (V - V_2)G_2$$

(1.6.8)

$$I_3 = (V_3 - V)G_3$$

$$I_4 = (V - V_4)G_4$$

These equations can be inserted into (1.6.7) to get

$$-(V_1-V)G_1+(V-V_2)G_2-(V_3-V)G_3(V-V_4)G_4=0$$

We can open the brackets and rewrite in the form

$$V(G_1+G_2+G_3+G_4)-V_1G_1-V_2G_2-V_3G_3-V_4G_4=0$$

(1.6.9)

Example 2.

In this example we take the same network, but assume that the voltage V is larger then all the other indicated voltages. Due to this assumptions, the directions of the currents must change. They will be as indicated in Fig. 1.6.4. KCL for the central node provides

$$I_1 + I_2 + I_3 + I_4 = 0$$

(1.6.10)

The currents through the branches are

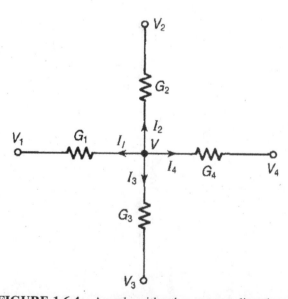

FIGURE 1.6.4   A node with other current directions.

$$I_1 = (V - V_1)G_1$$

$$I_2 = (V - V_2)G_2 \qquad (1.6.11)$$

$$I_3 = (V - V_3)G_3$$

$$I_4 = (V - V_4)G_4$$

Inserting these currents into (1.6.10) results in

$$(V - V_1)G_1 + (V - V_2)G_2 + (V - V_3)G_3 + (V - V_4)G_4 = 0$$

Rearranging terms we get

$$V(G_1 + G_2 + G_3 + G_4) - V_1G_1 - V_2G_2 - V_3G_3 - V_4G_4 = 0 \qquad (1.6.12)$$

The result is interesting, in both cases we get *exactly* the same equation. This is an important result, because it clearly does not matter which voltage we think is higher and which is lower. All we have to do is correctly relate the currents in terms of the voltages and conductances. We will make use of this fact in the next chapter.

1.7. CONNECTIONS OF RESISTORS

The first application of Ohm's and Kirchhoff's laws comes in the study of series and parallel connections of resistors. The simplest case is one loop of resistors, for instance as shown in Fig. 1.7.1. The arrow indicates the direction of our walk around the loop and in this case also indicates the direction of the current, I, which will flow through the combination. Since the same current flows through all resistors, the + and − signs of the voltages across the individual resistors are as shown. Let the source be a *dc* source, $E = 22\ V$. From the Kirchhoff's voltage law we have

$$V_1 + V_2 + V_3 + V_4 - E = 0,$$

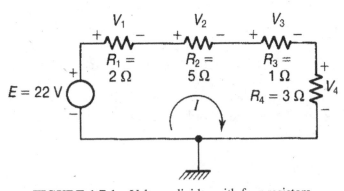

FIGURE 1.7.1 Voltage divider with four resistors.

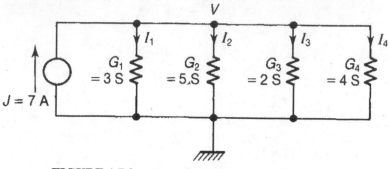

FIGURE 1.7.2   Current divider with four branches.

the minus sign before E coming from the fact that the circulating current enters the voltage source at its minus sign and thus has negative direction according to our rules. The voltage of the source is known and we transfer it on the other side of the equation:

$$V_1 + V_2 + V_3 + V_4 = E \tag{1.7.1}$$

Each of the resistors follows the Ohm's law:

$$V_1 = IR_1$$

$$V_2 = IR_2 \tag{1.7.2}$$

$$V_3 = IR_3$$

$$V_4 = IR_4$$

Substituting into Eq. (1.7.1)

$$IR_1 + IR_2 + IR_3 + IR_4 = E$$

or

$$I\,(R_1 + R_2 + R_3 + R_4) = E \tag{1.7.3}$$

We can introduce the notation

$$R_{tot} = R_1 + R_2 + R_3 + R_4 \tag{1.7.4}$$

and obtain the overall resistance of the combination, in our case $R_{tot} = 11\Omega$. The voltage of the source is $22V$ and the equation is solved for the current

$$I = \frac{22}{11} = 2\,A\,.$$

Returning to (1.7.2), we can now obtain the voltage across each resistor: $V_1 = 4\ V$, $V_2 = 10$, $V_3 = 2\ V$ and $V_4 = 6\ V$. The sum of these individual voltages must be equal to the voltage of the voltage source.

Since the voltage of the source was divided into several partial voltages, the network in Fig. 1.7.1 is also called a *voltage divider*. We say that the resistors are connected in *series*. We will also remember that the overall resistance of the resistors connected in series is equal to the sum of the individual resistances. In other words, we do not have to go through all the steps derived above. All we have to do is add the resistances of resistors connected in series.

A different situation is shown in Fig. 1.7.2 where we have one current source and several resistors connected *in parallel*. We would like to know the voltage common to all these resistors and the currents flowing through them. It is obvious that the same voltage, V, appears across all the resistors and the sum of the currents flowing through them must be equal to the current delivered by the current source. The Kirchhoff current law states that the sum of currents at the upper node must be equal to zero:

$$I_1 + I_2 + I_3 + I_4 - J = 0, \tag{1.7.5}$$

where the negative sign for the current source comes from the fact that its current flows *into* the node. As always, known values are transferred to the right hand side of the equation with a change of sign:

$$I_1 + I_2 + I_3 + I_4 = J \tag{1.7.6}$$

The current through each resistor can be expressed in terms of its conductance:

$$I_1 = VG_1$$

$$I_2 = VG_2 \tag{1.7.7}$$

$$I_3 = VG_3$$

$$I_4 = VG_4$$

Inserting these values into (1.7.5)

$$VG_1 + VG_2 + VG_3 + VG_4 = J$$

or

$$V(G_1 + G_2 + G_3 + G_4) = J. \tag{1.7.8}$$

The total conductance is

$$G_{tot} = G_1 + G_2 + G_3 + G_4$$

and the equation simplifies into

$$V = \frac{J}{G_{tot}} = \frac{7}{14} = 0.5$$

Going back to the individual equations (1.7.6), we calculate the currents through all resistors as $I_1 = 1.5\ A$, $I_2 = 2.5\ A$, $I_3 = 1\ A$ and $I_4 = 2\ A$. The sum is equal $7\ A$. The current of the source was divided into individual currents of the resistors and for this reason the network is also called the *current divider*.

The steps explained above can be summarized into two simple rules:

(a)   If all resistors are in series, add their resistances to obtain the overall resistance.
(b)   If all resistors are in parallel, add their conductances to obtain the overall conductance.

The rules help us to find the overall resistance of more complicated networks without the need to calculate the currents. We give several examples with increasing complexity.

Example 1.

In the network Fig. 1.7.3 find the current and power delivered into the network by the source. On the right, the two resistors are in parallel and thus their conductances are added to get $G = \frac{3}{4}$ or $R = \frac{4}{3}$. This is redrawn in the middle of the figure. Now we have two resistors in series and we can add their resistances to obtain the overall resistance $R_{tot} = 1 + \frac{4}{3} = \frac{7}{3}$. The network is again redrawn on the right. We now have a source $E = 14\ V$, a resistance $R_{tot} = \frac{7}{3}$ and thus the current flowing from the source is $I = E/R = 6\ A$. The power delivered into the combination is $P = E \times I = 14 \times 6 = 84\ W$.

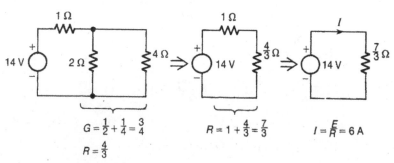

FIGURE 1.7.3   Calculating equivalent resistance.

Example 2.

Find the voltage across the current source in Fig. 1.7.4. Also find the power delivered by the source.

In the solution we first add the resistances of the two resistors on the right, $R = 3 + 9 = 12\Omega$. This is redrawn in the middle. Now we have two resistors in parallel and we must add their conductances, $G = \dfrac{1}{2} + \dfrac{1}{12} = \dfrac{7}{12}$. The network can be redrawn again, shown on the right, where we have a resistance $R = \dfrac{12}{7}$. The volt age across the source (and also across the equivalent resistor) is $V = R \times J = 12\ V$ and the power delivered into the network is $P = J \times V = 7 \times 12 = 84\ W$.

Example 3.

Find the equivalent resistance of the combination in Fig. 1.7.5.

Starting from the right we first add $R_3 + R_4 = R' = 3\Omega$. We now have two resistors in parallel and we must add their conductances. This is done in the middle of the figure. Their resistance is $R'' = \dfrac{6}{5}$. Finally, the resistance of R_1 is added to R'' to obtain the total resistance $R_{tot} = \dfrac{11}{5}\Omega$.

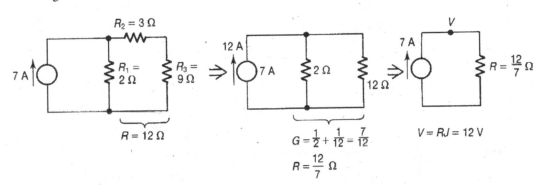

FIGURE 1.7.4 Calculating equivalent resistance.

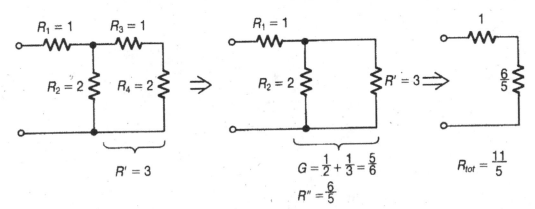

FIGURE 1.7.5 Calculating equivalent resistance.

Example 4.

Find the equivalent resistance of the combination in Fig. 1.7.6.

The steps are all indicated in the figure, with the result $R_{tot} = \dfrac{292}{141}\,\Omega.$

Example 5.

We will use the voltage divider to introduce the concept of an internal resistance of a *non-ideal* voltage source. In Fig. 1.7.7 is one ideal voltage source and two resistors. This is a simple voltage divider but consider the voltage source with the resistor R_1 as a model of a nonideal source, for instance a battery. The current flowing from the source will be

$$I = \frac{E}{R_1 + R_2}$$

and the voltage across R_2 will be

$$V = IR_2 = \frac{ER_2}{R_1 + R_2}$$

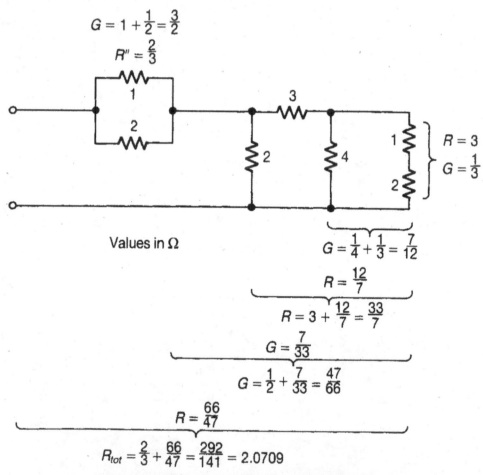

FIGURE 1.7.6   Calculating equivalent resistance.

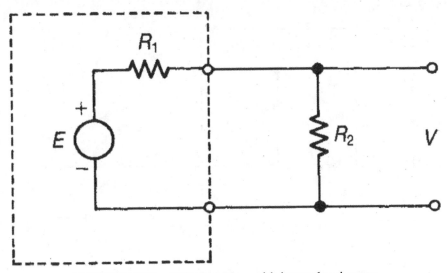

FIGURE 1.7.7   Voltage source with internal resistance.

If R_2 is very large, $R_2 \to \infty$, then there will be no current and the voltage V will be equal to E. If we replace R_2 by a short circuit, then the voltage will be zero. Any finite value of R_2 will make the output voltage $0 < V < E$. We can also calculate the power delivered by the source into the resistor R_2. It will be equal to

$$P = I \times V = \frac{E}{R_1 + R_2} \times \frac{ER_2}{R_1 + R_2} \times E^2 \frac{R_2}{(R_1 + R_2)^2}$$

If we select $R_2 = 0$, then the power consumed in it is zero. If we select $R_2 \to \infty$ (the same as if we disconnect R_2 from the network), then the power consumed by it will also be zero. For all other cases there will be some power consumed in R_2. We could plot such curve for a number of values and discover that the largest power is delivered when $R_2 = R_1$, in which case the voltage $V = \dfrac{E}{2}$.

Example 6.

A nonideal current source can be modeled as in Fig. 1.7.8. Using the KCL, we have

$$V(G_1 + G_2) = J$$

or

$$V = \frac{J}{G_1 + G_2}$$

The current flowing through G_2 is

$$I = VG_2 = \frac{JG_2}{G_1 + G_2}$$

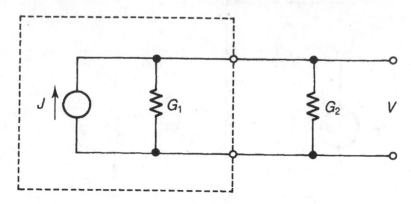

FIGURE 1.7.8   Current source with internal conductance.

and the power consumed (or as we say dissipated) in G_2 is

$$P = V \times I = J^2 \frac{G_2}{(G_1 + G_2)^2}$$

For the two extreme values of G_2, namely zero and infinity, there will be no power consumed in G_2. In all other cases the power will be nonzero and maximum transfer of power will occur for If $G_1 = G_2$.

PROBLEMS CHAPTER 1

P.1.1   Find equivalent resistances of the networks.

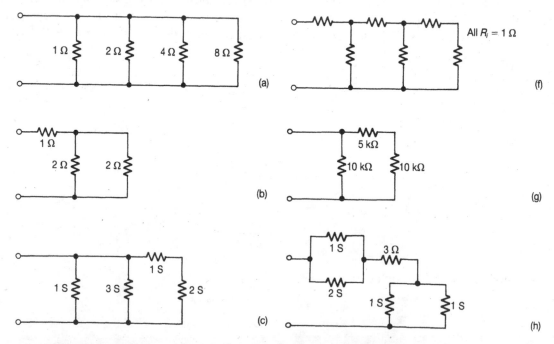

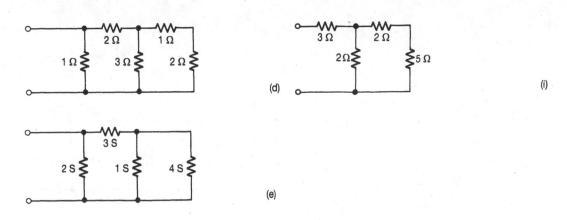

(d)

(i)

(e)

P.1.2 Find equivalent resistances.

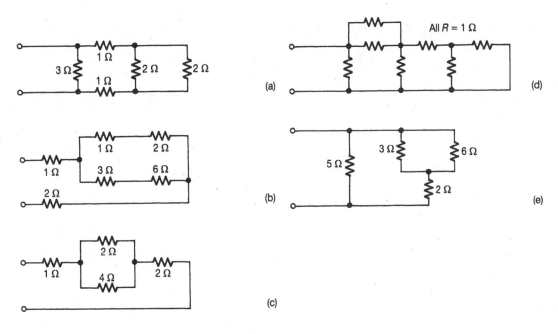

(a)

(d)

(b)

(e)

(c)

P.1.3 Find equivalent resistances.

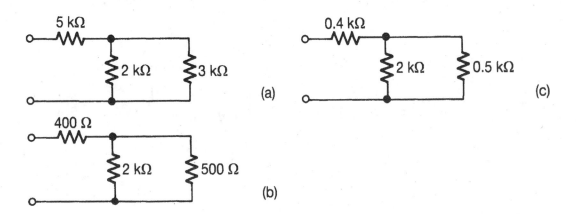

(a)

(c)

(b)

P.1.4   Find equivalent sources for the given combinations. Are a and b possible?

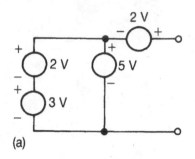

(a)

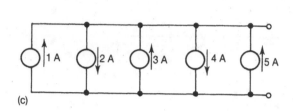

(c)

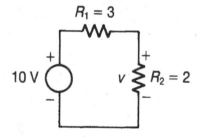

(b)

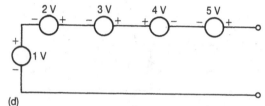

(d)

P.1.5   Find the voltage across the resistor R_2.

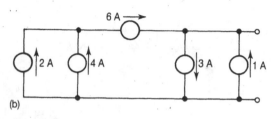

P.1.6   Find the voltages across the resistors and currents flowing through them.

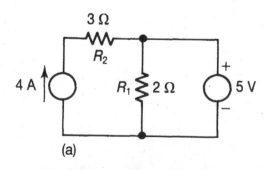

(a)

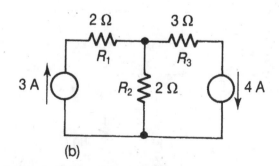

(b)

P.1.7   Find the current flowing through the resistor.

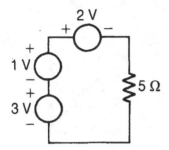

P.1.8   Find the nodal voltage V.

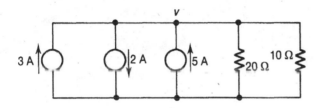

P.1.9   Find the current flowing through the resistor.

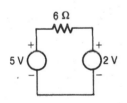

P.1.10   What current is flowing through the resistor? Which source delivers and which consumes power?

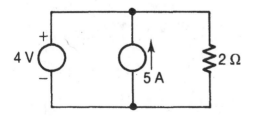

P.1.11 Find the powers delivered and onsumed in the three elements.

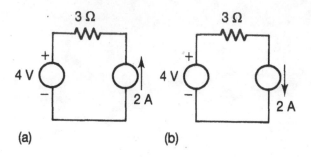

(a)   (b)

P.1.12 Find the powers delivered and consumed by the three elements.

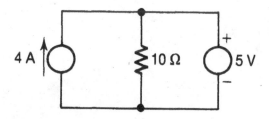

P.1.13 Find the power consumed in the resistor.

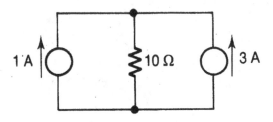

P.1.14 Find the voltages, currents and powers delivered and consumed.

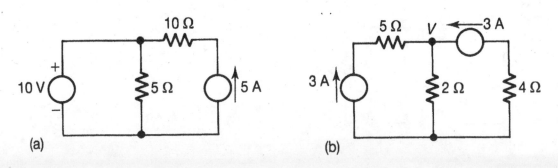

(a)   (b)

P.1.15 Find the voltage across the combination and the powers delivered or consumed by all elements.

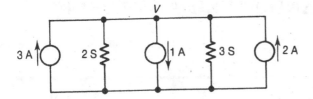

P.1.16 Find the voltages across the elements and powers delivered or consumed by each element.

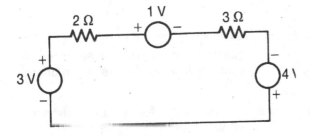

P.1.17 Find the voltages across the resistors, currents flowing through them and powers consumed in them.

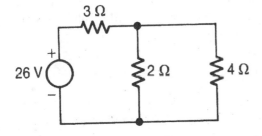

2 NODAL AND MESH ANALYSIS

INTRODUCTION

The voltage and current dividers explained in the previous chapter are the first useful networks, but they are needed only rarely. In most situations we need voltages at several (or all) nodes or currents through the elements. Such problems lead to systems of simultaneous equations which must be solved.

We will explain two classical methods how to write equations for networks: the nodal and the mesh formulations. The nodal formulation is based on the use of the Kirchhoff's current law and is suitable for networks with current sources. The mesh method uses the Kirchhoff's voltage law and is easy to apply for networks with voltage sources.

The equations which we set up have to be solved and practical methods how to do it will be given in the next chapter. Our main interest in this brief chapter is to learn how to write the two types of equations and master enough skills so that we will be able to write them by inspection.

2.1. NODAL ANALYSIS

Nodal analysis is a way of writing network equations by considering the Kirchhoff current law, KCL, at all ungrounded nodes of the network. In this chapter we will deal with only two types of elements: independent current sources and resistors, given in terms of their conductance values.

We will start with two simple situations. Consider first the network in Fig. 2.1.1, with one source and five resistors. We can make a good estimate which way the currents will flow through the resistors; in fact we have marked them in the network.

Consider next the same network, but attach another source on the right, Fig. 2.1.2. Suddenly we have a dilemma: we cannot guess which way we should mark directions of the currents in the resistors. The currents will obviously depend on the current values of the sources as well as on the values of the resistors. If $J_2 >> J_1$ it is likely that in the floating resistors (those not connected to ground) the currents will flow from right to left. If we have a larger network with several sources, nobody will be able to guess the directions. To avoid problems in assigning the directions, we must develop a unified method which works for any values of resistors and any number of sources. Such method is the subject of this section.

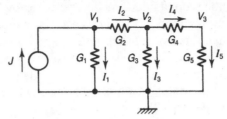

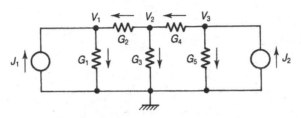

FIGURE 2.1.1 Network with estimated current directions.

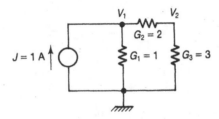

FIGURE 2.1.2 Network with estimated current directions.

FIGURE 2.1.3 Network with one current source.

Let us now turn our attention to Fig. 2.1.3. The bottom line is the reference node (ground), always assumed to have zero voltage. Assign voltages to the nodes which are not grounded. Normally we use subscripted variables, V_1 and V_2.

The *KCL* states that the sum of currents leaving a node is zero. Consider the first node. We do not know the voltage V_1, but if we *assume* that it is the largest voltage in the network, then the currents through all resistors connected to the node must flow away from the node. Since the first resistor is connected to ground, its current will be, as follows from the Ohm's law, $I_{G1} = G_1 V_1$. The voltage across the second resistor is equal to $V_1 - V_2$ and using the Ohm's law, the current will be $I_{G2} = (V_1 - V_2)G_2$. Note the way of writing. By giving first V_1 and by subtracting from it V_2 we indicate that we consider V_1 to be more positive than V_2. Currents flowing away from the node are considered to be positive by our definition. The current from the source must be taken negatively, because if flows into the node. The KCL equation is

$$G_1 V_1 + (V_1 - V_2)G_2 - J = 0 \qquad (2.1.1)$$

Let us now move to the second node. As before, we know nothing about the voltages and we can still make the assumption that now it is V_2 which is the highest voltage in the network. Once we make this assumption, all currents through the resistors must flow *away* from node 2. The voltage across the second resistor is now $V_2 - V_1$ (because of our assumption we must put first the voltage of the second node and subtract the voltage of the first node). The current through this resistor will be $I_{G2} = (V_2 - V_1)G_2$. The current through the third resistor must flow down and $I_{G3} = G_3 V_2$. Since there is no source at node two, the *KCL* for the second node is

$$G_3 V_2 + (V_2 - V_1)G_2 = 0 \tag{2.1.2}$$

Rearrange the terms in both equations and transfer the independent source to the right, with a change of sign. We get a set of two equations

$$(G_1 + G_2)V_1 - G_2 V_2 = J$$

$$-G_2 V_1 + (G_2 + G_3)V_2 = 0 \tag{2.1.3}$$

Numerical values simplify them to

$$3V_1 - 2V_2 = 1$$

$$-2V_1 + 5V_2 = 0 \tag{2.1.4}$$

At this moment we do not know yet a good method for the solution of systems of equations but the reader certainly knows elimination. The result of such solution is $V_1 = \dfrac{5}{11}$ and $V_2 = \dfrac{2}{11}$, as can be easily checked by inserting these values into the two equations. We can also calculate the currents through the elements and their directions:

$$I_{G1} = \frac{5}{11} A, \; I_{G2} = (\frac{5}{11} - \frac{2}{11})2 = \frac{6}{11} A, \text{ and } I_{G3} = \frac{2}{11}3 = \frac{6}{11} A.$$

The network in Fig. 2.1.4 has two current sources and we wish to calculate the nodal voltages, the currents and their directions. As before, consider the first node as the most positive one; the KCL is

$$G_1 V_1 + G_2(V_1 - V_2) - J_1 = 0$$

or rearranging the terms

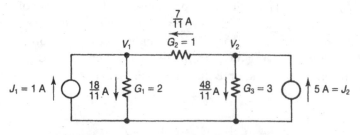

FIGURE 2.1.4   Network with two current sources.

$$(G_1 + G_2)V_1 - G_2V_2 = J_1$$

and numerically

$$3V_1 - 1V_2 = 1 \qquad (2.1.5)$$

Moving to the next node and considering it as the most positive one we obtain the KCL equation

$$G_3V_2 + G_2(V_2 - V_1) - J_2 = 0$$

or rearranging the terms

$$-G_2V_1 + (G_2 + G_3)V_2 = J_2$$

and numerically

$$-V_1 + 4V_2 = 5 \qquad (2.1.6)$$

In this case we will use elimination to solve. From Eq. (2.1.5) get

$$V_2 = 3V_1 - 1 \qquad (2.1.7)$$

and insert into Eq. (2.1.6) with the result

$$V_1 = \frac{9}{11}V.$$

Inserting this result into Eq. (2.1.7)

$$V_2 = \frac{16}{11}V \qquad (2.1.9)$$

The currents are $I_{G1} = V_1G_1 = \frac{18}{11}A$ and $I_{G2} = (V_1 - V_2)G_2 = -\frac{7}{11}A$. The minus sign in the last case indicates that the current actually flows from node 2 to node 1. For the third resistor we get $I_{G3} = V_2G_3 = \frac{48}{11}A$. The resulting current directions are indicated in the network by arrows. Taking absolute values of the currents and the directions indicated in the figure we confirm that the sum of currents at each node is indeed zero.

Example 1.

Consider Fig. 2.1.5, write the KCL equations for the two nodes and collect terms multiplied by the same voltages.

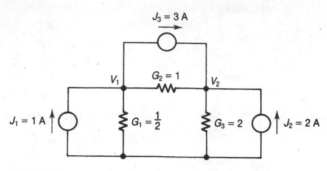

FIGURE 2.1.5   Network with three current sources.

Keeping first the general notation we get for the first node

$$(G_1 + G_2)V_1 - G_2V_2 = J_1 - J_3$$

and for the second node

$$-G_2V_1 + (G_2 + G_3)V_2 = J_2 + J_3$$

Using numerical values, the equations become

$$\frac{3}{2}V_1 - V_2 = -2$$

$$-V_2 + 3V_2 = 5$$

We will use one more network, Fig. 2.1.6, to write the KCL equations. This network is already so complicated that even a skilled person cannot see which way the currents will flow. Following the rules stated above, we first assume that node 1 is the most positive one.

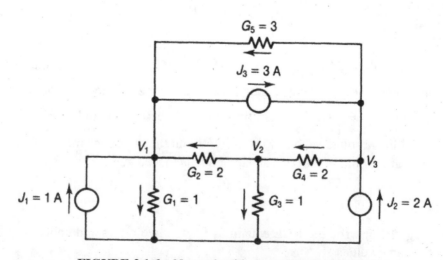

FIGURE 2.1.6   Network with three current sources.

All currents through the conductances must leave this node: $I_{G1} = G_1 V_1$, $I_{G2} = G_2(V_1 - V_2)$, $I_{G5} = G_5(V_1 - V_3)$, and the KCL for the first node is

$$G_1 V_1 + G_2 (V_1 - V_2) + G_5 (V_1 - V_3) - J_1 + J_3 = 0 \qquad (2.1.10)$$

Going to node 2, we assume that it is now this node which is most positive. The currents leaving the second node will be $I_{G2} = G_2 (V_2 - V_1)$, $I_{G3} = G_3 V_2, I_{G4} = G_4 (V_2 - V_3)$, and the sum of currents for the second node is

$$G_2 (V_2 - V_1) + G_3 V_2 + G_4 (V_2 - V_3) = 0 \qquad (2.1.11)$$

Going to node 3 we make the assumption that now this node is the most positive one. The currents of the conductances leaving the third node will be: $I_{G4} = G_4(V_3 - V_2)$, $I_{G5} = G_5(V_3 - V_1)$, and the KCL at node 3 is

$$G_4(V_3 - V_2) + G_5(V_3 - V_1) - J_2 - J_3 = 0 \qquad (2.1.12)$$

If we rearrange the terms and transfer the independent sources on the right side, we get the system

$$\begin{aligned}
(G_1 + G_2 + G_5)V_1 - G_2 V_2 - G_5 V_3 &= J_1 - J_3 \\
-G_2 V_1 + (G_2 + G_3 + G_4)V_2 - G_4 V_3 &= 0 \\
-G_5 V_1 - G_4 V_2 + (G_4 + G_5)V_3 &= J_2 + J_3
\end{aligned} \qquad (2.1.13)$$

Insertion of numerical values leads to

$$\begin{aligned}
6V_1 - 2V_2 - 3 &= -2 \\
-2V_1 + 5V_2 - 2V_3 &= 0 \\
-3V_1 - 2V_2 + 5V_3 &= 5
\end{aligned} \qquad (2.1.14)$$

We state the result of the solution of this system: $V_1 = \dfrac{53}{37}, V_2 = \dfrac{58}{37},$ and $V_3 = \dfrac{92}{37}$.

The currents through the resistors are: $I_{G1} = V_1 G_1 = \dfrac{53}{37} A$, $I_{G2}(V_1 - V_2)G_2 = -\dfrac{10}{37}$, $I_{G3} = V_2 G_3 = \dfrac{58}{37}$, $I_{G4} = (V_2 - V_3)G_4 = -\dfrac{68}{37}$, $I_{G5}(V_1 - V_3)G_5 = -\dfrac{117}{37}$.

In this case the currents through all floating resistors actually flow from right to left. The directions of the currents are drawn in the network. Taking absolute values of the currents *and* the marked directions we confirm that the sum of currents at each node is indeed equal to zero.

Let us return to the set of Eq. (2.1.13) because they give us the opportunity to find a general rule how to write the equations. Consider the first node: The sum of conductances connected to the first node is multiplied by the nodal voltage of the first node. Subtracted from this sum is the product of the same conductances, multiplied by the voltage of the other node to which they are onnected. Check the other equations to convince yourself that the same applies for the other nodes as well.

We are now in the position to formulate a general rule for writing CL at any node, i:

1. Multiply the sum of all conductances connected to node i by V_i.
2. Subtract from this sum the same conductances, each multiplied by the voltage of the other node where the element is connected.
3. Since the voltage of ground is defined to be zero, no subtraction occurs or grounded resistors.
4. The current of an independent current source flowing *into* node i is taken positive *on the right side of the equation*. The sign changes if the current flows in the opposite direction.

Using the above rules and the numerical values of the above examples, set up the equations directly, without first defining the currents.

We have not attempted to solve the systems of three equations at this moment. It is a fairly laborious process and elimination is not a very good way to proceed. Our next chapter will give the most convenient methods to solve small systems of linear equations; here we take one more example.

Example 2.

Consider the network in Fig. 2.1.7. Using the rules stated above write the nodal equations by inspection. The set of equations is:

$$\left(G_1 + G_2 + G_6\right)V_1 - G_2V_2 - G_6V_3 = J_1 - J_4$$
$$-G_2V_1 + \left(G_2 + G_3 + G_4\right)V_2 - G_4V_3 = J_2 - J_3$$
$$-G_6V_1 - G_4V_2 + \left(G_4 + G_5 + G_6\right)V_3 = J_3 + J_4 + J_5$$

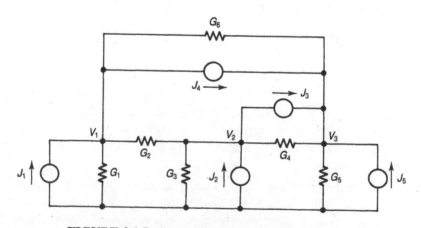

FIGURE 2.1.7   Network with five current sources.

2.2. MESH ANALYSIS

Mesh analysis is a way of writing network equations by applying the Kirchhoff voltage law, *KVL*: the sum of voltages around a loop is zero. In this section we will deal with the principle of mesh analysis and only two types of elements: independent voltage sources and resistors, this time given by their resistance values. The method has some unpleasant limitations, as we will see. It is never used in computer applications. Instead, the nodal formulation is developed into *Modified Nodal Formulation*, explained in Chapter 17.

We start with the network in Fig. 2.2.1. The voltage source delivers currents into the network and in this simple case we can easily guess which sides of the elements will be marked with the + sign and which with the – sign; this is shown in the figure. The way we walk around a loop is indicated by an arrow. We have two such arrows in the "windows," or "meshes," marked as 1 and 2. These two cases are not the only ones we can select. Another loop can be closed around the whole network, also indicated in the figure. Using the KVL we can write the following equations:

Mesh one:

$$V_{R1} + V_{R2} - E = 0$$

Here we must subtract the voltage of the voltage source because we are walking through the source from – to +.

Mesh two:

$$-V_{R2} + V_{R3} + V_{R4} = 0$$

The negative sign comes from the fact that the current I_2 flows through $R2$ from – to +. The loop around the network:

$$V_{R1} + V_{R3} + V_{R4} - E = 0$$

We can think of the arrows in the meshes or around the whole network as directions of currents. Such circulating currents do not actually exist but the notion helps; they are the first of several abstractions we will make.

In larger networks we can find many loops around which we could walk. In Fig. 2.2.2 we have marked six possible loops: three in the meshes, two more circling two meshes and

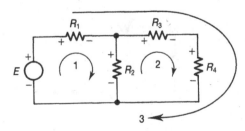

FIGURE 2.2.1   Possible loops on a simple network.

one around the whole network. In still larger networks this could become very complicated and we must accept certain simplifications. The first and most important one is that the network must be *planar*. A planar network is such that we can draw its diagram on paper without any connection crossing some other connection. The two networks we considered so far were clearly planar. The network in Fig. 2.2.3a seems not to be planar, but we can redraw it as shown in Fig. 2.2.3b and discover that it actually is planar. The network in Fig. 2.2.4 is not planar because some line will cross another line no matter how we redraw the network.

Why did we use so much space to discuss planar networks? Because only planar networks can be analyzed by the method we will discuss here. In every planar network we can identify the meshes and draw directions for the voltage summations. The directions around the meshes are arbitrary but it is advisable to select one direction and use it everywhere. We will use the clockwise direction.

Consider the network in Fig. 2.2.5; it has two meshes, one voltage source and four resistors. The arrows in the meshes indicate the direction in which we will collect the voltages. If we take the directions as currents, then the product of a current and a resistance

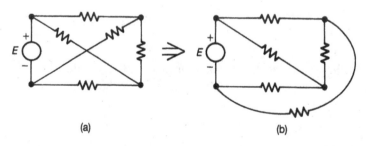

FIGURE 2.2.2   Possible loops on another network.

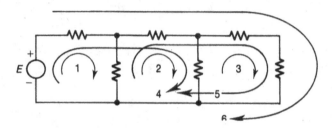

FIGURE 2.2.3   Is the network planner? (a) Seemingly nonplanner. (b) Redrawn for planarity.

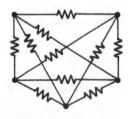

FIGURE 2.1.4   Nonplanar network.

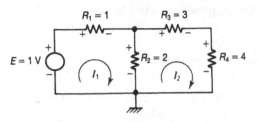

FIGURE 2.2.5 Network with one voltage source.

gives the voltage across the resistor, which is exactly what we need for the summations required by the KVL. There will also be cases where two currents in opposite directions will flow through a resistor; in our case it is resistor R_2. When we are summing the voltages around the first mesh, the current I_1 has positive direction while the current I_2 has negative direction. Thus the voltage which contributes to the first mesh will be $R_2(I_1 - I_2)$. When we move to the second loop and walk around it, it will be the current I_2 which will have positive direction and the current I_1 will have to be subtracted. This is similar to the situation we had in section 1 where we always considered the node for which we were writing the equations as the most positive one.

In Fig. 2.2.5 I_1 flows through the resistor R_1 and the voltage across it must be $V_{R1} = I_1 R_1$, with the + sign on the left side of R_1. As explained above, the voltage across the second resistor will be $V_{R2} = R_2(I_1 - I_2)$. The voltage of the voltage source must be subtracted *on the left side of the equation* because we are walking through it from − to +. Thus the sum of voltages around the first mesh is

$$R_1 I_1 + R_2 (I_1 - I_2) - E = 0 \tag{2.2.1}$$

We proceed similarly for the second mesh where we are summing the voltages in the direction of the second arrow. The voltage across R_2 is now $V_{R2} = R_2(I_2 - I_1)$, the voltage across resistor R_3 is $V_{R3} = R_3 I_2$ and the voltage across the fourth resistor is $V_{R4} = R_4 I_2$. The second equation becomes

$$R_2 (I_2 - I_1) + R_3 I_2 + R_4 I_3 = 0 \tag{2.2.2}$$

Transferring the known E in the first equation to the right and collecting terms multiplied by the same currents we get

$$\begin{aligned}(R_1 + R_2) I_1 - R_2 I_2 &= E \\ -R_2 I_1 + (R_2 + R_3 + R_4) I_2 &= 0\end{aligned} \tag{2.2.3}$$

Substituting the resistor values we simplify to the form

$$\begin{aligned}3I_1 - 2I_2 &= 1 \\ -2I_1 + 9I_2 &= 0\end{aligned}$$

This is a system of two equations in two unknowns, I_1 and I_2, and we give its solution:
$I_1 = \dfrac{9}{23}$ and $I_2 = \dfrac{2}{23}$.

The second network we consider is in Fig. 2.2.6. In the first mesh the voltage across R_1 is $V_{R1} = 2I_1$, the voltage across the second resistor is $V_{R2} = 4(I_1 - I_2)$, similarly as in the above problem. The first mesh equation is $2I_1 + 4(I_1 - I_2) - 2 = 0$. The second equation is written similarly as above. The contributions are $V_{R2} = 4(I_2 - I_1)$, $V_{R3} = I_2$ and the voltage of the voltage source is added, because the current flows through it from + to −: $4(I_2 - I_1) + 1I_2 + 4 = 0$. Rearranging terms multiplied by the same currents and transferring the constant terms on the right side of the equations we get the final form

$$6I_2 - 4I_2 = 2$$
$$-4I_1 + 5I_2 = -4$$

Similarly as in the nodal formulation, we can now try to find some general rule how to write these equations by inspection. Refer to (2.2.3). In the first equation, when we consider mesh 1, we multiply by I_1 the sum of all resistors in mesh 1. If a resistor is in two meshes, then we subtract its resistance, multiplied by the current of the other mesh. When we move to the mesh 2, a similar pattern appears. The current I_2 multiplies the sum of the resistors in the second mesh. If a resistor is shared by two meshes, its resistance, multiplied by the current of the other mesh, is subtracted. For the voltage source we see that if the circulating current flows into the mesh from the + terminal then its voltage it appears *on the right side of the equation* with a positive sign.

We summarize the steps in the following rule for any mesh i:

1. Draw all mesh currents with clockwise directions.

2. Multiply the sum of all resistances in mesh i by I_i.

3. If an opposite current of mesh j also flows through the resistor, subtract the product of its resistance and I_j.

4. The voltage of an independent voltage source appears on the *right side of the equation* with positive sign if the ith equation current flows into the mesh from the + terminal of the source.

Apply the above rules for the network in Fig. 2.2.7. For the first mesh

$$\left(R_1 + R_2\right)I_1 - R_2 I_2 = E_1 + E_2 + E_3$$

For the second mesh

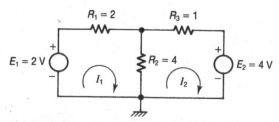

FIGURE 2.2.6   Network with two voltage sources.

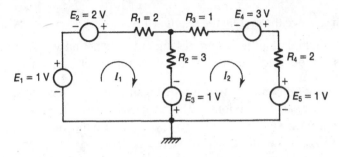

FIGURE 2.2.7   Network with four voltage sources.

$$-R_2 I_1 + \left(R_2 + R_3 + R_4\right) I_2 = -E_3 + E_4 - E_5$$

Inserting numerical values

$$5I_1 - 3I_2 = 4$$
$$-3I_1 + 6I_2 = 1$$

Example 1.

Use the above rule to write the equations for the network in Fig. 2.2.8.

$$\left(R_1 + R_2\right) I_1 - R_2 I_2 - R_1 I_3 = E_1 - E_2$$
$$-R_2 I_1 + \left(R_2 + R_3\right) I_2 - R_3 I_3 \qquad\qquad (2.2.4)$$
$$-R_1 I_1 - R_3 I_2 + \left(R_1 + R_3 + R_4\right) I_3 = -E_4$$

Inserting numerical values

$$3I_1 - 2I_2 - I_3 = -1$$
$$-2I_1 + 5I_2 - 3I_3 = -1$$
$$-I_1 - 3I_2 + 8I_3 = -4$$

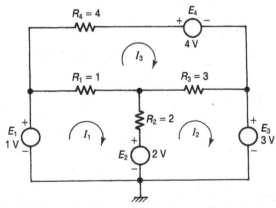

FIGURE 2.2.8   Network with four voltage sources.

For those who wish to solve the system the result is $I_1 = -\dfrac{94}{44}$ A, $I_2 = -\dfrac{86}{44}$ A and $I_3 = -\dfrac{66}{44}$ A. The negative signs mean that all three circulating currents actually flow in opposite directions. The currents can be used to find the voltage across each resistor.

A word of caution: Our rule for writing the mesh equations assumes that the circulating currents all have the same directions, either clockwise or counterclockwise. Only then it is possible to perform the steps as easily as above, because the current of the neighboring mesh flows always in opposite direction. If we select arbitrary directions, we will have to check where we must add the currents and where we must subtract them.

Example 2.

Apply the mesh method to the network in Fig. 2.2.9 by writing the equations directly: The equations are:

$$\left(R_1 + R_2 + R_3\right)I_1 - R_3I_2 - R_1I_4 = E_1 + E_2 - E_3$$

$$-R_3I_1 + \left(R_3 + R_4 + R_6\right)I_2 - R_6I_3 - R_4I_4 = -E_6 + E_7$$

$$-R_6I_2 + \left(R_6 + R_7 + R_8\right)I_3 - R_7I_4 = -E_7 + E_8$$

$$-R_1I_1 - R_4I_2 - R_7I_3 + \left(R_1 + R_4 + R_5 + R_7\right)I_4 = E_3 - E_4 + E_5 + E_6 - E_8$$

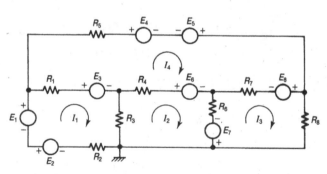

FIGURE 2.2.9   Network with seven voltage sources.

PROBLEMS CHAPTER 2

P.2.1   Find the indicated voltage V using one mesh formulation.

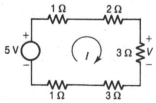

P.2.2   Find the current I using one mesh equation.

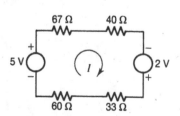

P.2.3 Find the voltage *V* using one nodal equation.

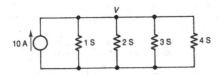

P.2.4 Find the voltage V using one nodal equation.

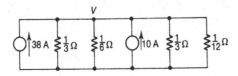

P.2.5 Write the mesh equations andfind the voltage *V*.

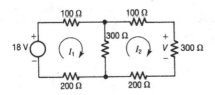

P.2.6 Connecting an element in series with a current source is not typical because the current forced through the element is known and for the rest of the network the voltages are the same as if the element was missing. The only difference is the voltage formed across the current source. Use this information, apply nodal formulation and calculate the current *I*. Also calculate the voltage across the current source.

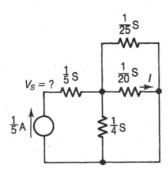

P.2.7 Find the indicated *V* and *I*. Use simplifications of the network to get easier to the result.

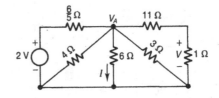

P.2.8 Calculate the indicated voltages V_1 and V_2 using any method or simplifications.

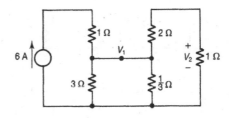

P.2.9 Use mesh formulation to obtain the indicated currents.

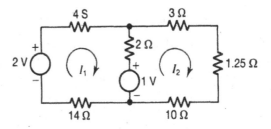

P.2.10 Use mesh formulation to calculate the currents.

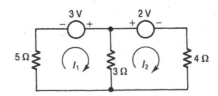

P.2.11 Calculate the voltage V using nodal formulation.

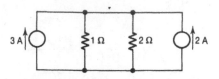

P.2.12 Calculate the nodal voltages V_1 and V_2.

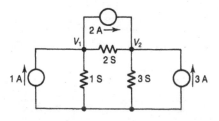

P.2.13 Use mesh formulation to get the currents. Also calculate the nodal voltages V_1 and V_2.

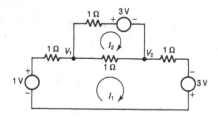

P.2.14 Calculate V_1 and V_2.

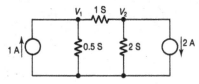

P.2.15 Calculate V_1 and V_2.

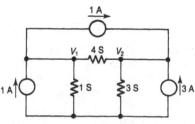

P.2.16 Calculate the nodal voltages V_1 and V_2 and V_3.

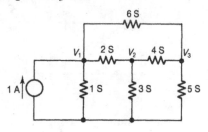

P.2.17 Obtain the voltages across the resistors.

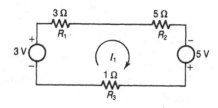

P.2.18 Obtain the voltages across the resistors.

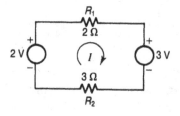

P.2.19 Calculate the currents and the voltage V.

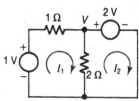

P.2.20 Calculate the currents and the voltage V.

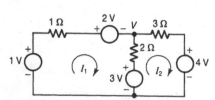

P.2.21 Calculate the currents and the voltages V_1 and V_2.

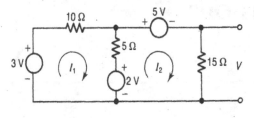

P.2.22 Calculate the currents, the voltages and the powers either delivered or consumed in all elements. Confirm hat the sum of powers delivered is equal to the sum of powers consumed

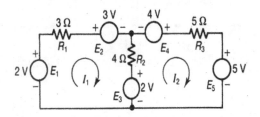

MATRIX METHODS

INTRODUCTION

Solutions of linear networks lead to systems of linear equations and the reader probably knows the elimination method for their solution: a variable is separated in one of the equations, substituted into the remaining ones and the process is repeated. Although conceptually simple, this method is confusing and leads to unpleasant fractions.

In order to learn and understand network theory, the student must do small solutions by hand, and we concluded that this can be bests done by matrix algebra. Very little of matrix algebra theory is actually needed in this book and can be easily taught in one hour. The student will have to learn how to write systems of linear equations in matrix form and how to calculate the product of a matrix by a vector. Cramer's rule, suitable for small systems, is used for the solutions. Its main advantage is that it does not lead to intermediate fractions, like the elimination method. Almost all problems in this book lead to two by two or three by three systems and the student has to learn how to evaluate determinants of these systems. Actually, in the first year, all the theoretical derivations can be skipped and the students can be taught only the mechanics of the Cramer's rule. We explain how to evaluate the determinant of a fourth order matrix by hand, but this can also be skipped at the beginning. Also skipped can be the inversion of matrices, given here for completness.

Some readers may already have learned the basics of matrix algebra. These readers will find the review presented in this book useful.

3.1. LINEAR EQUATIONS IN MATRIX FORM

In mathematics, the unknowns are usually denoted by x_1, x_2, etc. As an example, we may have a system

$$3x_1 - 2x_2 = 1$$
$$-2x_1 + 9x_2 - 4x_3 = 0 \qquad (3.1.1)$$
$$-4x_2 + 9x_3 = 3$$

Suppose that we apply elimination and separate x_1 from the first equation

$$x_1 = \frac{1}{3} + \frac{2}{3}x_2 \qquad (3.1.2)$$

We can now replace x_1 in the remaining two equations by the right hand side of Eq. (3.1.2) but here we have encountered something unpleasant: *the method invariably leads to fractions*. If we use a calculator to divide, we will almost certainly lose some digits in repunching the numbers and the final result will not be exact. In fact, it is usually better not to use the calculator and keep the fractions as long as possible, because they are *exact*. Division should be done only in the final result.

The above Eq. (3.2.1) can also be conveniently written in matrix form

$$\begin{bmatrix} 3 & -2 & 0 \\ -2 & 9 & -4 \\ 0 & -4 & 9 \end{bmatrix} \begin{bmatrix} x_1 \\ x_2 \\ x_3 \end{bmatrix} = \begin{bmatrix} 1 \\ 0 \\ 3 \end{bmatrix}$$

Standard mathematical notation uses bold letters for vectors and matrices and thus the equation could be expressed by

$$\mathbf{Ax} = \mathbf{b}$$

Here **A** is the system matrix (known), **x** is the unknown vector and **b** is the right hand side vector (known). We are looking for the best practical methods to solve such systems.

The above problem is still relatively simple because all entries of the matrix are numbers. In the solution of networks we often do not know the numerical values of the elements. All we know are the types of elements (resistors, capacitors etc.) and we wish to determine the values later so that the network behaves in a certain desired manner. This means that we must first get the solution in terms of the element *variables* and only later we try to select some particular values in order to achieve what is required.

This so called *symbolic* solution is a much more difficult problem. As an example, we give the simple resistive network in Fig. 3.1.1. In nodal formulation the equations are

$$(G_1 + G_2)V_1 + G_2 V_2 = 1$$
$$- G_2 V_1 + (G_2 + G_3 + G_4)V_2 - G_4 V_3 = 0$$
$$- G_4 V_2 + (G_4 + G_5)V_3 = 3$$

where V_1, V_2 and V_3 are the nodal voltages with respect to ground. If we insert the element values written in Fig. 3.1.1, we get the numerical example given above.

Consider this set of equations and the elimination method: separate V_1 from the first equation

$$V_1 = \frac{1}{G_1 + G_2} + \frac{G_2}{G_1 + G_2} V_2$$

and substitute for V_1 in the remaining equation: The fractions will stay with us through the whole solution process.

Such division can be avoided if we use matrix algebra and rewrite the equations in matrix form:

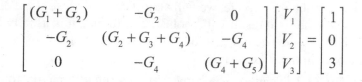

$$\begin{bmatrix} (G_1 + G_2) & -G_2 & 0 \\ -G_2 & (G_2 + G_3 + G_4) & -G_4 \\ 0 & -G_4 & (G_4 + G_5) \end{bmatrix} \begin{bmatrix} V_1 \\ V_2 \\ V_3 \end{bmatrix} = \begin{bmatrix} 1 \\ 0 \\ 3 \end{bmatrix}$$

Summarizing, we have presented two typical problems encountered in the solution of networks. The first and easier one has numbers in all entries of the matrix. Such case is usually solved by numerical methods, most likely by a computer. The second, theoretically much more important, keeps either all or some elements as variables (letters). Such problem can be solved only for relatively small networks and the solution is done by hand. Let us mention here that programs which can do this type of symbolic analysis are also available. In the following we explain appropriate methods for both these cases.

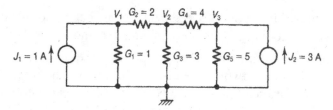

FIGURE 3.1.1   Network with three nodes.

Example 1.

Write the Eq. (2.1.14) in matrix form.

$$\begin{bmatrix} 6 & -2 & -3 \\ -2 & 5 & -2 \\ -3 & -2 & 5 \end{bmatrix} \begin{bmatrix} V_1 \\ V_2 \\ V_3 \end{bmatrix} = \begin{bmatrix} -2 \\ 0 \\ 5 \end{bmatrix}$$

Example 2.

Write the Eq. (2.1.13) in matrix form.

$$\begin{bmatrix} G_1 + G_2 + G_5 & -G_2 & -G_5 \\ -G_2 & G_2 + G_3 + G_4 & -G_4 \\ -G_5 & -G_4 & G_4 + G_5 \end{bmatrix} \begin{bmatrix} V_1 \\ V_2 \\ V_3 \end{bmatrix} = \begin{bmatrix} J_1 - J_3 \\ 0 \\ J_2 - J_3 \end{bmatrix}$$

3.2. DETERMINANTS

This section will summarize some rules for the evaluation of determinants. You will learn how to evaluate determinants but all theoretical details are left to specialized mathematical books.

Matrices may be rectangular or square but determinants are defined only for square matrices. Using standard notation with two subscripts, a 3 × 3 matrix **A** is written as

$$A = \begin{bmatrix} a_{11} & a_{12} & a_{13} \\ a_{21} & a_{22} & a_{23} \\ a_{31} & a_{32} & a_{33} \end{bmatrix} \tag{3.2.1}$$

The first subscript refers to rows and increases going down from the top, the second subscript refers to columns and increases from left to right.

The concept of a transpose matrix is also important. Columns of the transpose matrix, A^t, are the rows of the original matrix, **A**, and vice versa. Sometimes it is easier to think about the process of transposition as if the matrix was rotated along an axis formed by its main diagonal (a line connecting the top left and bottom right entry). The transpose of matrix in Eq. (3.2.1) is

$$A^t = \begin{bmatrix} a_{11} & a_{21} & a_{31} \\ a_{12} & a_{22} & a_{32} \\ a_{13} & a_{23} & a_{33} \end{bmatrix} \tag{3.2.2}$$

Many possible ways of calculating the value of a determinant exist; some are advantageous in certain situations, some are not. We will concentrate on those best suited *for solutions by hand*. First we will give methods for the evaluation of 2 × 2 and 3 × 3 determinants. In order to simplify things, let us use for each entry a different letter without a subscript. The determinant of a 2 × 2 matrix is

$$\det A = |A| = \begin{vmatrix} a & b \\ c & d \end{vmatrix} = ad - bc \tag{3.2.3}$$

The determinant of a 3 × 3 matrix is

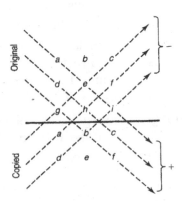

FIGURE 3.2.1   Evaluation of a 3 × 3 determinant.

$$|A| = \begin{vmatrix} a & b & c \\ d & c & f \\ g & h & i \end{vmatrix} = (aei + dhc + gbf) - (gec + dbi + ahf) \qquad (3.2.4)$$

The easiest way to remember the steps for the 3×3 determinant is to copy the first two lines under the determinant as shown in Fig. 3.2.1. Products along the lines pointing down are added, products along the lines pointing up are subtracted. Study carefully which entries are involved in the products because it is advantageous to memorize the method without copying the additional lines. We feel that this method is the *best* one for hand evaluation of a 3×3 determinant. Its main advantage is that *no fractions are involved*. Note that the method applies only to 3×3 determinants. It cannot be extended to determinants of other sizes. Hand evaluations of larger determinants will be discussed in section 5.

Having mastered the first two cases, we can now turn our attention to some useful rules:

1. If all entries of one row (column) are zero, then $|A| = 0$.
2. If two rows (columns) are equal, then $|A| = 0$.
3. If two rows (columns) are interchanged, then the determinant changes its sign.
4. If any row (column) is multiplied by a nonzero coefficient α, then the whole determinant is multiplied by this coefficient. In other words,

$$\alpha \begin{vmatrix} a_{11} & a_{12} \\ a_{12} & a_{22} \end{vmatrix} = \begin{vmatrix} \alpha a_{11} & a_{12} \\ \alpha a_{12} & a_{22} \end{vmatrix} = \begin{vmatrix} \alpha a_{11} & \alpha a_{12} \\ a_{12} & a_{23} \end{vmatrix} \qquad (3.2.5)$$

5. If we add to any row (column) some other row (column) multiplied by a nonzero coefficient α, the determinant does not change. For instance,

$$\begin{vmatrix} a_{11} & a_{12} \\ a_{12} & a_{22} \end{vmatrix} = \begin{vmatrix} a_{11} & a_{12} + \alpha a_{11} \\ a_{12} & a_{22} + \alpha a_{21} \end{vmatrix}$$

6. The determinant of the transpose matrix is equal to the determinant of the original matrix, $|A| = |A^t|$.

It is good to remember the above rules because sometimes we may obtain the determinant by simply inspecting the matrix or by performing some simple operation.

Let us now summarize our recommendations. For symbolic calculations (with letters instead of numbers) we recommend the above two rules for determinants of matrices 2×2 and 3×3. For larger matrices we recommend applying the Laplace expansion, covered in section 5. The main advantage of these steps lies in the fact that no fractions are involved.

It is more difficult to give unified recommendations for numerical solutions. Sometimes, inspection and application of the above rules may help. For 2×2 and 3×3 cases, it may still be easiest to use the above rules. Larger numerical problems are usually solved by numerical methods and by a computer. Modern calculators can also handle smaller numerical systems of equations.

Example 1.

Find the determinant of the matrix

$$A = \begin{bmatrix} 3 & -2 \\ -2 & 5 \end{bmatrix}$$

Using Eq. (3.2.3) we find $|A| = 15 - 4 = 11$.

Example 2.

Find the determinant of the matrix

$$A = \begin{bmatrix} G_1 + G_2 & -G_2 \\ -G_2 & G_2 + G_3 \end{bmatrix}$$

Using Eq. (3.2.3) we get

$$|A| = (G_1 + G_2)(G_2 + G_3) - G_2^2 = G_1 G_2 + G_1 G_3 + G_2 G_3$$

Example 3.

Find the determinant of the matrix

$$A = \begin{bmatrix} 6 & -2 & -3 \\ -2 & 5 & -2 \\ -3 & -2 & 5 \end{bmatrix}$$

Using Eq. (3.2.4) or Fig. 3.2.1

$$|A| = [(6)(5)(5) + (-2)(-2)(-3) + (-3)(-2)(-2)]$$

$$- [(-3)5(-3) + (-2)(-2)5 + 6(-2)(-2)] = 126 - 89 = 37$$

3.3. MATRIX INVERSION

Although we will not need matrix inversion to solve network problems, it is important to know some of the definitions and how to obtain the inverse theoretically.

First of all, matrix inversion is defined for square matrices only. Let us have an invertible square matrix A; then its inverse, A^{-1}, is a matrix satisfying

$$AA^{-1} = A^{-1}A = 1 \tag{3.3.1}$$

Here 1 is the *unit matrix*, having ones along the main diagonal and zeros everywhere else, like in the 3×3 matrix below:

$$1 = \begin{bmatrix} 1 & 0 & 0 \\ 0 & 1 & 0 \\ 0 & 0 & 1 \end{bmatrix}$$

In (3.3.1), the sequence of the matrices is not important; *in all other situations it is.*

Inversion of a matrix requires evaluation of its determinant and calculation of a matrix which is called adjoint, *adj* **A**. Then

$$A^{-1} = \frac{adj\, A}{|A|} \tag{3.3.2}$$

We already know how to calculate the determinant; all we need are the steps how to obtain the adjoint matrix. We give these steps as a set of rules by considering a 3 × 3 example. Let the matrix be

$$A = \begin{bmatrix} a_{11} & a_{12} & a_{13} \\ a_{21} & a_{22} & a_{23} \\ a_{31} & a_{32} & a_{33} \end{bmatrix} \tag{3.3.3}$$

Step 1.

Write the transpose of this matrix

$$A' = \begin{bmatrix} a_{11} & a_{21} & a_{31} \\ a_{12} & a_{22} & a_{32} \\ a_{13} & a_{23} & a_{33} \end{bmatrix} \tag{3.3.4}$$

Step 2.

Cross out one row and one column. The lines intersect at a given entry of a matrix; to be specific, let us cross out the second row and third column of **A'**. Copy the remaining entries of the matrix into a matrix which has the size one less than the original matrix and calculate the determinant. In our case we must evaluate

$$\begin{vmatrix} a_{11} & a_{21} \\ a_{13} & a_{23} \end{vmatrix} = a_{11}a_{23} - a_{13}a_{21}$$

Step 3.

Attach to the result a sign which is equal to $(-1)^{i+j}$ where i is the subscript of the row and j is the subscript of the column which were crossed out. In our case, the determinant will be multiplied by $(-1)^{2+3} = -1$ and the result is placed as one entry of the inverse matrix into the very same position we crossed out on the transpose matrix. In our example, this is the entry in second row and third column.

Step 4.

Repeat steps 2 and 3 for all other entries of the matrix.

We will take one example of inversion using

$$A = \begin{bmatrix} 1 & 1 & 0 \\ 1 & 0 & 1 \\ 1 & 2 & 2 \end{bmatrix}$$

In step 1 we transpose the matrix

$$A' = \begin{bmatrix} 1 & 1 & 1 \\ 1 & 0 & 2 \\ 0 & 1 & 2 \end{bmatrix}$$

In step 2 we cross out the first row and first column of A'. This asks for the evaluation of the determinant

$$\begin{vmatrix} 0 & 2 \\ 1 & 2 \end{vmatrix} = -2$$

In step 3 we find that this value is multiplied by +1. The result will be the (1,1) entry of the adjoint matrix. We now repeat steps 2 and 3 for the other positions. Crossing out row one and column two leads to the evaluation of the determinant

$$\begin{vmatrix} 1 & 2 \\ 0 & 2 \end{vmatrix} = 2$$

and this must be multiplied by −1. Similarly, for the (1,3) position we evaluate the determinant

$$\begin{vmatrix} 1 & 0 \\ 0 & 1 \end{vmatrix} = 1$$

and this result is multiplied by +1. Continuing similarly for the other positions we get the adjoint matrix

$$adj\, A = \begin{bmatrix} -2 & -2 & 1 \\ -1 & 2 & -1 \\ 2 & -1 & -1 \end{bmatrix}$$

We must still calculate the determinant of the original (or transpose) matrix; using the method of the previous section we find its value to be −3. Each entry of the adjoint matrix is divided by this number. The resulting inverse matrix is

$$A^{-1} = \begin{bmatrix} \dfrac{2}{3} & \dfrac{2}{3} & \dfrac{-1}{3} \\ \dfrac{1}{3} & \dfrac{-2}{3} & \dfrac{1}{3} \\ \dfrac{-2}{3} & \dfrac{1}{3} & \dfrac{1}{3} \end{bmatrix}$$

Check the correctness of the result by multiplying A and A^{-1}; you must get a unit matrix.

As can be seen, inversion is a rather laborious process and is rarely needed. For instance, in theoretical mathematics, the solution of the system $Ax = b$ is often described by simply writing $x = A^{-1}b$. Although this is formally correct, inversion in this case is *never* used. In smaller problems we may use Cramer's rule, explained in the next section.

In larger numerical problems we may use the well known Gaussian elimination, or other available numerical methods.

There is still one theoretical concept we must introduce. If the determinant of the matrix **A** happens to be zero, then such a matrix *does not have an inverse* and we say that the matrix is singular. The situation is extremely rare in cases when the matrix describes a practical network. Should it happen to you, then better check twice that you have not made a mistake somewhere.

If the above steps are applied to a 2 × 2 matrix,

$$A = \begin{bmatrix} a & b \\ c & d \end{bmatrix} \tag{3.3.5}$$

then the inverse is

$$A^{-1} = \frac{1}{ad - bc} \begin{bmatrix} d & -b \\ -c & a \end{bmatrix} \tag{3.3.6}$$

The entries on the main diagonal are interchanged and minus sign is attached to the other two entries. It is useful to memorize this formula. It is the only practical case when it is advantageous to use matrix inversion to solve a system of linear equations.

Example 1.

Solve the system

$$2x_1 + 3x_2 = 2$$
$$-1x_1 + 4x_3 = 3$$

by writing it as a matrix equation and by inverting the matrix.

The system is

$$\begin{bmatrix} 2 & 3 \\ -1 & 4 \end{bmatrix} \begin{bmatrix} x_1 \\ x_2 \end{bmatrix} = \begin{bmatrix} 2 \\ 3 \end{bmatrix}$$

Using matrix inversion

$$\begin{bmatrix} x_1 \\ x_2 \end{bmatrix} = \frac{1}{8+3} \begin{bmatrix} 4 & -3 \\ 1 & 2 \end{bmatrix} \begin{bmatrix} 2 \\ 3 \end{bmatrix} = \frac{1}{11} \begin{bmatrix} 8-9 \\ 2+6 \end{bmatrix} = \begin{bmatrix} -1/11 \\ 8/11 \end{bmatrix}$$

3.4. CRAMER'S RULE

Cramer's rule is probably the best method for hand solutions of equations. It is especially advantageous if only one variable is needed or if we have a matrix with symbols instead of numbers. It should be stressed that Cramer's rule is not useful for programming.

We will derive Cramer's rule by using a 3 × 3 example. Consider the linear equations

$$a_{11}x_1 + a_{12}x_2 + a_{13}x_3 = b_1$$
$$a_{21}x_1 + a_{22}x_2 + a_{23}x_3 = b_2 \qquad (3.4.1)$$
$$a_{31}x_1 + a_{32}x_2 + a_{33}x_3 = b_3$$

or, in matrix form

$$\mathbf{Ax = b} \qquad (3.4.2)$$

The determinant of the system matrix is

$$|\mathbf{A}| = \begin{vmatrix} a_{11} & a_{21} & a_{31} \\ a_{21} & a_{22} & a_{32} \\ a_{31} & a_{32} & a_{33} \end{vmatrix} \qquad (3.4.3)$$

Using rule 4. from section 3.2, we know that multiplication of any column (row) by α means that the determinant is multiplied by a. Suppose that we multiply the third column by x_3. The result will be

$$x_3|\mathbf{A}| = \begin{vmatrix} a_{11} & a_{21} & a_{31}x_3 \\ a_{21} & a_{22} & a_{23}x_3 \\ a_{31} & a_{32} & a_{33}x_3 \end{vmatrix}$$

According to rule 5. of section 3.2, we can multiply any column by some α, add it to any other column and the determinant will not change. We choose to multiply column 1 by x_1 and column 2 by x_2 and add them to the third column:

$$x_3|\mathbf{A}| = \begin{vmatrix} a_{11} & a_{21} & a_{11}x_1 + a_{12}x_2 + a_{13}x_3 \\ a_{21} & a_{22} & a_{21}x_1 + a_{22}x_2 + a_{23}x_3 \\ a_{31} & a_{32} & a_{31}x_1 + a_{32}x_2 + a_{33}x_3 \end{vmatrix}$$

Compare the third column with the left side of (3.4.1): they are identical. We can thus replace the column by the right side of (3.4.1):

$$x_3|\mathbf{A}| = \begin{vmatrix} a_{11} & a_{21} & b_1 \\ a_{21} & a_{23} & b_2 \\ a_{31} & a_{32} & b_3 \end{vmatrix}$$

However, this is the determinant of a matrix similar to **A**, but with the third column replaced by the right hand side. Let us denote this matrix by $\mathbf{A}_3$ and obtain

$$x_3|\mathbf{A}| = |\mathbf{A}_3|$$

or

$$x_3 = \frac{|\mathbf{A}_3|}{|\mathbf{A}|}$$

In the general case of a $n \times n$ system, we find the solution of the kth variable x_k by

(a) calculating the determinant of the system matrix, $|\mathbf{A}|$, and
(b) calculating the determinant of a similar matrix $|\mathbf{A}_k|$ in which the kth column was replaced by the right hand side vector. Then

$$x_k = \frac{|\mathbf{A}_k|}{|\mathbf{A}|} \tag{3.4.4}$$

Consider

$$x_1 + 2x_2 + x_3 = 1$$

$$-x_1 + 3x_2 + 2x_3 = 0$$

$$2x_1 - 2x_2 + x_3 = 2$$

The determinant of the system, D, is

$$D = |\mathbf{A}| = \begin{vmatrix} 1 & 2 & 1 \\ -1 & 3 & 2 \\ 2 & -2 & 1 \end{vmatrix} = 13$$

Replacing the first column by the right hand side we get the first numerator of (3.4.4)

$$N_1 = |\mathbf{A}_1| = \begin{vmatrix} 1 & 2 & 1 \\ 0 & 3 & 2 \\ 2 & -2 & 1 \end{vmatrix} = 9$$

Replacing the second column by the right hand side we get

$$N_2 = |\mathbf{A}_2| = \begin{vmatrix} 1 & 1 & 1 \\ -1 & 0 & 2 \\ 2 & 2 & 1 \end{vmatrix} = -1$$

Replacing the third column by the right hand side

$$N_3 = |\mathbf{A}_3| = \begin{vmatrix} 1 & 2 & 1 \\ -1 & 3 & 0 \\ 2 & -2 & 2 \end{vmatrix} = 6$$

The solutions are

$$x_1 = \frac{N_1}{D} = \frac{|\mathbf{A}_1|}{|\mathbf{A}|} = \frac{9}{13}$$

$$x_2 = \frac{N_2}{D} = \frac{|\mathbf{A}_2|}{|\mathbf{A}|} = \frac{1}{13}$$

$$x_3 = \frac{N_3}{D} = \frac{|\mathbf{A}_3|}{|\mathbf{A}|} = \frac{6}{13}$$

Example 1.

Solve the system

$$\begin{bmatrix} 3 & 2 \\ 2 & 4 \end{bmatrix} \begin{bmatrix} x_1 \\ x_2 \end{bmatrix} = \begin{bmatrix} 2 \\ -2 \end{bmatrix}$$

Using Cramer's rule.

The determinant of the system is $D = 12 - 4 = 8$. The first numerator is

$$N_1 = \begin{vmatrix} 2 & 2 \\ -2 & 4 \end{vmatrix} = 8 + 4 = 12$$

The second numerator is

$$N_2 = \begin{vmatrix} 3 & 2 \\ 2 & -2 \end{vmatrix} = -6 - 4 = -10$$

The solution is $x_1 = \dfrac{N_1}{D} = \dfrac{12}{8}$ and $x_2 = \dfrac{N_2}{D} = -\dfrac{10}{8}$.

Example 2.

Using Cramer's rule solve the system

$$2x_1 + x_3 = 1$$

$$3x_1 + 2x_2 - 3x_3 = 2$$

$$x_1 - x_2 = 0$$

In matrix form these equations are

$$\begin{bmatrix} 2 & 0 & 1 \\ 3 & 2 & -3 \\ 1 & -1 & 0 \end{bmatrix} \begin{bmatrix} x_1 \\ x_2 \\ x_3 \end{bmatrix} = \begin{bmatrix} 1 \\ 2 \\ 0 \end{bmatrix}$$

The denominator is

$$D = \begin{vmatrix} 2 & 0 & 1 \\ 3 & 2 & -3 \\ 1 & -1 & 0 \end{vmatrix} = -3 - 2 - 6 = -11$$

The first numerator is

$$N_1 = \begin{vmatrix} 1 & 0 & 1 \\ 2 & 2 & -3 \\ 0 & -1 & 0 \end{vmatrix} = -2 - 3 = -5$$

The second numerator is

$$N_2 = \begin{bmatrix} 2 & 1 & 1 \\ 3 & 2 & -3 \\ 1 & 0 & 0 \end{bmatrix} = -3 - 2 = -5$$

The third numerator is

$$N_3 = \begin{bmatrix} 2 & 0 & 1 \\ 3 & 2 & 2 \\ 1 & -1 & 0 \end{bmatrix} = -3 - 2 + 4 = -1$$

and the solutions are $x_1 = \dfrac{N_1}{D} = \dfrac{5}{11}$; $x_2 = \dfrac{N_2}{D} = \dfrac{5}{11}$; $x_3 = \dfrac{N_3}{D} = \dfrac{1}{11}$.

The above steps are valid for determinants of three equations. For fourth order systems we must use expansion, explained in the next section.

3.5. EXPANSION OF DETERMINANTS

Hand evaluation of determinants 4×4 is best done by the Laplace expansion. It is especially convenient if we work with variables (letters) because no division is involved. Cramer's rule is not suitable for larger systems; a computer and other methods would be used in such cases.

Let us start with a numerical example

$$|A| = \begin{vmatrix} 2 & 3 & -2 & 4 \\ 3 & -2 & 1 & 2 \\ 3 & 2 & 3 & 4 \\ -2 & 4 & 0 & 5 \end{vmatrix} \tag{3.5.1}$$

Expansion of the determinant can be done along *any row or column* and proper signs have to be attached to the results, as explained below. In the example we will use expansion along the first column. Apply the following steps. Cross out the first row and column; take the number at the intersection of these two lines, in our case 2, and copy the remaining determinant less the crossed out lines:

$$2 \begin{vmatrix} -2 & 1 & 2 \\ 2 & 3 & 4 \\ 4 & 0 & 5 \end{vmatrix} = 2 \cdot (-48)$$

Return to the original determinant, go down to the next entry (first column, second row) and cross out the first column and second row. Write the number at the intersection and copy the remaining eterminant, less the crossed out lines:

$$-3 \begin{vmatrix} 3 & -2 & 4 \\ 2 & 3 & 4 \\ 4 & 0 & 5 \end{vmatrix} = -3 \cdot (-15)$$

We will return specially to the sign in front of this expression. The third expansion about the first column crosses out the first column and third row:

$$3 \begin{vmatrix} 3 & -2 & 4 \\ -2 & 1 & 2 \\ 4 & 0 & 5 \end{vmatrix} = 3 \cdot (-37)$$

and the last case crosses out the first column and fourth row:

$$-(-2) \begin{vmatrix} 3 & -2 & 4 \\ -2 & 1 & 2 \\ 2 & 3 & 4 \end{vmatrix} = 2 \cdot (-62)$$

Again, we will speak about the sign in front of this in a few moments. The determinant is

$$|A| = 2 \cdot (-48) - 3 \cdot (-15) + 3 \cdot (-37) + 2 \cdot (-62) = -286$$

Now about the signs in front of the expressions. Theoretically, each is multiplied by $(-1)^{i+j}$ where i and j are the subscripts of the term lying on the intersection of the lines we used for crossing out. Going down in the first column, the sequence will alternate signs, $+, -, +, -$, and this was used above.

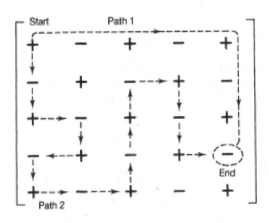

FIGURE 3.5.1 Determining sign for a subdeterminant.

The following mnemonic rule gives the same result and is much easier to remember. Start in the upper left corner with + sign. Walk to the entry under consideration in any sequence of horizontal and vertical moves, in each step always changing the sign. When we reach the desired entry, the sign is correct. This is shown schematically in Figure 3.5.1 where one way is straightforward and one quite complicated. Nevertheless, the sign is always the same one.

It is advisable to look for any row or column which has the largest number of zeros, and expand the determinant along this row or column. In such a case some partial determinants would be multiplied by zero and we would not calculate them at all. Looking at (3.5.1) we see that expansion about the third column would have saved us one determinant. Using the above rule for determining the signs, the expansion will be

$$|A| = -2\begin{vmatrix} 3 & -2 & 2 \\ 3 & 2 & 4 \\ -2 & 4 & 5 \end{vmatrix} - 1\begin{vmatrix} 2 & 3 & 4 \\ 3 & 2 & 4 \\ -2 & 4 & 5 \end{vmatrix} + 3\begin{vmatrix} 2 & 3 & 4 \\ 3 & -2 & 2 \\ -2 & 4 & 5 \end{vmatrix}$$

$$= -2 \cdot (60) - 1 \cdot (-17) + 3 \cdot (-61) = -286$$

The reader should also note that addition or subtraction of columns may considerably simplify the expansion. Using the above example and adding the first row to the last row will create a zero in the (4,1) position. Similarly, subtracting the second row from the third row will create a zero at the (3,1) position and the determinant will be

$$|A| = \begin{vmatrix} 2 & 3 & -2 & 4 \\ 3 & -2 & 1 & 2 \\ 0 & 4 & 2 & 2 \\ 0 & 7 & -2 & 9 \end{vmatrix} = 2\begin{vmatrix} -2 & 1 & 2 \\ 4 & 2 & 2 \\ 7 & -2 & 9 \end{vmatrix} - 3\begin{vmatrix} 3 & -2 & 4 \\ 4 & 2 & 2 \\ 7 & -2 & 9 \end{vmatrix}$$

$$= 2 \cdot (-110) - 3 \cdot (22) = -286$$

Instead of doing the expansion about the column we can achieve the same by doing the expansion about any row.

Summarizing this section we state that Laplace expansion of determinants is best suited for evaluations of determinants having letters instead of numbers, a frequent case in network solutions. Numerical solutions can be handled by many methods, but if done by hand, then Laplace expansion and Cramer's rule still retain one big advantage: there are no intermediate fractions, only products. Only he last step of Cramer's rule requires a division and this is done without any loss of accuracy.

Example 1.

Evaluate the determinant of the matrix in which not all entries are numbers:

$$\begin{bmatrix} 1 & 2 & 0 & 0 \\ 3 & 0 & A & 1 \\ 1 & B & 0 & 2 \\ 0 & 1 & 0 & 3 \end{bmatrix}$$

Expanding about the third column this reduces to the evaluation of

$$-A\begin{vmatrix} 1 & 2 & 0 \\ 1 & B & 2 \\ 0 & 1 & 3 \end{vmatrix} = 8A - 3AB$$

Less advantageous is the expansion about the first row which leads to

$$1\begin{vmatrix} 0 & A & 1 \\ B & 0 & 2 \\ 1 & 0 & 3 \end{vmatrix} - 2\begin{vmatrix} 3 & A & 1 \\ 1 & 0 & 2 \\ 0 & 0 & 3 \end{vmatrix} = 1[2A - 3AB] - 2[-3A] = 8A - 3AB$$

Another possibility is to expand about the fourth row. This leads to the evaluation of

$$1\begin{vmatrix} 1 & 0 & 0 \\ 3 & A & 1 \\ 1 & 0 & 2 \end{vmatrix} + 3\begin{vmatrix} 1 & 2 & 0 \\ 2 & 0 & A \\ 3 & B & 0 \end{vmatrix} = 2A + 3(2A - AB) = 8A - 3AB$$

3.6. SOLUTIONS OF RESISTIVE NETWORKS

Resistive linear networks are the simplest practical problems and we will go through several examples.

Example 1.

Consider the network in Fig. 3.6.1. Using mesh formulation and matrix algebra find the two mesh currents. Also find the equivalent resistance of the combination by calculating the ratio $\dfrac{E}{I_1}$. The two equations are:

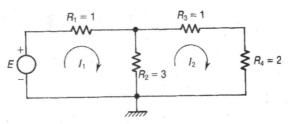

FIGURE 3.6.1   Network with two meshes.

$$4I_1 - 3I_2 = E$$
$$-3I_1 + 6I_2 = 0$$

In matrix form:

$$\begin{vmatrix} 4 & -3 \\ -3 & 6 \end{vmatrix} = \begin{bmatrix} I_1 \\ I_2 \end{bmatrix} = \begin{bmatrix} E \\ 0 \end{bmatrix}$$

The determinant is

$$D = 24 - 9 = 15$$

Using Cramer's rule we find the numerator for I_1

$$N_1 = \begin{vmatrix} E & -3 \\ 0 & 6 \end{vmatrix} = 6E$$

and for I_2

$$N_2 = \begin{vmatrix} 4 & E \\ -3 & 0 \end{vmatrix} = 3E$$

The currents are $I_1 = \dfrac{6}{15}E$ and $I_2 = \dfrac{3}{15}E$. The *current* I_1 flows from the source into the network. By taking the ratio $\dfrac{I_1}{E} = \dfrac{6}{15}$ we will get the conductance of the combination. The inverse is the resistance. Check this result by finding the overall conductance (resistance) of the network as a parallel/series combination of resistors (as explained in Chapter 1).

Example 2.

The network in Fig. 3.6.2 is suitable for nodal equations. Find the nodal voltages and the overall resistance of the network, as viewed from the source.

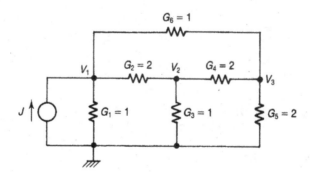

FIGURE 3.6.2   Network with three nodes.

The equations are:

$$4V_1 - 2V_2 - 1V_3 = J$$

$$-2V_1 + 5V_2 - 2V_3 = 0$$

$$-1V_1 - 2V_2 + 5V_3 = 0$$

In matrix form

$$\begin{bmatrix} 4 & -2 & -1 \\ -2 & 5 & -2 \\ -1 & -2 & 5 \end{bmatrix} \begin{bmatrix} V_1 \\ V_2 \\ V_3 \end{bmatrix} = \begin{bmatrix} J \\ 0 \\ 0 \end{bmatrix}$$

The determinant of the system is $D = 51$. Using Cramer's rule, we evaluate the determinants for the numerators:

$$N_1 = \begin{vmatrix} J & -2 & -1 \\ 0 & 5 & -2 \\ 0 & -2 & 5 \end{vmatrix} = 21J$$

$$N_2 = \begin{vmatrix} 4 & J & -1 \\ -2 & 0 & -2 \\ -1 & 0 & 5 \end{vmatrix} = 12J$$

$$N_3 = \begin{vmatrix} 4 & -2 & J \\ -2 & 5 & 0 \\ -1 & -2 & 0 \end{vmatrix} = 9J$$

The results are $V_1 = \dfrac{21}{51}J$, $V_2 = \dfrac{12}{51}J$ and $V_3 = \dfrac{9}{51}J$. The overall equivalent resistance of the network, as viewed from the source, is equal to $\dfrac{V_1}{J} = \dfrac{21}{51}$.

Example 3.

Consider the network in Fig. 3.6.3 where all conductances are $G_i = 1$. Find the voltage V_1 and the equivalent resistance of the network, seen from the terminals of the source.

The nodal equations are:

$$3V_1 - 1V_2 - 1V_3 + 0V_4 = J$$

$$-1V_1 + 4V_2 - 1V_3 - 1V_4 = 0$$

$$-1V_1 - 1V_2 + 4V_3 - 1V_4 = 0$$

$$0V_1 - 1V_2 - 1V_3 + 3V_3 = 0$$

or in matrix form

$$\begin{bmatrix} 3 & -1 & -1 & 0 \\ -1 & 4 & -1 & -1 \\ -1 & -1 & 4 & -1 \\ 0 & -1 & -1 & 3 \end{bmatrix} \begin{bmatrix} V_1 \\ V_2 \\ V_3 \\ V_4 \end{bmatrix} = \begin{bmatrix} J \\ 0 \\ 0 \\ 0 \end{bmatrix}$$

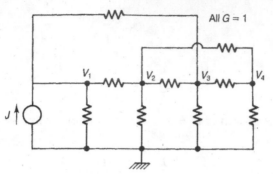

FIGURE 3.6.3   Network with four nodes

In this example we find only the voltage V_1. The determinant of the system matrix is obtained by expansion about the first column:

$$D = 3\begin{vmatrix} 4 & -1 & -1 \\ -1 & 4 & -1 \\ -1 & -1 & 3 \end{vmatrix} + 1\begin{vmatrix} -1 & -1 & 0 \\ -1 & 4 & -1 \\ -1 & -1 & 3 \end{vmatrix} - 1\begin{vmatrix} -1 & -1 & 0 \\ 4 & -1 & -1 \\ -1 & -1 & 3 \end{vmatrix}$$

$$= 105 - 15 - 15 = 75$$

The numerator is found by replacing the first column of the system matrix by the right hand side and by expanding the determinant about the first column. Since only one entry in this column is nonzero, this expansion is much simpler:

$$N_1 = \begin{vmatrix} J & -1 & -1 & 0 \\ 0 & 4 & -1 & -1 \\ 0 & -1 & 4 & -1 \\ 0 & -1 & -1 & 3 \end{vmatrix} = J\begin{vmatrix} 4 & -1 & -1 \\ -1 & 4 & -1 \\ -1 & -1 & 3 \end{vmatrix} = 35J$$

The resulting voltage is $V_1 = \dfrac{35}{75}J$ and the input resistance of the network is $\dfrac{V_1}{J} = R_{in} = \dfrac{35}{75}$.

Example 4.

The network in Fig. 3.6.4 is suitable for nodal equations. Find all nodal voltages and the equivalent resistance of the network. All conductances are equal to $G_i = 1$. Write directly the matrix equation.

$$\begin{bmatrix} 3 & -1 & -1 \\ -1 & 3 & -1 \\ -1 & -1 & 3 \end{bmatrix}\begin{bmatrix} V_1 \\ V_2 \\ V_3 \end{bmatrix} = \begin{bmatrix} 1 \\ 0 \\ 0 \end{bmatrix}$$

Determinant of the matrix is $D = 16$, $N_1 = 8$, $N_2 = 4$, $N_3 = 4$. The voltages are $V_1 = \dfrac{1}{2}, V_2 = \dfrac{1}{4}$ and $V_3 = \dfrac{1}{4}$. The input resistance is $\dfrac{V_1}{J} = \dfrac{V_1}{1} = R_{in} = \dfrac{1}{2}$.

Example 5.

The network in Fig. 3.6.5 has two voltage sources and is suitable for mesh equations. Write the mesh equations, the matrix equation, and calculate the voltage V_{out} indicated in the figure.

The equations are

$$5I_1 - 4I_2 - I_3 = 2$$

$$-4I_1 + 8I_2 = -3$$

$$-1I_1 + 3I_3 = 3$$

In matrix form:

$$\begin{bmatrix} 5 & -4 & -1 \\ -4 & 8 & 0 \\ -1 & 0 & 3 \end{bmatrix} \begin{bmatrix} I_1 \\ I_2 \\ I_3 \end{bmatrix} = \begin{bmatrix} 2 \\ -3 \\ 3 \end{bmatrix}$$

We need the current I_2 to be able to calculate the voltage V_{out}. The system matrix determinant is $D = 64$. For the second mesh current we must find

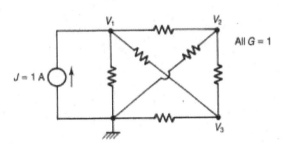

FIGURE 3.6.4   Network with three nodes.

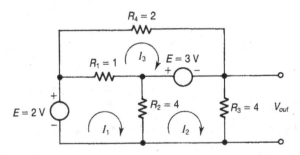

FIGURE 3.6.5 Network with three meshers.

$$N_2 = \begin{vmatrix} 5 & 2 & -1 \\ -4 & -3 & 0 \\ -1 & 3 & 3 \end{vmatrix} = -6$$

The second mesh current is $I_2 = \dfrac{N_2}{D} = -\dfrac{6}{64}$. This current flows through the resistor $R_3 = 4\Omega$ and thus the output voltage is

$$V_{out} R_3 I_2 = -4 \cdot \frac{6}{64} = -\frac{3}{8} V$$

Example 6.

Use mesh equations for the network in Fig. 3.6.6, find the current I_1 and the equivalent resistance seen from the source. All resistors have unit values, $R_i = 1\Omega$.

The system matrix is:

$$\begin{bmatrix} 2 & -1 & -1 & 0 \\ -1 & 3 & -1 & -1 \\ -1 & -1 & 3 & -1 \\ 0 & -1 & -1 & 3 \end{bmatrix} \begin{bmatrix} I_1 \\ I_2 \\ I_3 \\ I_4 \end{bmatrix} = \begin{bmatrix} 1 \\ 0 \\ 0 \\ 0 \end{bmatrix}$$

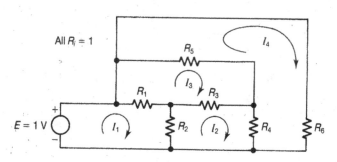

FIGURE 3.6.6   Network with four meshes

The determinant, found by Laplace expansion about the first column, is equal to $D = 8$. Replacing the first column by the right hand side we find $N_1 = 16$ and thus $I_1 = \dfrac{16}{8} = 2A$

The input resistance $R_{in} = \dfrac{E}{I_1} = 0.5\Omega$.

PROBLEMS CHAPTER 3

P.3.1 Find the solutions of the following problems by using the Cramer's rule. For systems larger than 3×3 use expansion of determinants described in Section 3.5.

$$x_1 + 2x_2 + x_3 = 1$$
$$2x_1 + 6x_2 + 2x_3 = 0 \qquad \text{(a)}$$
$$3x_1 + 7x_2 + 6x_3 = -2$$

$$2x_1 + x_2 + 5x_3 + x_4 = 5$$
$$x_1 + x_2 - 3x_3 - 4x_4 = -1 \qquad \text{(b)}$$
$$3x_1 + 6x_2 - 2x_3 + x_4 = 8$$
$$2x_1 + 2x_2 + 2x_3 - 3x_4 = 2$$

$$2x_1 + 0x_2 + 3x_3 + 0x_4 = 1$$
$$0x_1 + 3x_2 + 0x_3 + 1x_4 = 2 \qquad \text{(c)}$$
$$6x_1 + 6x_2 + 10x_3 + 4x_4 = 0$$
$$2x_1 + 0x_2 + 6x_3 + 8x_4 = 3$$

$$2x_1 + 3x_2 - 1x_3 = 1$$
$$4x_1 - 2x_2 + 4x_3 = 0 \qquad \text{(d)}$$
$$0x_1 + x_2 - x_3 = 0$$

$$3x_1 - 1x_2 - 1x_3 + 0x_4 = 5$$
$$-1x_1 + 3x_2 - 1x_3 - 1x_4 = -10 \qquad \text{(e)}$$
$$-1x_1 - 1x_2 + 3x_3 - 1x_4 = 10$$
$$0x_1 - 1x_2 - 1x_3 + 3x_4 = -5$$

$$1x_1 + 0x_2 + 1x_3 + 0x_4 = 0$$
$$2x_1 + 2x_2 + 2x_3 + 4x_4 = 0 \qquad \text{(f)}$$
$$0x_1 + 1x_2 + 3x_3 + 11x_4 = 0$$
$$1x_1 + 0x_2 - 1x_3 - 5x_4 = 3$$

P.3.2 Find the voltage V across the last resistor.

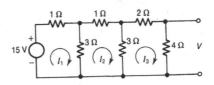

P.3.3 Find the indicated current I.

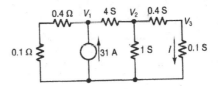

P.3.4 Use nodal formulation to obtain the three nodal voltages.

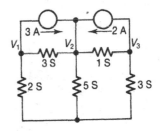

P.3.5 Use mesh equation, calculate the currents and the voltages V_1 and V_2.

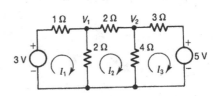

P.3.6 Use mesh equations to obtain the three currents.

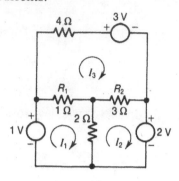

P.3.7 Use nodal formulation to obtain the three nodal voltages.

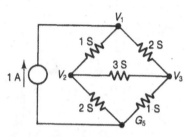

P.3.8 Use nodal formulation to obtain the three nodal voltages.

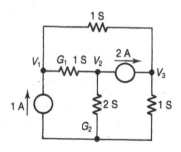

P.3.9 Use nodal formulation to obtain the three nodal voltages.

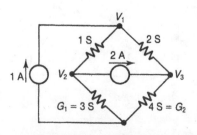

P.3.10 Find the three currents.

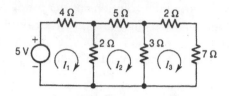

P.3.11 Find the three currents.

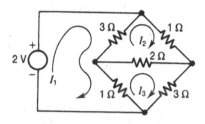

P.3.12 Set up the nodal equations and write them in matrix form. Use determinant expansion to obtain the nodal voltages.

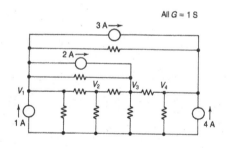

P.3.13 Do the same as in P.3.12. All G_i = 1S.

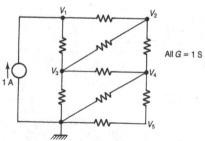

P.3.14 Use mesh equations to set up the matrix equation and solve.

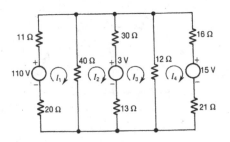

P.3.15 Use nodal equations to solve. Find the nodal voltages and also calculate the indicated current I.

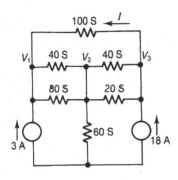

P.3.16 Connection of the current source on the left is not typical but since the same current flows from the source as through the resistor on the left, one can first replace the first resistor by a short circuit. This will not influence the distribution of voltages and currents in the remaining network. Find the nodal voltages this way first. Afterwards find the voltage across the current source on the right.

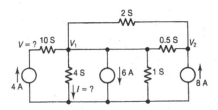

P.3.17 Calculate the the current I_2 and the voltage across the resistor R_4 by keeping this resistor as a variable.

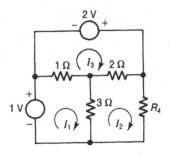

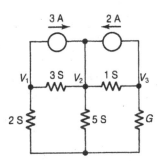

P.3.18 This network was already solved in P.3.4 but now keep the conductance on the right as a variable, G. Calculate the voltage V_3. Afterwards substitute $G = 3S$; you should get the same result as in P.3.4.

P.3.19 In problem P.3.6 keep the resistors R_1 and R_2 as variables and obtain the solution. Afterwards substitute the same values; you should obtain the same result.

P.3.20 In problem P.3.7 keep G_5 as variable and get the result. Afterwards substitute the same value as there and check that you have the same result.

P.3.21 In problem P.3.8 keep G_1 and G_2 as variables and find V_3.

P.3.22   In problem P.3.9 keep G_1 and G_2 as variables and find the voltage V_1.

P.3.23   Keep the conductance G as variable and find the voltage V_2. Afterwards get such value of G that $V_2 = 2V$.

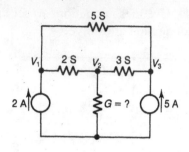

4 DEPENDENT SOURCES

INTRODUCTION

Since we now have the two fundamental equation formulation methods and also the means to solve them, we can turn our attention to the dependent sources. They are extremely important elements because the properties of semiconductors are modeled by them.

An independent voltage source, as we defined it in Chapter 1, delivers a voltage of E volts and this voltage does not depend on anything inside the network. The same was true about the independent current source which delivers a current of J amperes, independent of any external influences.

If we make the value of the voltage (or current) delivered by the source dependent on some other voltage (or current) in the network, we will have a dependent source. There are altogether four possibilities:

1. Voltage controlled current source, VC.
2. Current controlled voltage source, CV.
3. Current controlled current source, CC.
4. Voltage controlled voltage source, VV.

The elements were already introduced in Section 1.6. In the abbreviations the first letter refers to the controlling variable and the second to the type of source.

We will now study all these controlled sources in more detail and show how we use them in the nodal or mesh formulations.

4.1. VOLTAGE CONTROLLED CURRENT SOURCE

The voltage controlled current source, VC, is an element which can easily be incorporated into the nodal equations and we consider it first. The element is shown in Fig. 4.1.1. The terminals on the left are called the *input* or *controlling* terminals, those on the right the *output* or *controlled* terminals. Voltages are measured with respect to ground and *positive* directions of currents are taken as flowing *away* from the terminals. Since no current flows into the element at the input terminals

$$I_i = I_j = 0. \tag{4.1.1}$$

The currents at the output terminals, k and l, are

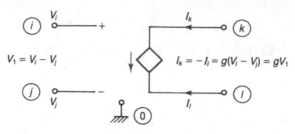

FIGURE 4.1.1   Voltage-controlled current source, VC.

$$I_k = g(V_i - V_j)$$
$$I_l = -g(V_i - V_j). \qquad (4.1.2)$$

The constant g is called the *transconductance* and is measured in Siemens, S. The amount of current delivered by the source at the controlled terminals depends on the transconductance value and on the voltage difference $V_i - V_j$. If this difference is zero, the element does not deliver any current and behaves as an open circuit. Recall here that independent current source not delivering current also behaves as an open circuit.

Consider the simple network in Fig. 4.1.2. The VC in the network has its controlling terminals between the nodes 1 and 0, its controlled terminals between nodes 0 and 2. The sequence must be given from 0 to 2, because of the direction of the current. All the necessary information is already in the figure. Identifying the voltages with those in the diagram of the VC we have $V_i = V_1$, $V_j = 0$, $V_k = 0$, $V_l = V_2$ and $I_l = -gV_1$.

The nodal equations are written in exactly the same way as we have done in section 2.1. Consider node 1 and assume that it has the highest voltage with respect to ground. Having made this assumption we must accept the consequence that all currents through the resistors flow *away* from the node. The Kirchhoff current law states that the sum of these currents must be equal to zero:

$$G_1 V_1 + G_2(V_1 - V_2) - J = 0$$

When considering node 2 we assume that the voltage of the second node is now the highest one. This means that all currents through the resistors must flow away from this node. The current gV_1 flows into node 2 and has to be taken negatively in the KCL:

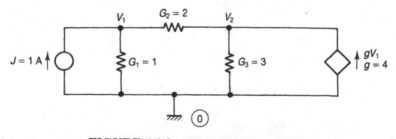

FIGURE 4.1.2   Network with one VC.

$$G_3V_2 + G_2(V_2 - V_1) - gV_1 = 0$$

There is no difference in writing equations for the dependent or independent current sources. Since the transconductance is multiplied by some unknown voltage, the term must stay on the left side of the equations. The independent source, which is known, is transferred to the right side with a change of sign. Collecting terms multiplied by the same voltage

$$(G_1 + G_2)V_1 - G_2V_2 = J$$

$$-(G_2 + g)V_1 + (G_2 + G_3)V_2 = 0.$$

If we use the values indicated in Fig. 4.1.2, the system of equations will be

$$3V_1 - 2V_2 = 1$$

$$-6V_1 + 5V_2 = 0$$

Using matrix algebra

$$\begin{bmatrix} 3 & -2 \\ 6 & 5 \end{bmatrix} \begin{bmatrix} V_1 \\ V_2 \end{bmatrix} = \begin{bmatrix} 1 \\ 0 \end{bmatrix}$$

and the solution is $V_1 = \dfrac{5}{3}V$, $V_2 = 2V$. If we need the currents through the three conductances, they are:

$$I_{G1} = G_1V_1 = \frac{5}{3}A$$

$$I_{G2} = 2(V_1 - V_2) = 2(\frac{5}{3} - 2) = -\frac{2}{3}A$$

$$I_{G3} = G_3V_2 = 6A$$

The sign of I_{G2} indicates that the current actually flows from node 2 to node 1.

Example 1.

Write the nodal equations for the network in Fig. 4.1.3 and find the nodal voltages. The equations are

$$3V_1 - 2V_2 + 4(V_1 - V_2) = J$$

$$-2V_1 + 3V_2 = 0$$

In matrix form

$$\begin{bmatrix} 7 & -6 \\ -2 & 3 \end{bmatrix} \begin{bmatrix} V_1 \\ V_2 \end{bmatrix} = \begin{bmatrix} 1 \\ 0 \end{bmatrix}$$

4.1

4.2

4.3

4.4

4.5

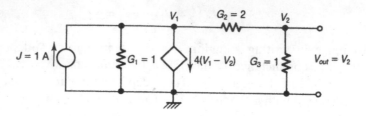

FIGURE 4.1.3   Network with one *VC*.

The determinant $D = 21 - 12 = 9$,

$$N_1 = \begin{vmatrix} 1 & -6 \\ 0 & 3 \end{vmatrix} = 3$$

$$N_2 = \begin{vmatrix} 7 & 1 \\ -2 & 0 \end{vmatrix} = 2$$

and the solutions are $V_1 = \dfrac{3}{9}$ and $V_2 = \dfrac{2}{9}$.

Example 2.

The network in Fig. 4.1.4 has two independent and two voltage controlled current sources. Find the nodal voltages.

The equations are

$$3V_1 - 2V_2 - g_2 V_2 = 1$$

$$-2V_1 + 5V_2 + g_1(V_1 - V_2) = 2$$

We have intentionally retained the symbols for the controlled sources. Inserting their values

$$\begin{bmatrix} 3 & -4 \\ 0 & 3 \end{bmatrix} \begin{bmatrix} V_1 \\ V_2 \end{bmatrix} = \begin{bmatrix} 1 \\ 2 \end{bmatrix}$$

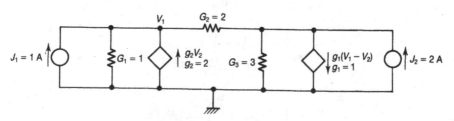

FIGURE 4.1.4   Network with two *VC*.

The system determinant is $D = 9$, $N_1 = 11$ and $N_2 = 6$. The solutions are $V_1 = \dfrac{3}{9}$ and $V_2 = \dfrac{6}{9}$.

Example 3.

Do the same for Fig. 4.1.4 but retain all elements as variables.

The equations are

$$(G_1 + G_2)V_1 - G_2V_2 - g_2V_2 = J_1$$

$$- G_2V_1 + (G_2 + G_3)V_2 + g_1(V_1 - V_2) = J_2$$

or in matrix form

$$\begin{bmatrix} G_1 + G_2 & -g_2 - G_2 \\ g_1 - G_2 & G_2 + G_3 - g_1 \end{bmatrix} \begin{bmatrix} V_1 \\ V_2 \end{bmatrix} = \begin{bmatrix} J_1 \\ J_2 \end{bmatrix}$$

When we perform the necessary steps, we get

$$D = G_1G_2 + G_1G_3 + G_2G_3 - G_1g_1 - G_2g_2 + g_1g_2$$

$$N_1 = J_1(G_2 + G_3 - g_1) + J_2(g_2 + G_2)$$

$$N_2 = J_1(G_2 - g_1) + J_2(G_1 + G_2)$$

Use numerical values to find the nodal voltages.

A network with three nodes is in Fig. 4.1.5. Using the same rules as above we write for node 1

$$(G_1 + G_2 + G_6)V_1 - G_2V_2 - G_6V_3 - g_1(V_2 - V_3) = J_1$$

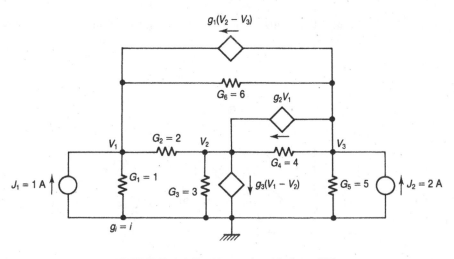

FIGURE 4.1.5 Network with three *VC*.

for node 2

$$-G_2V_1 + (G_2 + G_3 + G_4)V_2 - G_4V_3 - g_2V_1 + g_3(V_1 - V_2) = 0$$

and for node 3

$$-G_6V_1 - G_4V_2 + (G_4 + G_5 + G_6)V_3 + g_1(V_2 - V_3) + g_2V_1 = J_2$$

The transconductances of the dependent sources must stay on the left side, the independent sources are transferred on the right side of the equations. Collect the terms multiplied by the same voltages,

$$(G_1 + G_2 + G_6)V_1 - (G_2 + g_1)V_2 - (g_1 - G_6)V_3 = J_1$$

$$(g_3 - G_2 - g_2)V_1 + (G_2 + G_3 + G_4 - g_3)V_2 - G_4 V_3 = 0$$

$$(g_2 - G_6)V_1 + (g_1 - G_4)V_2 + (G_4 + G_5 + G_6 - g_1)V_3 = J_2$$

In matrix form:

$$\begin{bmatrix} (G_1 + G_2 + G_6) & -G_2 - g_1 & g_1 - G_6 \\ g_3 - G_2 - g_2 & G_2 + G_3 + G_4 - g_3 & -G_4 \\ g_2 - G_6 & g_1 - G_4 & G_4 + G_5 + G_6 - g_1 \end{bmatrix} \begin{bmatrix} V_1 \\ V_2 \\ V_3 \end{bmatrix} = \begin{bmatrix} J_1 \\ 0 \\ J_2 \end{bmatrix}$$

If we substitute the values and solve, we get $V_1 = \dfrac{156}{423} = 0.3688\,V$, $V_2 = \dfrac{112}{423} = 0.2648\,V$ and $V_3 = \dfrac{129}{423} = 0.3050V$.

Example 4.

Write the nodal equations and the matrix equation for the network in Fig. 4.1.6. Use the numerical values and find the nodal voltages.

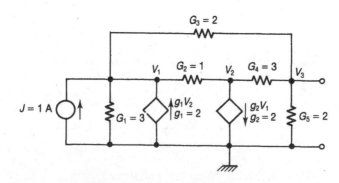

FIGURE 4.1.6   Network with two *VC*.

The equations are

$$6V_1 - 1V_2 - 2V_3 - g_1V_2 = J$$

$$-1V_1 + 4V_2 - 3V_3 + g_2V_1 = 0$$

$$-2V_1 - 3V_2 + 7V_3 = 0.$$

Inserting the values for g_1 and g_2 we get

$$\begin{bmatrix} 6 & -3 & -2 \\ 1 & 4 & -3 \\ -2 & -3 & 7 \end{bmatrix} \begin{bmatrix} V_1 \\ V_2 \\ V_3 \end{bmatrix} = \begin{bmatrix} 1 \\ 0 \\ 0 \end{bmatrix}$$

The system determinant is $D = 107$ and the numerators are $N_1 = 19$, $N_2 = -1$ and $N_3 = 5$. The voltages are $V_1 = \dfrac{19}{107}$, $V_2 = \dfrac{-1}{107}$ and $V_3 = \dfrac{5}{107}$.

4.2. CURRENT CONTROLLED VOLTAGE SOURCE

The current controlled voltage source, CV, is the second of the four possible controlled sources. It is easy to incorporate into the mesh equations and is shown in Fig. 4.2.1. The input terminals on the left are connected by a short circuit and a current I flows through this connection from node i to node j. It is obvious that the voltages of the two input terminals are equal,

$$V_i - V_j = 0 \tag{4.2.1}$$

and the currents flowing away from the terminals i and j are

$$I_i = I$$
$$I_j = -I \tag{4.2.2}$$

A voltage source is connected between the output terminals k and l, its voltage depending on the amount of current flowing through the short circuit of the input terminals. Because

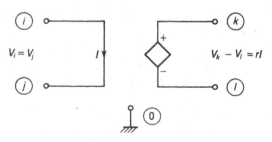

FIGURE 4.2.1 Current-controlled voltage source, CV.

the voltages are measured with respect to ground, the equation describing the output properties of the CV is

$$V_k - V_l = rI \tag{4.2.3}$$

The constant r is called the *transresistance* and is measured in ohms. If no current flows through the controlling short circuit, the voltage difference of the controlled terminals is zero and the dependent voltage source behaves as a short circuit. We recall that an independent voltage source delivering zero voltage also behaves as a short circuit.

Consider the network in Fig. 4.2.2. Because all sources, dependent and independent, are voltages, we can use the mesh method. The current I in the first mesh controls the voltage of the CV on the right. In the first step we realize that $I = I_1$ and we write this into the Figure. The KVL equation for the first mesh is

$$R_1 I_1 + R_2(I_1 - I_2) - E = 0$$

and for the second

$$R_2(I_2 - I_1) + R_3 I_2 - rI_1 = 0$$

The dependent source is handled exactly the same way as an independent source, with the only difference that it will stay on the left side. Rearranging the terms and transferring the independent voltage E to the right side we get

$$(R_1 + R_2)I_1 - R_2 I_2 = E$$

$$(-R_2 - r)I_1 + (R_2 + R_3)I_2 = 0 \tag{4.2.4}$$

If we use numerical values for the elements, the equations become

$$3I_1 - 2I_2 = 1$$

$$-4I_1 + 5I_2 = 0$$

or in matrix form

$$\begin{bmatrix} 3 & -2 \\ -4 & 5 \end{bmatrix} \begin{bmatrix} I_1 \\ I_2 \end{bmatrix} = \begin{bmatrix} 1 \\ 0 \end{bmatrix}$$

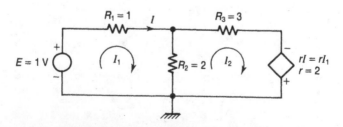

FIGURE 4.2.2   Network with one *CV*.

Using matrix algebra we find $I_1 = \dfrac{5}{7}$ and $I_2 = \dfrac{4}{7}$.

Example 1.

For the network in Fig. 4.2.3 write the mesh equations, calculate the mesh currents and the output voltage.

The controlling current, indicated in the short circuit in the middle of the figure, is equal to $I = I_1 - I_2$ (because both I and I_1 are in the same direction while I_2 has opposite direction). The mesh equations are

$$3I_1 = E$$

$$3I_2 - 3I = 0$$

If we now substitute for I, we get the equations

$$3I_1 = E$$

$$-3I_1 + 6I_2 = 0$$

or

$$\begin{bmatrix} 3 & 0 \\ -3 & 6 \end{bmatrix} \begin{bmatrix} I_1 \\ I_2 \end{bmatrix} = \begin{bmatrix} 1 \\ 0 \end{bmatrix}$$

The result is $D = 18$, $N_1 = 6$, $N_2 = 3$ and the currents are $I_1 = \dfrac{6}{18}$ and $I_2 = \dfrac{3}{18}$. The output voltage is taken across the dependent source and is equal to $V_{out} = -3(I_1 - I_2) = -\dfrac{1}{2}$.

Example 2.

In Fig. 4.2.4 calculate the voltage across the resistor R_2 using the mesh equations. We first note that $I = I_1$.

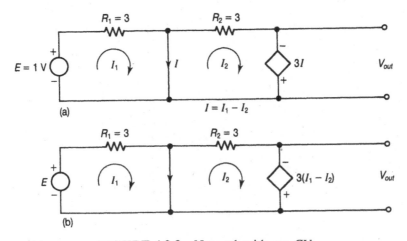

FIGURE 4.2.3   Network with one *CV*.

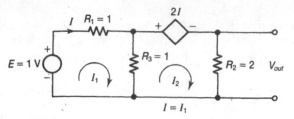

FIGURE 4.2.4   Network with one *CV.*

The equations are

$$(R_1 + R_3)I_1 - R_3 I_2 = E$$

$$- R_3 I_1 + (R_2 + R_3)I_2 + 2I = 0$$

Inserting numerical values

$$\begin{bmatrix} 2 & -1 \\ 1 & 3 \end{bmatrix} \begin{bmatrix} I_1 \\ I_2 \end{bmatrix} = \begin{bmatrix} 1 \\ 0 \end{bmatrix}$$

The solutions are $D = 7$, $N_1 = 3$, $N_2 = -1$ and the voltage across R_2 is $V_{out} = -\dfrac{2}{7}$.

Example 3.

The network in Fig. 4.2.5 has three meshes. Calculate the indicated voltage V_{out}. In the first step we realize that the controlling current is equal to $I = I_2 - I_3$. In addition we note that all we need to calculate is the current I_3 and add the voltage across the resistor R_4 to the voltage of the independent source E_2. The mesh equations are

$$1I_1 + 2I = 2$$

$$- 2I + 2I_2 - I_3 = 0$$

$$- I_2 + 2I_3 = -1$$

Inserting for the current I we get the system

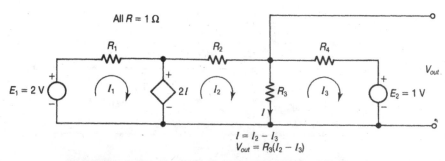

$$I = I_2 - I_3$$
$$V_{out} = R_3(I_2 - I_3)$$

FIGURE 4.2.5   Network with two sources and one *CV.*

$$\begin{bmatrix} 1 & 2 & -2 \\ 0 & 0 & 1 \\ 0 & -1 & 2 \end{bmatrix} \begin{bmatrix} I_1 \\ I_2 \\ I_3 \end{bmatrix} = \begin{bmatrix} 2 \\ 0 \\ -1 \end{bmatrix}$$

The third numerator is obtained as

$$N_3 = \begin{vmatrix} 1 & 2 & 2 \\ 0 & 0 & 0 \\ 0 & -1 & -1 \end{vmatrix}$$

Since the second and third column are identical, the determinant is zero and I_3 is zero. The output voltage will be equal to the voltage $E_2 = 1$ V.

Example 4.

Consider Fig. 4.2.6 and find the voltage V shown in the figure.

For the solution we first realize that the controlling current $I = -I_2$. The mesh equations are

$$3I_1 - 2I_2 + 3I = 2$$

$$-2I_1 + 5I_2 - 3I = -4$$

Substituting for I and inserting numerical values

$$\begin{bmatrix} 3 & -5 \\ -2 & 8 \end{bmatrix} \begin{bmatrix} I_1 \\ I_2 \end{bmatrix} = \begin{bmatrix} 2 \\ -4 \end{bmatrix}$$

The results are $D = 14$, $N_1 = -4$, $N_2 = -8$ and the currents are $I_1 = -\frac{4}{14}$, $I_2 = -\frac{8}{14}$. The voltage V can be calculated in several ways; the simplest is to take the voltage of E_1 and subtract the voltage across the resistor R_1. The result is $V = E_1 - 1I_1 = 2 + \frac{4}{14} = \frac{32}{14} V$. We can obtain the same result by taking $V = E_2 + R_3 I_2$.

A more complicated network is in Fig. 4.2.7a. We will show that a small amount of preprocessing makes the problem as simple as the above ones. The network has two

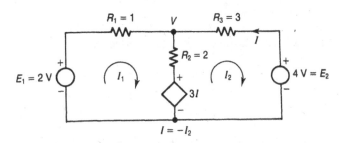

FIGURE 4.2.6 Network with two sources and one CV.

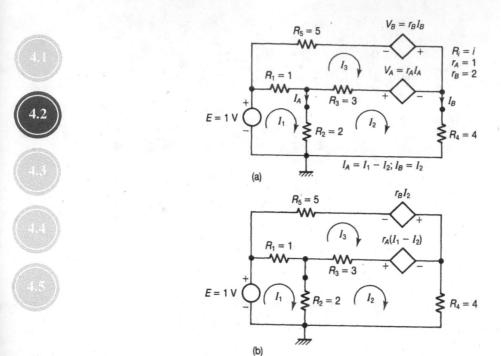

(a)

(b)

FIGURE 4.2.7 Netwok with two *CV:* (a) Original. (b) with information transferred to the dependent source.

current controlled voltage sources; the controlling currents are denoted by I_A and I_B, the controlled voltage sources by V_A and V_B. First we realize that $I_A = I_1 - I_2$ and $I_B = I_2$; we write these values to the controlled sources, as shown in Fig. 4.2.7 b. Once this has been done, all the necessary information has been transferred to the dependent sources and we can write the mesh equations. For the first mesh

$$(R_1 + R_2)I_1 - R_2I_2 - R_1I_3 = E$$

For the second

$$- R_2I_1 + (R_2 + R_3 + R_4)I_2 - R_3I_3 + r_A(I_1 - I_2) = 0$$

and for the third

$$- R_1I_1 - R_3I_2 + (R_1 + R_3 + R_5)I_3 - r_A(I_1 - I_2) - r_BI_2 = 0$$

The terms multiplied by the same currents are collected:

$$(R_1 + R_2)I_1 - R_2I_2 - R_1I_3 = E$$

$$(-R_2 + r_A)I_1 + (R_2 + R_3 + R_4 - r_A)I_2 - R_3I_3 = 0$$

$$(-R_1 - r_A)I_1 + (- R_3 + r_A - r_B)I_2 + (R_1 + R_3 + R_5)I_3 = 0$$

and the equations are now in their final form. If we substitute numerical values ($R_i = i$, $r_A = 1$, $r_B = 2$, $E = 1$), we will get the system

$$3I_1 - 2I_2 - I_3 = 1$$

$$-I_1 + 8I_2 - 3I_3 = 0$$

$$-2I_1 - 4I_2 + 9I_3 = 0$$

In matrix form

$$\begin{bmatrix} 3 & -2 & -1 \\ -1 & 8 & -3 \\ -2 & -4 & 9 \end{bmatrix} \begin{bmatrix} I_1 \\ I_2 \\ I_3 \end{bmatrix} = \begin{bmatrix} 1 \\ 0 \\ 0 \end{bmatrix}$$

The solution is $I_1 = \dfrac{60}{130}$, $I_2 = \dfrac{15}{130}$ and $I_3 = \dfrac{20}{130}$.

4.3. CURRENT CONTROLLED CURRENT SOURCE

The current controlled current source, CC, is shown in Fig. 4.3.1. The input terminals on the left are connected by a short circuit which senses the amount of current, I, flowing through it. The voltages at both input terminals are equal,

$$V_i - V_j = 0 \qquad (4.3.1)$$

The currents flowing away from the terminals i and j are

$$I_i = I \qquad (4.3.2)$$

$$I_j = -I$$

The output of the CC is formed by a current source connected between terminals k and l. The dependent source current is proportional to the current I of the input short circuit. The proportionality constant is usually denoted as α; it is dimensionless. The currents at the output terminals are

$$I_k = \alpha I \qquad (4.3.3)$$

$$I_l = -\alpha I$$

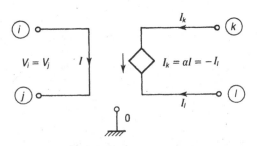

FIGURE 4.3.1 Current-controlled current source, CC.

If no current flows through the controlling short circuit, the element does not deliver any current at its output, and behaves as the open circuit between the terminals k and l. Because the output of the CC is a current, it can be handled by the nodal formulation if we do some preprocessing steps.

Consider the network in Fig. 4.3.2a. The input terminals of the CC are indicated by the short circuit marked by the arrow and by I_A. The output of the CC is current $\alpha\, I_A$. The controlling current I_A is equal to the current flowing through the conductance G_2. We can express this current as $I_A = G_2(V_1 - V_2)$ and write it into the network, see Fig. 4.3.2b. We are now ready to write the nodal equations. For node 1 the KCL is

$$G_1V_1 + G_2(V_1 - V_2) - J = 0$$

and for node two

$$G_2(V_2 - V_1) + G_3V_2 + \alpha\, G_2(V_1 - V_2) = 0$$

The equations are rearranged

$$(G_1 + G_2)V_1 - G_2V_2 = J$$

$$-G_2(1 - \alpha)V_1 + [G_3 + G_2(1 - \alpha)]V_2 = 0$$

and written in matrix form

$$\begin{bmatrix} (G_1 + G_2) & -G_2 \\ -G_2(1-\alpha) & [G_3 + G_2(1-\alpha)] \end{bmatrix} \begin{bmatrix} V_1 \\ V_2 \end{bmatrix} = \begin{bmatrix} J \\ 0 \end{bmatrix}$$

Using Cramer's rule we find the determinant of the matrix

$$D = G_1G_2(1 - \alpha) + G_1G_3 + G_2G_3$$

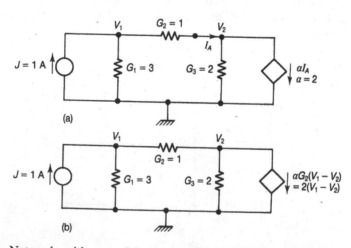

FIGURE 4.3.2 Network with one CC (a) Original. (b)With information transferred to the dependent source.

At this moment, it is worth noting one important property of determinants. Intermediate steps lead to two terms $G_2^2(1-\alpha)$, one with + and one with − sign; they cancel. We will remember that as long as we give every element a different subscript (or name), then it is *never* possible that the result would contain some element raised to power more than one. Should such a term remain in your result, you can be sure that you have made a mistake somewhere.

Using Cramer's rule we find the numerator

$$N_1 = J[G_3 + G_2(1-\alpha)]$$

and the voltage V_1 is equal to

$$V_1 = \frac{N_1}{D} = \frac{J[G_3 + G_2(1-\alpha)]}{G_1 G_2(1-\alpha) + G_1 G_3 + G_2 G_3}$$

We find similarly

$$N_2 = JG_2(1-\alpha)$$

and the voltage is

$$V_2 = \frac{N_2}{D} = \frac{JG_3(1-\alpha)}{G_1 G_2(1-\alpha) + G_1 G_3 + G_2 G_3}$$

Example 1.

Find the nodal voltages for the network in Fig. 4.3.3.

In this problem we first realize that the controlling current $I = G_3 V_2 = 4V_2$. We can now write the nodal equations

$$(G_1 + G_2)V_1 - G_2 V_2 - 12V_2 = J_1$$

$$-G_2 V_1 + (G_2 + G_3)V_2 = J_2$$

Using numerical values

$$\begin{bmatrix} 3 & -13 \\ -1 & 5 \end{bmatrix} \begin{bmatrix} V_1 \\ V_2 \end{bmatrix} = \begin{bmatrix} 1 \\ 3 \end{bmatrix}$$

and the results are $D = 2$, $N_1 = 44$, $N_2 = 10$. The nodal voltages are $V_1 = 22$ and $V_2 = 5$.

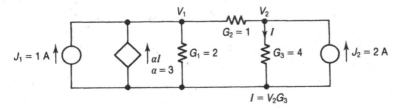

FIGURE 4.3.3 Network with two sources and one *CC*.

Example 2.

Find the nodal voltages for the network in Fig. 4.3.4.

In the first step we realize that the controlling current $I = (V_1 - V_2)G_2$. The nodal equations are

$$(G_1 + G_2)V_1 - G_2V_2 + \alpha I = J_1$$

$$- G_2V_1 + (G_2 + G_3)V_2 - \alpha I = J_2$$

Inserting numerical values

$$\begin{bmatrix} 9 & -6 \\ -6 & 10 \end{bmatrix} \begin{bmatrix} V_1 \\ V_2 \end{bmatrix} = \begin{bmatrix} 2 \\ 3 \end{bmatrix}$$

The results are $D = 54$, $N_1 = 38$, $N_2 = 39$ and the voltages are $V_1 = \dfrac{58}{54}$ and $V_2 = \dfrac{39}{54}$.

The next network in Fig. 4.3.5a has three nodes and three current controlled current sources. All conductances have, for simplicity, their values equal to the subscript. The controlling currents are denoted I_A, I_B, I_C and are marked in the figure by arrows. The dependent sources are marked as 1 I_A, 2 I_B and 3 I_C. Looking at the figure we see that the controlling currents are

$$I_A = G_3V_2 = 3V_2$$

$$I_B = G_4(V_2 - V_3) = 4V_2 - 4V_3$$

$$I_C = G_2(V_1 - V_2) = 2V_1 - 2V_2$$

The network is redrawn in Fig. 4.3.5b. In this figure the controlling currents are omitted because the information has been transferred to the controlled sources. All sources in the network are current sources and we use nodal formulation. For the first node:

$$3V_1 - 2V_2 - (6V_1 - 6V_2) = 10$$

For the second node:

$$- 2V_1 + 9V_2 - 4V_3 + (8V_2 - 8V_3) = 0$$

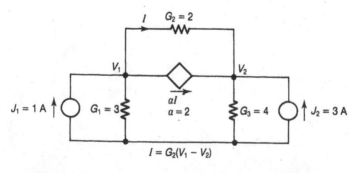

FIGURE 4.3.4   Network with two sources and one CC.

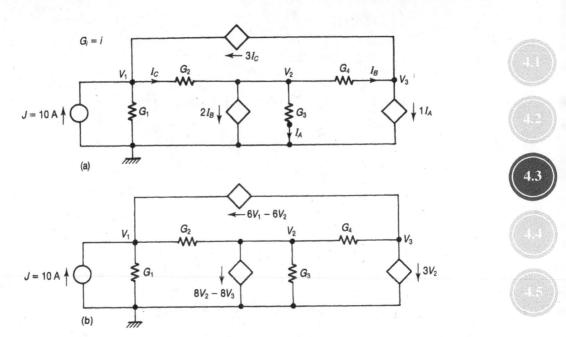

FIGURE 4.3.5 Network with three CC.(a) Original (b) With information transferred to the dependent sources

and for the third node:

$$-4V_2 + 4V_3 + (3V_2) + (6V_1 - 6V_2) = 0$$

In matrix form

$$\begin{bmatrix} -3 & 4 & 0 \\ -2 & 17 & -12 \\ 6 & -7 & 4 \end{bmatrix} \begin{bmatrix} V_1 \\ V_2 \\ V_3 \end{bmatrix} = \begin{bmatrix} 10 \\ 0 \\ 0 \end{bmatrix}$$

The determinant of the system is $D = -208$, the numerators are $N_1 = -160$, $N_2 = -640$ and $N_3 = -880$. This makes the voltages $V_1 = \dfrac{N_1}{D} = 0.7692\ V$, $V_2 = \dfrac{F_2}{D} = 3.0769\ V$ and $V_3 = \dfrac{N_3}{D} = 4.2308\ V$.

Example 3.

Find nodal voltages for the network in Fig. 4.3.6a.

The network has one current controlled current source and one voltage controlled current source. First we realize that the current I is given by the expression $I = G_2(V_1 - V_2)$. Although it is not necessary to redraw the network, we will do so for clarification, as shown in Fig. 4.3.6 b. As before, the controlling current is now omitted because the information has been transferred to the dependent source. The equations are

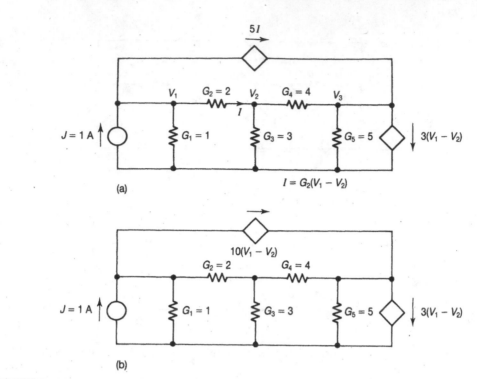

FIGURE 4.3.6   Network with one *CC* and one *VC:* (a) Original.(b)With information transferred to the dependent sources.

$$3V_1 - 2V_2 + 10V_1 - 10V_2 = 1$$

$$-2V_1 + 9V_2 - 4V_3 = 0$$

$$-4V_2 + 9V_3 - 10V_1 + 10V_2 + 3V_1 - 3V_2 = 0$$

Collecting terms

$$\begin{bmatrix} 13 & -12 & 0 \\ 2 & 9 & -4 \\ -7 & 3 & 9 \end{bmatrix} \begin{bmatrix} V_1 \\ V_2 \\ V_3 \end{bmatrix} = \begin{bmatrix} J \\ 0 \\ 0 \end{bmatrix}$$

The solutions are $D = 657$, $N_1 = 93$, $N_2 = 46$, $N_3 = 57$ and the voltages are the appropriate ratios.

4.4.   VOLTAGE CONTROLLED VOLTAGE SOURCE

The voltage controlled voltage source, *VV*, is shown in Fig. 4.4.1. Its input terminals sense the difference of the voltages at nodes *i* and *j* and the output is formed by a voltage source whose voltage is μ times the difference $V_i - V_j$. The amplification constant μ is dimensionless. The currents at the input terminals are zero,

$$I_i = I_j = 0 \qquad\qquad (4.4.1)$$

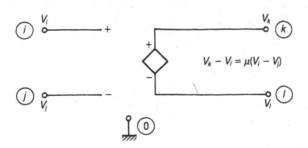

FIGURE 4.4.1 Voltage-controlled voltge source, VV.

The output voltage, taken between nodes k and l, is

$$V_k - V_l = \mu(V_i - V_j) \tag{4.4.2}$$

If the voltage difference $V_i - V_j = 0$, the output voltage is zero and the voltage source becomes a short circuit between terminals k and l. Because the VV is a special type of a voltage source, it can be handled by the mesh formulation if wc do some preprocessing steps.

A small example, Fig. 4.4.2, will clarify how we should proceed. The network has two meshes and we have already indicated the clockwise current directions by arrows. The voltage V_A across the resistor R_1 indicates the input terminals of the VV; this voltage, multiplied by μ, appears as a voltage source on the right of the figure. Since only the current I_1 flows through R_1, the controlling voltage is equal to $V_A = R_1 I_1$. We have redrawn the network in Fig. 4.4.2b and transferred this information to the dependent source. This preprocessing, written into the figure, helps us write the mesh equations directly by using the rules we learned in Chapter 2. For the first mesh we have

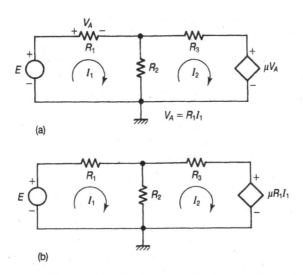

(a)

(b)

FIGURE 4.4.2 Network with one VV: (a) Original, (b) With information transferred to the dependent source

$$(R_1 + R_2)I_1 - R_2 I_2 = E$$

and for the second

$$- R_2 I_1 + (R_2 + R_3)I_2 + \mu R_1 I_1 = 0$$

As always, we put the equations into matrix form

$$\begin{bmatrix} R_1 + R_2 & -R_2 \\ \mu R_1 - R_2 & R_2 + R_3 \end{bmatrix} \begin{bmatrix} I_1 \\ I_2 \end{bmatrix} = \begin{bmatrix} E \\ 0 \end{bmatrix}$$

The determinant is

$$D = R_1 R_2 (1 + \mu) + R_1 R_3 + R_2 R_3$$

and the currents are

$$I_1 = \frac{E(R_2 + R_3)}{R_1 R_2 (1 + \mu) + R_1 R_3 + R_2 R_3}$$

$$I_2 = \frac{E(R_2 - \mu R_1)}{R_1 R_2 (1 + \mu) + R_1 R_3 + R_2 R_3}$$

Example 1.

The network in Fig. 4.4.3a has one voltage controlled voltage source with the controlling voltage taken across R_2. Find the output voltage by using the mesh method.

For the solution we realize that the controlling voltage $V = R_2(I_1 - I_2)$. To simplify the steps we transfer this information into a redrawn network Fig. 4.4.3b. The mesh equations are now

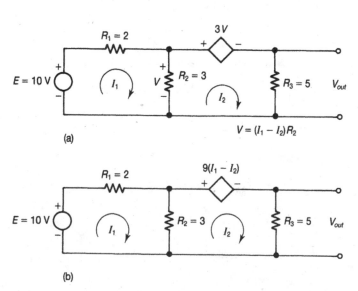

(a)

(b)

FIGURE 4.4.3   Network with one VV: (a) Original, (b) With information transferred to the dependent source.

$$5I_1 - 3I_2 = 10$$
$$-3I_1 + 8I_2 + 9I_1 - 9I_2 = 0$$

or

$$\begin{bmatrix} 5 & -3 \\ 6 & -1 \end{bmatrix} \begin{bmatrix} I_1 \\ I_2 \end{bmatrix} = \begin{bmatrix} 10 \\ 0 \end{bmatrix}$$

For the output voltage we need I_2. The determinant is $D = 13$, $N_2 = -60$ and the output voltage is $V_{out} = 5 \cdot (-\dfrac{60}{13}) = -\dfrac{300}{13}$.

Example 2.

The network in Fig. 4.4.4a has two voltage controlled voltage sources. Find the voltage at the output.

We first determine the expressions for the controlling voltages: $V_A = I_2 R_3 = 4I_2$, $V_B = R_1 I_1 = 2I_1$. The network is redrawn in Fig. 4.4.4b with the information transferred to the dependent sources. The mesh equations are

$$(R_1 + R_2)I_1 - R_2 I_2 + 12I_2 = E$$
$$- R_2 I_1 + (R_2 + R_3)I_2 + 4I_1 - 12I_2 = 0$$

Inserting numerical values

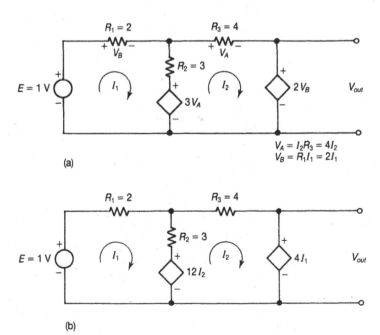

(a)

(b)

FIGURE 4.4.4 Network with one *VV*: (a) Original, (b) With information transferred to the dependent sources.

$$\begin{bmatrix} 5 & 9 \\ 1 & -5 \end{bmatrix} \begin{bmatrix} I_1 \\ I_2 \end{bmatrix} = \begin{bmatrix} 1 \\ 0 \end{bmatrix}$$

For the output voltage we need I_1. The determinant is $D = -34$, $N_1 = -5$ and the current is $I_1 = \dfrac{5}{34}$. The output voltage is $V_{out} = 4I_1 = \dfrac{20}{34}$.

A network with three meshes is in Fig. 4.4.5a. It has two voltage controlled voltage sources and the controlling voltages are V_A and V_B. In order to simplify insertions, the resistors are given values equal to their subscripts. The amplification constants were chosen to be $\mu_A = 2$ and $\mu_B = 3$. The mesh currents are indicated in the figure.

Similarly as in the above example, we first determine the controlling voltages in terms of the elements and the mesh currents. They are

$$V_A = R_4 I_3 = 4I_3$$

and

$$V_B = R_3(I_2 - I_3) = 3I_2 - 3I_3.$$

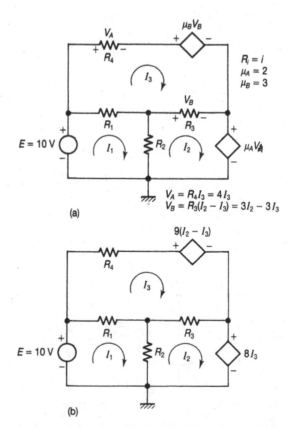

(a)

(b)

FIGURE 4.4.5 Network with two VV: (a) Original, (b) With information transferred to the dependent sources.

These values are transferred into Fig. 4.4.5b. The figure is now prepared for the mesh equations:

$$3I_1 - 2I_2 - I_3 = 10$$

$$-2I_1 + 5I_2 - 3I_3 + (8I_3) = 0$$

$$-I_1 - 3I_2 + 8I_3 + (9I_2 - 9I_3) = 0$$

The expressions in brackets indicate the dependent voltage sources. Rearranging the terms we get

$$\begin{bmatrix} 3 & -2 & -1 \\ -2 & 5 & 5 \\ -1 & 6 & -1 \end{bmatrix} \begin{bmatrix} I_1 \\ I_2 \\ I_3 \end{bmatrix} = \begin{bmatrix} 10 \\ 0 \\ 0 \end{bmatrix}$$

Cramer's rule provides $D = -84$, $N_1 = -350$, $N_2 = N_3 = -70$ and the currents are $I_1 = 4.16667$, $I_2 = I_3 = 0.83333$.

Example 3.

The network in Fig. 4.4.6a has three meshes, two independent voltage sources and two voltage controlled voltage sources. Find the nodal voltage V.

We first define the voltages V_A and V_B in terms of the mesh currents. They are

$$V_A = (I_2 - I_3)R_4 = 2I_2 - 2I_3$$

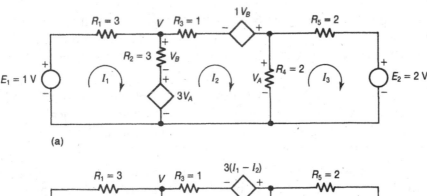

(a)

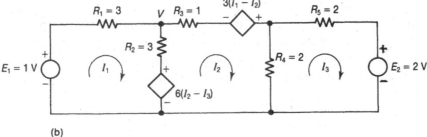

(b)

FIGURE 4.4.6 Network with two *VV:* (a) Original, (b) With information transferred to the dependent sources.

$$V_B = (I_1 - I_2)R_2 = 3I_1 - 3I_2$$

The values are transferred to the dependent sources, as given in Fig. 4.4.6b. The mesh equations are:

$$(R_1 + R_3)I_1 - R_3 I_2 + 6(I_2 - I_3) = 1$$

$$- R_2 I_1 + (R_2 + R_3 + R_4)I_2 - R_4 I_3 - 6(I_2 - I_3) - 3(I_1 - I_2) = 0$$

$$- R_4 I_2 + (R_4 + R_5)I_3 = -2$$

Inserting values:

$$\begin{bmatrix} 6 & 3 & -6 \\ -6 & 3 & 6 \\ 0 & -2 & 4 \end{bmatrix} \begin{bmatrix} I_1 \\ I_2 \\ I_3 \end{bmatrix} = \begin{bmatrix} 1 \\ 0 \\ -2 \end{bmatrix}$$

To obtain the indicated voltage we need I_1 and $V = E_1 - R_1 I_1$. The determinant of the system matrix is $D = 120$, $N_1 = -40$ and $I_1 = -\dfrac{1}{3}$. The voltage $V = 1 - 3 \cdot (-\dfrac{1}{3}) = 2$.

4.5. SUMMARY OF DEPENDENT SOURCES

This chapter explained all possible cases of dependent sources. They are very important because we use them to model properties of many practical devices, like transistors. We now summarize our development, point out the restrictions and indicate how the restrictions can be removed.

Our present knowledge shows that current sources (controlled or independent) are more suitable for nodal formulation, but only the voltage controlled current source can be used without any additional steps. If we wish to use nodal formulation for a network with current controlled current sources, we must first express the controlling currents in terms of nodal voltages.

Voltage sources (controlled or independent) are suitable for mesh formulation but only the current controlled voltage sources can be used without modifications. If we wish to use mesh formulation for a network with a voltage controlled voltage sources, we must first express the controlling voltage in terms of mesh currents.

These restrictions are quite serious because we cannot easily analyze networks with a mixture of current and voltage sources. There are two ways out of this problem. The simpler one, to be explained in the next chapter, applies some transformations and redrawings of the networks. A better one, but requiring more knowledge, uses a method which is neither nodal nor mesh. It is called the modified nodal formulation and is explained later.

PROBLEMS CHAPTER 4

P.4.1   Use mesh equations to find the currents.

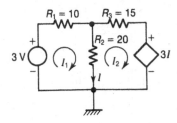

P.4.2   Use mesh equations to find the currents.

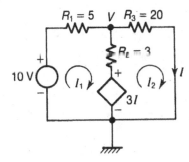

P.4.3   Use mesh equations to find the currents.

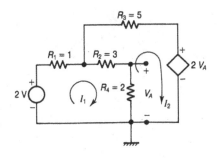

P.4.4   Use mesh equations to find the currents.

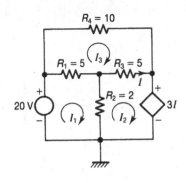

P.4.5   Use nodal formulation to get the nodal voltages.

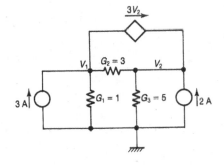

P.4.6   Use nodal formulation to get the nodal voltages.

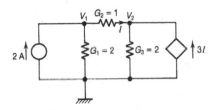

P.4.7   Use mesh formulation to get the currents.

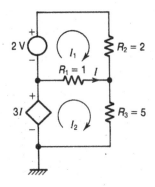

P.4.8 Use mesh equations to get the currents.

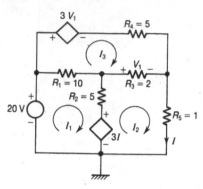

P.4.9 Use nodal equations to get the nodal voltages.

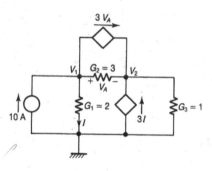

P.4.10 Use nodal equations to get the nodal voltages.

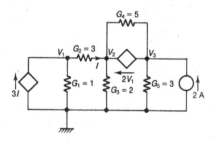

P.4.11 Use nodal equations to get the output voltage V_{out}.

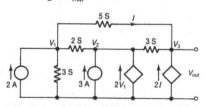

P.4.12 Find the output voltage V_{out}.

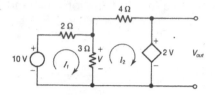

P.4.13 Find the output voltage V_{out}.

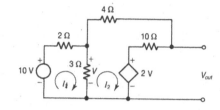

P.4.14 Find the output voltage V_{out}.

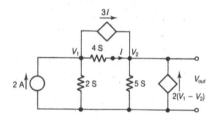

P.4.15 Find the output voltage V_{out}.

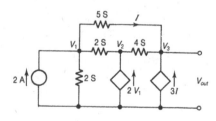

P.4.16 Find the output voltage V_{out}.

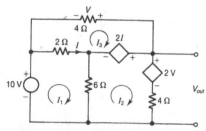

P.4.17 Find the two nodal voltages.

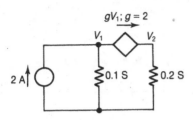

P.4.18 Find the two nodal voltages.

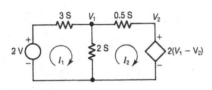

P.4.19 Find the two nodal voltages.

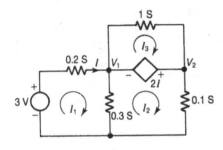

P.4.20 Find the two nodal voltages.

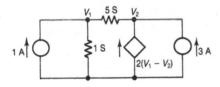

P.4.21 Find the two nodal voltages.

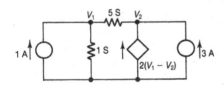

P.4.22 Find the output voltage.

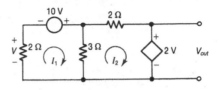

P.4.23 Find the output voltage.

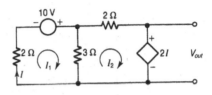

P.4.24 Calculate the output voltage.

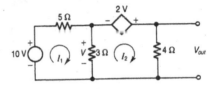

P.4.25 Calculate the output voltage.

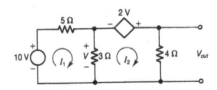

P.4.26 Calculate the output voltage.

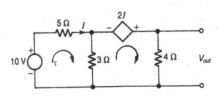

P.4.27   Calculate the output voltage.

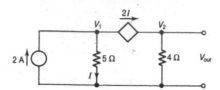

P.4.28   Calculate the output voltage.

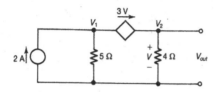

P.4.29   Calculate the output voltage.

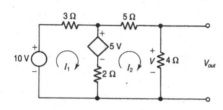

P.4.30   Calculate the output voltage.

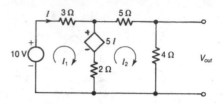

P.4.31   Calculate the output voltage.

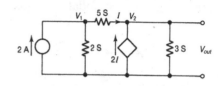

P.4.32   Calculate the output voltage.

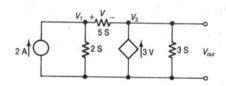

5 NETWORK TRANSFORMATIONS

INTRODUCTION

We now know most of the elements which can be used in resistive networks. We also know two analysis methods (nodal and mesh) and how to write the network equations, but there are cases where our knowledge is not sufficient. We still do not know how to handle situations when we have a mixture of voltage and current sources.

This chapter explains several important concepts: transformation of sources, splitting of sources and transformation of networks by the Thevenin and Norton theorems. It also demonstrates the law of superposition, valid for any linear network, and the concept of input and output resistance.

Until now, we have treated the nodal and mesh methods as equally important; they are not. The last section of this chapter gives important recommendations which we follow in the rest of this book.

5.1. TRANSFORMATION OF SOURCES

All the previous examples had only one type of source. If they were current sources, we used nodal formulation. If they were voltage sources, we used mesh formulation. This situation does not occur in most practical cases and we must find a way to incorporate all kinds of sources (dependent or independent) into any of the two formulations.

One step which helps in this direction is the transformation of voltage sources into current sources and vice versa. In this section we will assume that each voltage source is connected in series with a resistor or each current source in parallel with the same resistor. The two cases are sketched in Fig. 5.1.1. On the left is a voltage source with the resistor R_S in series; the combination is connected to a load resistor R_L. On the right is a current source with the same resistor R_S in parallel. We wish to find a transformation by which we can transform any of the two cases into the other one. In both cases, the same current must flow into the loading resistor, R_L, and the same voltage must appear across it.

Consider first the network on the left. It is a voltage divider and the voltage V is equal to

$$V = \frac{R_L}{R_S + R_L} E \qquad (5.1.1)$$

In the case of a current source, the two resistors are in parallel and we add their conductances to get the overall conductance,

$$G = \frac{1}{R_S} + \frac{1}{R_L} = \frac{R_S + R_L}{R_S R_L}. \qquad (5.1.2)$$

Using nodal formulation for just one node we get for the network on the right

$$\frac{R_S + R_L}{R_S R_L} V = J \qquad (5.1.3)$$

Because V must be equal in both situations, we substitute (5.1.1) into (5.1.3) and get $E = JR_S$. This looks very much like the Ohm's law but it actually expresses transformation of sources. We indicate this fact by writing the subscript EQ into the following two equations

$$E_{EQ} = J_{EQ} R_S$$

$$J_{EQ} = E_{EQ} G_S. \qquad (5.1.4)$$

but we will not use the subscripts later on. The transformation can be applied only in cases where a resistor is in series with the voltage source or in parallel with the current source.

We apply the transformation to the network in Fig. 5.1.2a. All resistor values are in ohms and for simplicity set equal to the subscript of the resistor. Fig. 5.1.2b shows the transformation of the voltage source into a current source and we use the nodal formulation:

$$\begin{bmatrix} G_1 + G_2 + G_3 & -G_3 \\ -G_3 & G_3 + G_4 \end{bmatrix} \begin{bmatrix} V_1 \\ V_2 \end{bmatrix} = \begin{bmatrix} J_{EQ} \\ J \end{bmatrix}$$

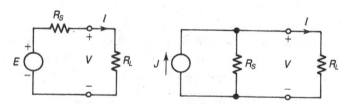

FIGURE 5.1.1   Transformation of sources.

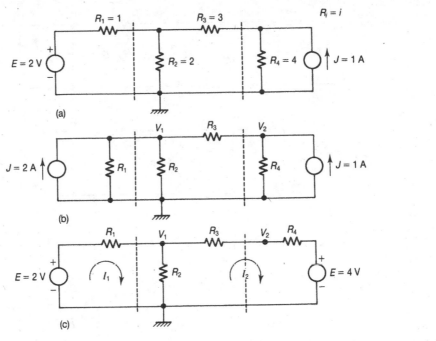

FIGURE 5.1.2 Network with two sources. (a) Original. (b) Transformed to current sources. (c) Transformed to voltage sources.

Substituting numerical values

$$\begin{bmatrix} 11/6 & -1/3 \\ -1/3 & 7/12 \end{bmatrix} \begin{bmatrix} V_1 \\ V_2 \end{bmatrix} = \begin{bmatrix} 2 \\ 1 \end{bmatrix}$$

The determinant of the matrix is $D = 23/24$ and the voltages are $V_1 = 36/23 = 1.5652$ and $V_2 = 60/23 = 2.6087$.

We could also transform the current source on the right of Fig. 5.1.2a into a voltage source and get the network in Fig. 5.1.2c. Use of the mesh method leads to

$$\begin{bmatrix} 3 & -2 \\ -2 & 9 \end{bmatrix} \begin{bmatrix} I_1 \\ I_2 \end{bmatrix} = \begin{bmatrix} 2 \\ -4 \end{bmatrix}$$

The solution is $I_1 = 10/23$ and $I_2 = -8/23$. The negative sign indicates that the current I_2 actually flows the other way. We can confirm that the two results are equivalent. For instance, we can calculate the voltage V_1 either as $V_1 = E - R_1 I_1$ or as $V_1 = R_2(I_1 - I_2)$. In both cases we get $V_1 = 1.5652$.

Example 1.

A somewhat more complicated situation is in Fig. 5.1.3a where we first transform the voltage source into the current source, Fig. 5.1.3b, and then use the nodal formulation. The matrix equation is

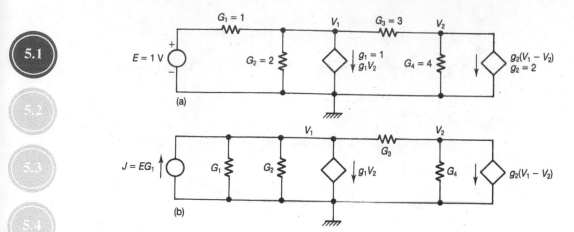

FIGURE 5.1.3   Network with dependent sources: (a) Original. (b) Transformed for nodal formulation.

$$\begin{bmatrix} G_1 + G_2 + G_3 & -G_3 + g_1 \\ -G_3 + g_2 & G_3 + G_4 - g_2 \end{bmatrix} \begin{bmatrix} V_1 \\ V_2 \end{bmatrix} = \begin{bmatrix} EG_1 \\ 0 \end{bmatrix}$$

Substituting numerical values

$$\begin{bmatrix} 6 & -2 \\ -1 & 5 \end{bmatrix} \begin{bmatrix} V_1 \\ V_2 \end{bmatrix} = \begin{bmatrix} 1 \\ 0 \end{bmatrix}$$

The determinant is $D = 28$ and the voltages are $V_1 = 5/28$, $V_2 = 1/28$.

The transformations can be applied equally to dependent and independent sources. We will use the network in Fig. 5.1.4a to demonstrate transformation of a dependent current source into a dependent voltage source. Using (5.1.4) we redraw the network as shown in Fig. 5.1.4b. Note that these transformations did not modify the part of the network where we sense the controlling current of the CC and the controlling voltage of the VV. After this transformation we realize that the controlling current of the CC is $I = I_1 - I_2$ and that the controlling voltage of the VV is $V_1 = R_2(I_1 - I_2)$. We write these values into the network in Fig. 5.1.4c. All the necessary information has now been transferred into the last figure for which we write the mesh equations. In matrix form they are

$$\begin{bmatrix} (R_1 + R_2) & -R_2 \\ (-R_2 + mR_2 - a\,R_4) & (R_2 - mR_2 + R_3 + R_4 + a\,R_4) \end{bmatrix} \begin{bmatrix} I_1 \\ I_2 \end{bmatrix} = \begin{bmatrix} E \\ 0 \end{bmatrix}$$

Substituting numerical values

$$\begin{bmatrix} 3 & -2 \\ -10 & 17 \end{bmatrix} \begin{bmatrix} I_1 \\ I_2 \end{bmatrix} = \begin{bmatrix} 2 \\ 0 \end{bmatrix}.$$

The result is $I_1 = 34/31$ and $I_2 = 20/31$.

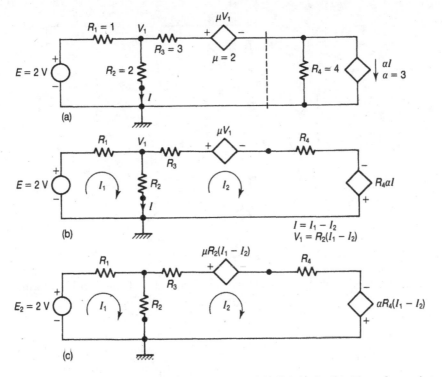

FIGURE 5.1.4 Network with dependent sources: (a) Original. (b) Transformed to voltage sources. (c) Final form for mesh formulation.

Example 2.

In this example, Fig. 5.1.5a, we transform the voltage sources into current sources. The VV on the right is transformed into a current source in parallel with G_5. The voltage source on the left is transformed into a current source in parallel with G_1. We note again that the transformations did not modify the part of the network where we sense the current of the VV and CC. We express the current I in terms of the voltage V_2, $I = G_4V_2$ and write it instead of I. All these transformations are transferred into Fig. 5.1.5b for which we write the nodal equations

$$(G_1 + G_2 + G_3 + G_5)V_1 - G_3V_2 + \alpha V_2G_4 - \mu V_2G_5 = E\,G_1$$

$$- G_3V_1 + (G_3 + G_4)V_2 = 0$$

or

$$\begin{bmatrix} (G_1 + G_2 + G_3 + G_5) & (-G_3 + \alpha G_4 - \mu G_5) \\ -G_3 & G_3 + G_4 \end{bmatrix} \begin{bmatrix} V_1 \\ V_2 \end{bmatrix} = \begin{bmatrix} EG_1 \\ 0 \end{bmatrix}$$

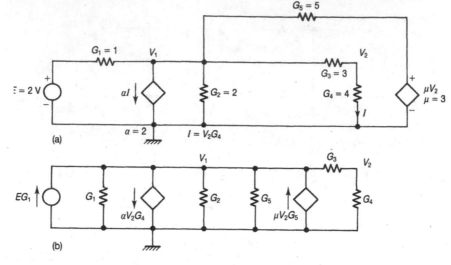

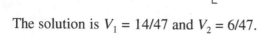

FIGURE 5.1.5   Network with voltage and current sources: (a) Original. (b) Transformed for nodal formulation.

Using numerical values

$$\begin{bmatrix} 11 & -10 \\ -3 & 7 \end{bmatrix}\begin{bmatrix} V_1 \\ V_2 \end{bmatrix} = \begin{bmatrix} 2 \\ 0 \end{bmatrix}.$$

The solution is $V_1 = 14/47$ and $V_2 = 6/47$.

Example 3.

Apply both transformations to the network in Fig. 5.1.6a. First we transform the voltage sources into current sources, as shown in Fig. 5.1.6b. Resistor values must be converted into conductances and the two nodal equations, in matrix form, are

$$\begin{bmatrix} 2 & -1/2 \\ -1/2 & 5/6 \end{bmatrix}\begin{bmatrix} V_1 \\ V_2 \end{bmatrix} = \begin{bmatrix} 15/2 \\ 2 \end{bmatrix}$$

The solutions are $D = 17/12$, $N_1 = 29/4$, $N_2 = 31/4$, $V_1 = 87/17$ and $V_2 = 93/17$.

Fig. 5.1.6c shows the transformation of the current source on the right into a voltage source. The mesh equations are

$$\begin{bmatrix} 3 & -1 \\ -1 & 6 \end{bmatrix}\begin{bmatrix} I_1 \\ I_2 \end{bmatrix} = \begin{bmatrix} 0 \\ -1 \end{bmatrix}$$

The solutions are $D = 17$, $N_1 = -1$, $N_2 = -3$ and the currents are $I_1 = -1/17$, $I_2 = -3/17$. To check correctness of both solutions we calculate the voltage V_1 by subtracting the voltage across the resistor R_1 from the voltage of the voltage source on the left: $V_1 = E_1 - R_1 I_1 = 87/17$. The voltage $V_2 = V_1 - R_3 I_2 = 93/17$.

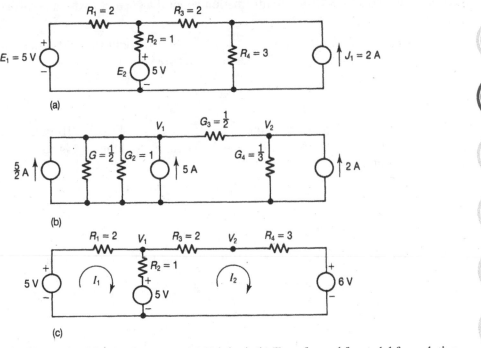

FIGURE 5.1.6   Network with mixed sources: (a) Original. (b) Transformed for nodal formulation. (c) Transformed for mesh formulation.

We now summarize the rules when we can apply the transformation. We will make the rules general by using the word *element* instead of only a *resistor*. As we will see later, the rules apply to capacitors and inductors as well.

1. The voltage source (dependent or independent) has one element in series with it.

2. The current source (dependent or independent) has one element in parallel with it.

3. Transformation of the part of the circuit where we sense the controlling voltage or current is not permitted without special care. This is explained in detail in Section 2.

4. Cases with more complicated connections are handled by source splitting, the subject of Section 3.

The next section covers another important transformation.

5.2.   TRANSFORMATIONS OF CONTROLLING TERMINALS

Transformations of sources are always possible, but sometimes this may simultaneously involve transformation of a voltage or current which controls another source. To be specific, consider the Fig. 5.2.1a where we wish to transform the voltage source into a current source. If we do so we loose all information about the current I which controls the current source.

A bit of preprocessing will solve the problem. All we have to do is determine the current I *before* the transformation. In the given network the controlling current is

$$I = G_1(E - V_1)$$

and we write this information to the current source, see Fig. 5.2.1b. Once this has been done we are free to transform the voltage source, as is done in Fig. 5.2.1b, and we can proceed with analysis:

$$(G_1 + G_2 + G_3)V_1 - V_2G_3 = EG_1$$

$$-G_3V_1 + (G_3 + G_4)V_2 - \alpha G_1(E - V_1) = 0$$

Transferring the independent source in the second equation to the right we arrive at

$$(G_1 + G_2 + G_3)V1 - G_2V_2 = EG_1$$

$$(\alpha G_1 - G_3)V_1 + (G_3 + G_4)V_2 = \alpha G_1 E$$

Inserting numerical values and writing in matrix form:

$$\begin{bmatrix} 6 & -1 \\ 3 & 3 \end{bmatrix}\begin{bmatrix} V_1 \\ V_2 \end{bmatrix} = \begin{bmatrix} 4 \\ 8 \end{bmatrix}$$

The solutions are: $D = 21$, $N_1 = 20$, $N_2 = 36$, $V_1 = 20/21$ and $V_2 = 36/21$.

Another case where special attention must be paid is the network in Fig. 5.2.2a with a voltage controlled current source. The controlling voltage is taken across R_1 and we wish

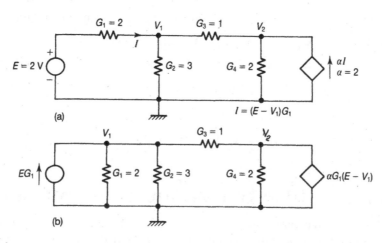

FIGURE 5.2.1   Transformation of controlling branch: (a) Given network. (b) Modified for nodal formulation.

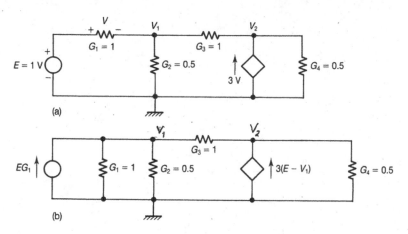

(a)

(b)

FIGURE 5.2.2   Transformation of controlling branch: (a) Given network. (b) Modified for nodal formulation.

to apply nodal formulation. The preprocessing first establishes that the controlling voltage is equal to

$$V = (E - V_1)G_1$$

The network is redrawn and the information transferred to the dependent current source. We are now free to transform the voltage source into a current source, see Fig. 5.2.2b. The nodal equations are

$$2.5V_1 - V_2 = 1$$

$$-V_1 + 1.5V_2 - 3(E - V_1) = 0$$

As always, the independent voltage source goes to the right and the system becomes

$$\begin{bmatrix} 2.5 & -1 \\ 2 & 1.5 \end{bmatrix} \begin{bmatrix} V_1 \\ V_2 \end{bmatrix} = \begin{bmatrix} 1 \\ 3 \end{bmatrix}$$

The solutions are $D = 5.75$, $N_1 = 4.5$ and $N_2 = 5.5$.

5.3.   SPLITTING OF SOURCES

Source transformations help us to modify networks if the voltage source is connected in series with one resistor or if the current source is in parallel with one resistor. This may not always be the case. Splitting of sources is the last step we need in order to prepare any network for analysis by either the nodal or mesh method.

We start with the example shown in Fig. 5.3.1a. The voltage source is connected to two resistors and we cannot immediately apply the transformation into a current source. We know that an ideal voltage source can deliver any current and that we normally do not connect voltage sources in parallel, but in this situation we intentionally replace one

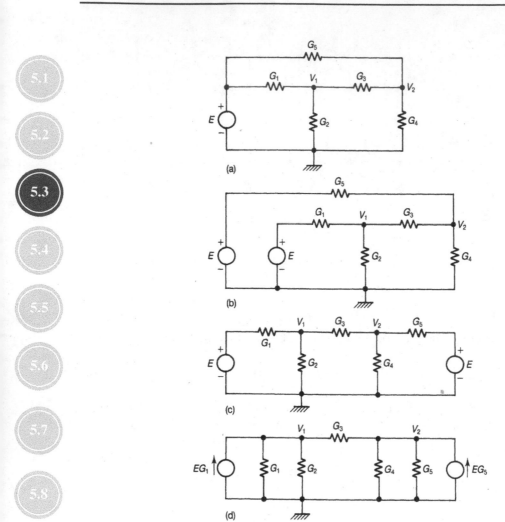

FIGURE 5.3.1   Example of voltage source splitting: (a) Original. (b) After splitting. (c) Redrawn. (d) Transformed for nodal formulation.

voltage source by two voltage sources with *the same voltage*. Once we have two such sources, we can separate them and still have the same voltages on the left side of G_1 and G_5. This separation is done in Fig. 5.3.1b and each conductance is in series with its own voltage source. The network is redrawn by taking G_5 to the right, as shown in Fig. 5.3.1c. We could apply mesh formulation to this network by converting conductances into resistances, but it would lead to a 3×3 system matrix. In the last step we transform the voltage sources into current sources and the network in Fig. 5.3.1d is prepared for nodal formulation

$$\begin{bmatrix} G_1 + G_2 + G_3 & -G_3 \\ -G_3 & G_3 + G_4 + G_5 \end{bmatrix} \begin{bmatrix} V_1 \\ V_2 \end{bmatrix} = \begin{bmatrix} E G_1 \\ E G_5 \end{bmatrix}$$

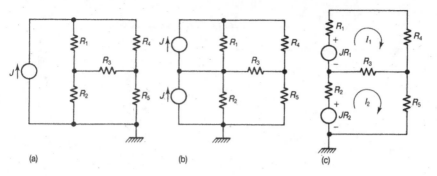

FIGURE 5.3.2 Example of current source splitting: (a) Original. (b) After splitting. (c) Transformed for mesh formulation.

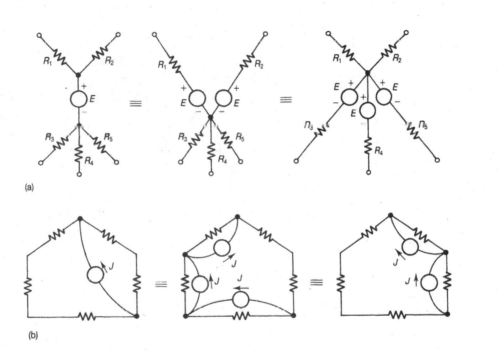

FIGURE 5.3.3   Splitting of sources: (a) Voltage sources. (b) Current sources.

The splitting in the above network was done as an exercise; we could have used the mesh method directly.

A similar splitting can also be applied to current sources. An example is in Fig. 5.3.2a. The current source can be split into two current sources in series if *both deliver the same current*. This has been done in Fig. 5.3.2b where now each source is connected in parallel with its own resistor. We could have used nodal formulation from the beginning but it would lead to a 3×3 system matrix. If we transform the current sources into voltage sources, as shown in Fig. 5.3.2c, we get a network suitable for mesh formulation and the system matrix will be only 2×2:

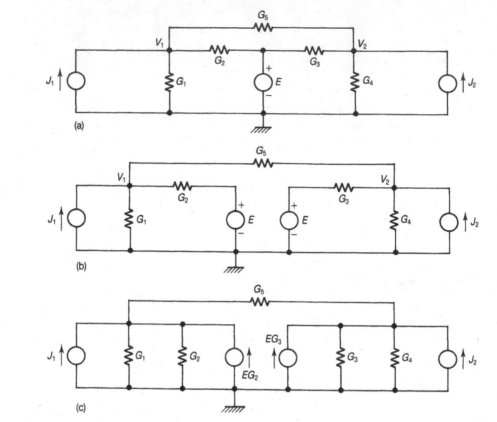

FIGURE 5.3.4 Example of voltage source splitting. (a) Original. (b) After splitting. (c) Transformed for nodal formulation.

$$\begin{bmatrix} R_1 + R_3 + R_4 & -R_3 \\ -R_3 & R_2 + R_3 + R_5 \end{bmatrix} \begin{bmatrix} I_1 \\ I_2 \end{bmatrix} = \begin{bmatrix} J\,R_1 \\ J\,R_2 \end{bmatrix}$$

More complicated situations can also occur. Fig. 5.3.3a indicates splitting of one voltage source into two or three voltage sources, each in series with a resistor. Fig. 5.3.3b shows splitting of one current source into two or three current sources, each in parallel with a resistor.

Fig. 5.3.4a shows a network with two current sources and one voltage source in the middle. We wish to prepare this network for nodal formulation. The voltage source is split into two voltage sources in Fig. 5.3.4b and each is transformed into a current source as shown in Fig. 5.3.4c. The nodal equations are:

$$\begin{bmatrix} G_1 + G_2 + G_5 & -G_5 \\ -G_5 & G_3 + G_4 + G_5 \end{bmatrix} \begin{bmatrix} V_1 \\ V_2 \end{bmatrix} = \begin{bmatrix} J_1 + E\,G_2 \\ J_2 + E\,G_3 \end{bmatrix}$$

Example 1.

Consider the network in Fig. 5.3.5 with one voltage source and one current source. Apply voltage splitting and current splitting to obtain the solution.

Figs. 5.3.6a,b,c show the steps for voltage splitting and transformation to current sources. It leads to the network 5.3.6 c and the nodal matrix equation is

$$\begin{bmatrix} 4 & 0 \\ 0 & 6 \end{bmatrix} \begin{bmatrix} V_1 \\ V_2 \end{bmatrix} = \begin{bmatrix} -1 \\ 4 \end{bmatrix}$$

with the solution $V_1 = -1/4$ and $V_2 = 2/3$.

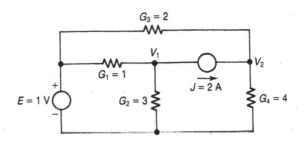

FIGURE 5.3.5   Network with a voltage and a current source.

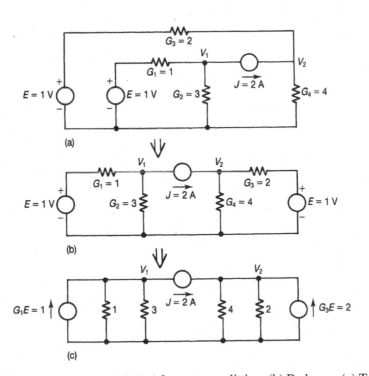

FIGURE 5.3.6   Network in Fig. 5.3.5:(a) After source splitting. (b) Redrawn. (c) Transformed for nodal formulation.

5.4. SUPERPOSITION PRINCIPLE

Linear networks obey the important law of superposition: If the network has several independent sources, then we can calculate the contribution of each of them individually, as if the others were not present, and add the results. In order to do it correctly, we must properly remove the sources which we are not considering at that moment: replace *independent* voltage sources by a short circuit and remove *independent* current sources from the network. We have stressed the words *independent*, because we must *never* remove the dependent sources. They stay in the network without any change.

At present we know only resistive networks and dc signals (voltages delivered by batteries) but the law is completely general and applies for any signals, as long as the network is linear.

Consider the network in Fig. 5.4.1. It has two independent current sources and one *VC*. The matrix equation is

$$\begin{bmatrix} G_1 + G_2 & -G_2 \\ -G_2 - g & G_2 + G_3 + g \end{bmatrix} \begin{bmatrix} V_1 \\ V_2 \end{bmatrix} = \begin{bmatrix} J_1 + J_2 \\ -J_2 \end{bmatrix}$$

(a) We solve first the case $J_2 = 0$:

$$\begin{bmatrix} 3 & -2 \\ -4 & 7 \end{bmatrix} \begin{bmatrix} V_1 \\ V_2 \end{bmatrix} = \begin{bmatrix} 1 \\ 0 \end{bmatrix}$$

The result is $V_1^a = \dfrac{7}{13}$ and $V_2^a = \dfrac{4}{13}$.

(b) In the next step we set $J_1 = 0$ and solve the system

$$\begin{bmatrix} 3 & -2 \\ -4 & 7 \end{bmatrix} \begin{bmatrix} V_1 \\ V_2 \end{bmatrix} = \begin{bmatrix} 2 \\ -2 \end{bmatrix}$$

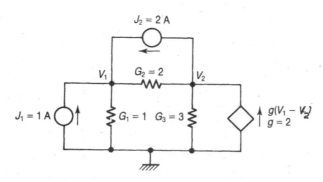

FIGURE 5.4.1   Demonstrating superposition principle.

with the result $V_1^b = \dfrac{10}{13}$ and $V_2^b = \dfrac{2}{13}$. Adding the results we get $V_1 = V_1^a + V_1^b = \dfrac{7}{13} + \dfrac{10}{13} = \dfrac{17}{13}$ and $V_2 = V_2^a + V_2^b = \dfrac{4}{13} + \dfrac{2}{13} = \dfrac{6}{13}$.

To show that the addition indeed gives a correct result, we now leave both sources in place and solve

$$\begin{bmatrix} 3 & -2 \\ -4 & 7 \end{bmatrix}\begin{bmatrix} V_1 \\ V_2 \end{bmatrix} = \begin{bmatrix} 3 \\ -2 \end{bmatrix}.$$

The two results are equal, as expected.

There are several important conclusions which we can draw from the above considerations and which are obvious only because we are using matrix algebra. The rules would be quite difficult to discover if we had used equation elimination.

1. Independent sources appear in the right hand column of the matrix equation.
2. Dependent sources appear in the system matrix.
3. Changing the nodes where the source is applied does not influence the matrix of the system.
4. The determinant of the system matrix (denominator of the solution) remains the same irrespective where we have the independent sources.

The fourth conclusion is perhaps the most important one, because it saves a lot of calculations.

Our next network in Fig. 5.4.2 is suitable for mesh formulation. It has one current controlled voltage source. The controlling current is indicated in the figure as I and we see that it is equal to the circulating current I_2. This information is transferred into the figure. The matrix equation is

$$\begin{bmatrix} R_1 + R_2 & -R_2 + r \\ -R_2 & R_2 + R_3 - r \end{bmatrix}\begin{bmatrix} I_1 \\ I_2 \end{bmatrix} = \begin{bmatrix} E_1 \\ -E_2 \end{bmatrix}.$$

Note again that the dependent source appears in the system matrix while the independent sources are in the right hand side vector only. We can set either E_1 or E_2 equal to zero, solve and add the results; this is left to the student as an exercise. If we use both sources simultaneously, we get $I_1 = \dfrac{3}{9}$ and $I_2 = -\dfrac{4}{9}$. We also note that setting $E_1 = 0$ is equivalent to short-circuiting this source; the same applies for E_2.

In a normal situation we solve the network with all its sources because this gives immediately the final result. We will see in the following two sections of this chapter that there may be sometimes special reasons why we should apply the sources one at a time. Even if we do so, we should still remember that the *system matrix does not change* and we must evaluate the system determinant only once.

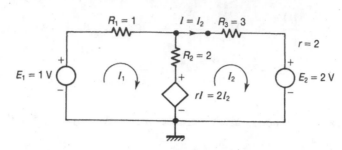

FIGURE 5.4.2   Demonstrating superposition principle.

5.5.   INPUT AND OUTPUT RESISTANCE

More complicated transformations require the concept of an input resistance (conductance) at various points of the network. The concept gives us additional information about the network and will be needed in the next section. The word *input* relates to any two terminals of the network where we can connect an independent source. The expression *input resistance* describes the ratio of the voltage across the input terminals, divided by the current flowing into the network through the same terminals. Sometimes we also use the expression *output resistance*; it is nothing but an input resistance calculated at the terminals where we normally take the output of the network.

Any pair of terminals where we can connect an independent source can be considered as an input terminal pair. The first example, shown in Fig. 5.5.1, deals with independent current sources. The network has three resistors, one voltage controlled current source and three independent current sources, not yet connected to the network.

If we connect *only one of the sources* and calculate the voltage across that particular source, we obtain the input resistance, R_{in}, as the ratio of the voltage across the source to the current delivered by the source.

Let us attach the first current source, J_1. Since one terminal of this source is grounded, the voltage across it will be V_1 and the input resistance will be $R_{in,1} = \dfrac{V_1}{J_1}$. If we disconnect the first source and attach instead the second source, J_2, we can calculate V_2, the voltage across this source, and obtain the input resistance at this second pair of terminals as the

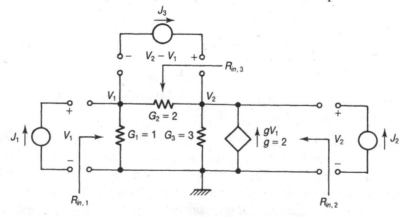

FIGURE 5.5.1   Input and output resistances, nodal formulation.

ratio $R_{in,2} = \dfrac{V_2}{J_2}$. Finally, if we attach the third current source, J_3, the voltage across it will be $V_2 - V_1$ and the input resistance at this third pair of terminals will be $R_{in,3} = \dfrac{V_2 - V_1}{J_3}$.

The important point is that we must *always remove all independent current sources except one* and calculate the voltage across the source which is left in the network. We can also proceed by preparing the equations with all independent sources simultaneously, but set always all except one equal to zero.

Let us now find the three input resistances. In all three cases, the network is the same and the dependent source, which is part of it, *must not be removed.* If all three independent sources are connected to the network, the system equation is

$$\begin{bmatrix} 3 & -2 \\ -4 & 5 \end{bmatrix}\begin{bmatrix} V_1 \\ V_2 \end{bmatrix} = \begin{bmatrix} J_1 - J_3 \\ J_2 + J_3 \end{bmatrix} \tag{5.5.1}$$

The determinant of the matrix is always the same, no matter which sources we remove from the right hand side: $D = 7$. Setting $J_2 = J_3 = 0$ we find the voltage across the first source to be

$$V_1 = \frac{5J_1}{7}$$

The input resistance, looking into the network from the left, is

$$R_{in,1} = \frac{V_1}{J_1} = \frac{5}{7} \tag{5.5.2}$$

If we keep only J_2, we must solve

$$\begin{bmatrix} 3 & -2 \\ -4 & 5 \end{bmatrix}\begin{bmatrix} V_1 \\ V_2 \end{bmatrix} = \begin{bmatrix} 0 \\ J_2 \end{bmatrix} \tag{5.5.3}$$

The solution is

$$V_2 = \frac{3J_2}{7}$$

and the input resistance at the terminals on the right is

$$R_{in,2} = \frac{3}{7} \tag{5.5.4}$$

Keeping only the third source, J_3, we must solve

$$\begin{bmatrix} 3 & -2 \\ -4 & 5 \end{bmatrix}\begin{bmatrix} V_1 \\ V_2 \end{bmatrix} = \begin{bmatrix} -J_3 \\ J_3 \end{bmatrix} \tag{5.5.5}$$

but now we need both $V_1 = \dfrac{-3J_3}{7}$ and $V_2 = \dfrac{-J_3}{7}$. The input resistance at the terminals on top is

$$R_{in,3} = \frac{V_2 - V_1}{J_3} = \frac{2}{7} \qquad (5.5.6)$$

If the network has independent voltage sources, we can also calculate the input resistance (or conductance) by evaluating the amount of current flowing from the source into the network. We will use the network Fig. 5.5.2. It has one current controlled voltage source. With both sources in place, the matrix equation is

$$\begin{bmatrix} R_1 + R_2 & -R_2 \\ -R_2 + r & R_2 + R_3 - r \end{bmatrix} \begin{bmatrix} I_1 \\ I_2 \end{bmatrix} = \begin{bmatrix} E_1 - E_2 \\ E_2 \end{bmatrix} \qquad (5.5.7)$$

The input resistance for the source E_1 is calculated by setting $E_2 = 0$ and by finding the mesh current I_1. Using numerical values

$$\begin{bmatrix} 3 & -2 \\ 0 & 3 \end{bmatrix} \begin{bmatrix} I_1 \\ I_2 \end{bmatrix} = \begin{bmatrix} E_1 \\ 0 \end{bmatrix} \qquad (5.5.8)$$

The determinant is $D = 9$ and the current I_1 is obtained by Cramer's rule

$$I_1 = \frac{3E_1}{9}$$

from which we get the input conductance

$$G_{in,1} = \frac{I_1}{E_1} = \frac{3}{9} \qquad (5.5.9)$$

The input resistance is, of course, the inverse.

In order to obtain the input conductance for the second source, we set $E_1 = 0$.

$$\begin{bmatrix} 3 & -2 \\ 0 & 3 \end{bmatrix} \begin{bmatrix} I_1 \\ I_2 \end{bmatrix} = \begin{bmatrix} -E_2 \\ E_2 \end{bmatrix} \qquad (5.5.10)$$

Since the current flowing into the network from this source is equal to $I_2 - I_1$, we must solve for both these currents. The system determinant is already known and using Cramer's rule we get $I_1 = -\frac{E_2}{9}$ and $I_2 = \frac{3E_2}{9}$.

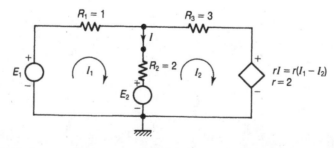

FIGURE 5.5.2   Input resistance for network with dependent sources

The current flowing into the network is

$$I_{in} = I_2 - I_1 = \frac{4E_2}{9}$$

and the input conductance is

$$G_{in} = \frac{I_{in}}{E_2} = \frac{I_2 - I_1}{E_2} = \frac{4}{9}$$

5.6. THEVENIN AND NORTON THEOREMS

Transformation of sources explained in section 1 of this chapter can be extended to more complicated situations. Consider Fig. 5.6.1 where we have one network with independent sources and one external network connected to it on the right. We wish to transform the left network into another one which will behave exactly equivalently, no matter how we change the network on the right. This can be achieved only if the network on the left can deliver the same current, I, while maintaining the same voltage, V, across the indicated terminals.

Two transformations exist, called the Thevenin and the Norton transformations. Schematically, both are given in Fig. 5.6.2. The upper part replaces the network which has independent sources by one voltage source and one resistor in series. This is called the Thevenin transformation. The lower part replaces the same network by a combination of a current source in parallel with the same resistor. This is called the Norton transformation. The two transformed sources on the right are also coupled by the equation which we have already derived in section 1 for the transformation of the voltage source into a current source.

To get the Thevenin transformation, we must perform two calculations, as we explain with the help of Fig. 5.6.1.

1. Disconnect the external network and calculate the voltage at the indicated terminals by taking into account all independent and dependent sources of the network on the left. The voltage we obtain, $V = E_{EQ}$, will be the value of the equivalent Thevenin voltage source.

2. Set all independent sources of the network on the left equal to zero (but keep all dependent sources) and attach one independent source at the terminals. If it is a voltage source, E, calculate the current which it is delivering, I. If it is a current source, J, calculate the voltage across it, V. The Thevenin equivalent resistance is

$$R_{EQ} = \frac{E}{I}$$

if we used a voltage source, or

$$R_{EQ} = \frac{V}{J}$$

if we used a current source.

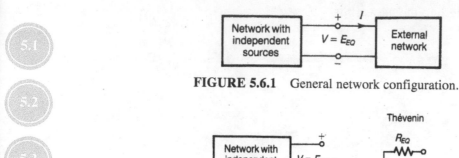

FIGURE 5.6.1   General network configuration.

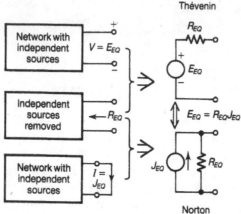

FIGURE 5.6.2   Thévenin and Norton equivalents.

The Thevenin transformation is particularly easy to obtain if the network has a form suitable for nodal equations. We will see it on the example in Fig. 5.6.3a. The network has one current source, $J_1 = 2\ A$, one voltage controlled current source and three resistors given by their conductance values. Output of the network is on the right. We have also indicated in the figure another (dotted) current source which we will use to calculate the output resistance. Keeping both sources in place we write

$$\begin{bmatrix} G_1 + G_2 & -G_2 \\ -G_2 - g & G_2 + G_3 \end{bmatrix} \begin{bmatrix} V_1 \\ V_2 \end{bmatrix} = \begin{bmatrix} J_1 \\ J_2 \end{bmatrix}$$

In order to obtain the output voltage, we first remove J_2 by setting it equal to zero. This is equivalent to removing the source from the network and making an open circuit in its place, as required by the rule which gets us the equivalent voltage source. Inserting numerical values

$$\begin{bmatrix} 3 & -2 \\ -4 & 5 \end{bmatrix} \begin{bmatrix} V_1 \\ V_2 \end{bmatrix} = \begin{bmatrix} 2 \\ 0 \end{bmatrix}$$

The determinant of the matrix is $D = 7$ and the solution is $V_2 = E_{EQ} = \dfrac{8}{7}$.

In the next step we remove the original independent source, J_1, by setting its value equal to zero, and connect the second source, $J_2 = 1$. This leads to

$$\begin{bmatrix} 3 & -2 \\ -4 & 5 \end{bmatrix} \begin{bmatrix} V_1 \\ V_2 \end{bmatrix} = \begin{bmatrix} 0 \\ J_2 \end{bmatrix}$$

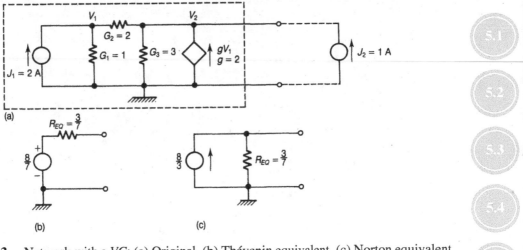

FIGURE 5.6.3 Network with a *VC*: (a) Original. (b) Thévenin equivalent. (c) Norton equivalent.

The only difference from the previous case is in the right hand side. We do not have to calculate the system determinant again since it is already known from the previous step. We find V_2 and the ratio $\dfrac{V_2}{J_1} = E_{EQ} = \dfrac{3}{7}$. The Thevenin equivalent of the network is in Fig. 5.6.3b. The Norton equivalent, if needed, is obtained by using equation (5.1.4); it is in Fig. 5.6.3c.

If the network is suitable for mesh equations, the Norton equivalent may be easier to obtain. Consider Fig. 5.6.4a and replace the output short circuit with another independent voltage source, to be later used for calculation of the output conductance. Note in particular the positions of the + and − signs with respect to the mesh current; its polarity must be such that it would deliver a current of the same direction into the network.

Keeping first both independent sources in the network, we write the mesh equations:

$$(R_1 + R_2)I_1 - R_2 I_2 + rI_2 = E_1$$

$$- R_2 I_1 + (R_2 + R_3)I_2 - rI_2 = E_2$$

and inserting numerical values

$$\begin{bmatrix} 3 & -1 \\ -2 & 4 \end{bmatrix} \begin{bmatrix} I_1 \\ I_2 \end{bmatrix} = \begin{bmatrix} E_1 \\ E_2 \end{bmatrix}$$

Set $E_2 = 0$ and calculate the current flowing through this source: $I_2 = J_{EQ}$. This requires the solution of

$$\begin{bmatrix} 3 & -1 \\ -2 & 4 \end{bmatrix} \begin{bmatrix} I_1 \\ I_2 \end{bmatrix} = \begin{bmatrix} 3 \\ 0 \end{bmatrix}$$

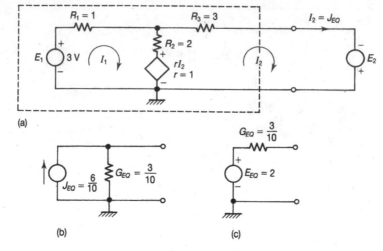

(a)

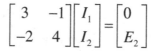

(b)   (c)

FIGURE 5.6.4   Network with a *CV*: (a) Original. (b) Norton equivalent. (c) Thévenin equivalent.

The determinant of the system is $D = 10$ and $J_{EQ} = \dfrac{6}{10}$. In the next step we set $E_1 = 0$ and keep E_2. We must solve the system

$$\begin{bmatrix} 3 & -1 \\ -2 & 4 \end{bmatrix}\begin{bmatrix} I_1 \\ I_2 \end{bmatrix} = \begin{bmatrix} 0 \\ E_2 \end{bmatrix}$$

and obtain $I_2 = \dfrac{3E_2}{10}$. The ratio $\dfrac{I_2}{E_2} = \dfrac{3}{10}$. is the output conductance. The Norton equivalent is in Fig. 5.6.4b the Thevenin in Fig. 5.6.4c.

Example 1.

Find the Thevenin equivalent for the network in Fig. 5.6.5a.

   We select nodal formulation and transform the voltage source into a current source, Simultaneously we must find, on the original network, the controlling current $I = G_1(E - V_1)$. The network is redrawn in figure b with another, auxiliary source on the right. The nodal equations with all sources kept in the equations as symbols are:

$$7V_1 - 2V_2 = 2E_1 + J_1$$

$$-2V_1 + 2V_2 - 4E_1 + 4V_1 = J_2$$

and

$$\begin{bmatrix} 7 & -2 \\ 2 & 2 \end{bmatrix}\begin{bmatrix} V_1 \\ V_2 \end{bmatrix} = \begin{bmatrix} 2E_1 + J_1 \\ 4E_2 + J_2 \end{bmatrix}$$

The determinant is $D = 18$.

(a) In the first step set $J_2 = 0$, insert values for the other sources and solve for V_2. The result is $N_2 = 20$ and $V_2 = E_{EQ} = \dfrac{20}{18}$.

(b) In the next step set $E_1 = 0$, $J_1 = 0$ and calculate again V_2. The result is now $N_2 = 7J_2$ and $\dfrac{V_2}{J_2} = R_{EQ} = \dfrac{7}{18}$.

The Thevenin and Norton equivalents are in Fig. 5.6.5c,d, respectively.

Let us summarize the steps: Both transformations require two solutions of the same network.

1. Calculate either the output voltage (Thevenin) or the output current (Norton) using all independent sources in the network.

2. Remove all independent sources, retain all dependent sources and calculate the resistance of the network from the output terminals.

3. If necessary, use (5.1.4) to get the desired form of the equivalent network.

In practical applications, it is useful to do the steps by first preparing the system matrix with all sources, even the auxiliary one for calculation of the resistance from the output terminals. This can save us a lot of difficulties:

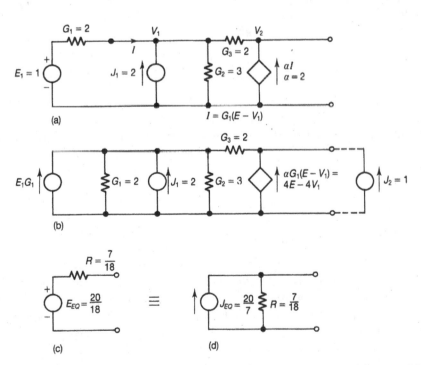

FIGURE 5.6.5 Network with a CC: (a) Original. (b) With the necessary sources. (c) Thévenin equivalent. (d) Norton equivalent.

1. We avoid mistakes by removing dependent sources.

2. Removal of the various independent sources is done simply by setting their values equal to zero in the right hand side vector.

3. We avoid unnecessary steps by realizing that the system determinant is the same for both solutions.

Although the two transformations are very practical because they simplify the source network, the resulting sources are equivalent only as far as the external network is concerned. This should be very clearly understood. The transformations are *not* equivalent in terms of power delivered by the sources. If there is no external network, the Thevenin equivalent does not consume any power, because there is no current flowing from the source. The Norton equivalent *does* draw power if it is disconnected from the external network.

5.7.   OPERATIONAL AMPLIFIERS

There exists one more ideal and widely used element which we have not covered yet. It is the ideal operational amplifier, OPAMP. Its application will be covered fully in Chapter 12, but the student will probably need some understanding earlier in his studies. For this reason we give here a simple, yet very efficient method for the analysis of networks with operational amplifiers.

An ideal operational amplifier is a special voltage controlled voltage source with two input terminals, i and j, as given in Fig. 5.7.1a. The voltages at these points are taken and their difference, $V_i - V_j$, is multiplied by the gain, A. Thus the output of the dependent voltage source, V_o, whose *one terminal is grounded* as shown, will be

$$V_o = A(V_i - V_j) \qquad (5.7.1)$$

The usual symbol for an operational amplifier, OPAMP, is in Fig. 5.7.1b. It is a common practice to mark the terminal i with a + sign and terminal j with – sign, as shown, but these signs do not mean that the first voltage is positive and the second negative. They only reflect the fact that in (5.7.1) the first one is taken positively and the second one is subtracted. For the same reason it is also an accepted practice to write (5.7.1) in the form

$$V_o = A(V_+ - V_-) \qquad (5.7.2)$$

which we will be using in the following.

The definition of the operational amplifier is not yet finished. In order to have an *ideal* OPAMP we assume that the gain is *infinitely large*. Infinity is always troublesome; we cannot display it and we cannot use it in calculators or in computers. To get out of this difficulty we recall from mathematics how to handle a variable which may become infinitely large. We first divide the equation by this variable and only afterwards let it grow without bound. Anything finite, divided by such variable, will become zero. Applying these steps to (5.7.2) we first obtain

$$\frac{V_0}{A} = V_+ - V_-$$ (5.7.3)

and then apply the limiting step. In this case the expression on the left becomes zero and

$$0 = V_+ - V_-$$

or, alternately

$$V_+ = V_-$$ (5.7.4)

This is a very important conclusion. It tells us that the voltage at the + terminal will be *exactly* equal to the voltage at the − terminal and *no other equation is needed*. In other words, although the ideal OPAMP is in principle a voltage controlled voltage source, we never use its equation.

There are some additional advantages: we can use nodal formulation. In addition, we need not even write any equation for the output node of the OPAMP. All the necessary information has been transferred to the statement that the input terminals are at the same potential. We will indicate that there is no need to write KCL at the output node by marking this node with a cross. We must still indicate the output voltage, V_{out}.

Several simple examples will clarify the procedure. If you are eager to learn more about networks with OPAMPS, turn to Chapter 12. Its initial sections will be relatively easy to follow; all you need is nodal formulation which we have already covered in detail.

Consider the network in Fig. 5.7.2a. The voltage source on the left is transformed into a current source by the Thevenin transformation, as shown in Fig. 5.7.2b. The output voltage, V_{out}, is marked, but the node is crossed out to indicate that we will not write the KCL there. The + terminal is grounded and is at zero potential; the − terminal must be, according to (5.7.4), at the same zero potential. This voltage is written at the terminal. Proceeding our standard way we write the current balance for the only node which does not have the cross:

$$V_-(G_1 + G_2) - G_2 V_{out} = EG_1$$ (5.7.5)

However, we know that the voltage V_- is zero and any finite expression multiplied by zero becomes zero. Thus the first term is zero and we did not even have to write the term. We could have written directly

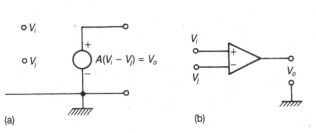

(a) (b)

FIGURE 5.7.1   Operational amplifier: (a) Equivalent network. (b) Symbol.

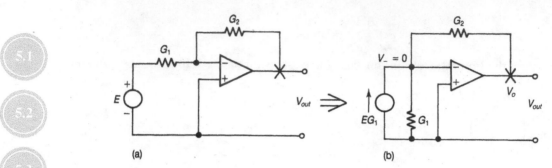

FIGURE 5.7.2   OPAMP inverting network: (a) Original. (b) After Thévenin transformation.

$$-G_2 V_{out} = EG_1$$

The solution is

$$V_{out} = -\frac{G_1}{G_2} E$$

Example 1.

Write the equations and find the output voltage for the network in Fig. 5.7.3. The voltage at the + terminal is equal to E and according to (5.7.4) the same voltage must be at the other input terminal. This we express by writing *the same* voltage at the - terminal, as done in Fig. 5.7.3. The output terminal is crossed out, but has the voltage V_{out}. We need to write the balance of currents at the – terminal only:

$$E(G_1 + G_2) - G_2 V_{out} = 0$$

The solution is

$$V_{out} = (1 + \frac{G_1}{G_2})E$$

Example 2.

Write the nodal equations for the network in Fig. 5.7.4a.

We first transform the source on the left into a current source. This leads to Fig. 5.7.4b. The output voltage, due to the short circuit, appears at the - terminal, and due to the fact that the element is an ideal OPAMP, the same voltage will be at the + terminal, where we write the voltage V_{out}. We have two nodes which are not crossed out and where we must write the KCL:

$$(G_1 + G_2 + G_3)V_1 - G_2 V_{out} - G_3 V_{out} = EG_1$$

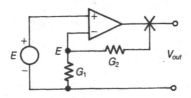

FIGURE 5.7.3 Noninverting OPAMP amplifier.

$$-G_2V_1 + (G_2 + G_4)V_{out} = 0$$

As usual, we put the equations into matrix form:

$$\begin{bmatrix} G_1 + G_2 + G_3 & -(G_2 + G_3) \\ -G_2 & G_2 + G_4 \end{bmatrix} \begin{bmatrix} V_1 \\ V_{out} \end{bmatrix} = \begin{bmatrix} EG_1 \\ 0 \end{bmatrix}$$

The determinant is $D = G_1(G_2 + G_4) + G_4(G_2 + G_3)$ and $N_o = G_1G_2E$. The output voltage is

$$V_{out} = \frac{G_1G_2}{G_1(G_2 + G_4) + G_4(G_2 + G_3)} E$$

Example 3.

Find the output voltage for the network in Fig. 5.7.5a.

Splitting the voltage source on the left into two voltage sources and transforming them into current sources leads to the network in Fig. 5.7.5b. V_{out} is marked, the node crossed out. The remaining two nodes *must be at the same potential*, which we express by writing the same voltage, V_1, at both. For these nodes we write separate nodal equations:

$$(G_2 + G_4)V_1 - G_4V_{out} = EG_2$$

$$(G_1 + G_3)V_1 = EG_1$$

In matrix form

$$\begin{bmatrix} G_2 + G_4 & -G_4 \\ G_1 + G_3 & 0 \end{bmatrix} = \begin{bmatrix} V_1 \\ V_{out} \end{bmatrix} \begin{bmatrix} EG_2 \\ EG_1 \end{bmatrix}$$

The determinant is $D = G_4(G_1 + G_3)$, for the numerator we must find the determinant

$$N_0 = \begin{vmatrix} G_2 + G_4 & EG_2 \\ G_1 + G_3 & EG_1 \end{vmatrix} = E(G_1G_4 - G_2G_3)$$

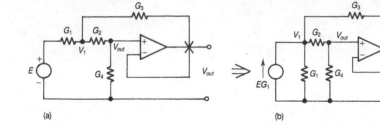

(a)   (b)

FIGURE 5.7.4 Network with OPAMP as an amplifier with unity gain: (a) Original. (b) Independent voltage source transformed into a current source.

and the solution is

$$V_{out} = \frac{G_1 G_4 - G_2 G_3}{G_4 (G_1 + G_3)} E.$$

All networks with ideal OPAMPs, modeled as in Fig. 5.7.1, can be solved by this method. The only condition is that one end of the dependent voltage source, see Fig. 5.7.1b, must be grounded. The method can be extended to *all* voltage sources, dependent or independent, as long as one of their terminals is grounded. Detailed explanations are in Chapter 12, but here we will use the method for the problems solved above.

Example 4.

Analyze the network in Fig. 5.7.2a by omitting the Thevenin transformation.

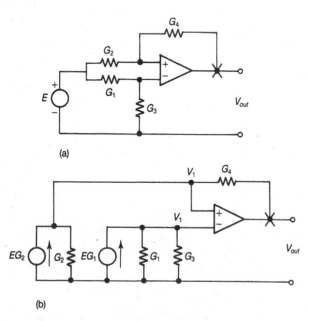

(a)

(b)

FIGURE 5.7.5 Simple network with OPAMP: (a) Original. (b) After Thévenin transformation.

The network is redrawn in Fig. 5.7.6. but now two nodes are crossed out, because they have voltage sources which have the other terminal grounded. The equation is written for only one node, omitting the term which would be multiplied by zero:

$$- G_2 V_{out} - G_1 E = 0$$

which leads to the same result as above.

Example 5.

Use the method for the network in Fig. 5.7.5.

The network is redrawn in Fig. 5.7.7. Crossed out are the two nodes with the voltage sources and the equations are written at the remaining nodes. They are the same as before,

$$(G_2 + G_4)V_1 - G_4 V_{out} = EG_2$$

$$(G_1 + G_3)V_1 = EG_1$$

Example 6.

Write the equations for the network in Fig. 5.7.8 by keeping all elements as variables. Put the equations into matrix form. Afterwards insert numerical values and solve for V_{out}. Let the input voltage be $E = 1V$.

The crosses are at two nodes and voltages are written at three nodes. We write the equations in the sequence of numbers in circles:

$$(G_1 + G_2 + G_3)V_1 - G_2 V_2 - G_3 V_{out} = EG_1$$

$$- G_2 V_1 + (G_2 + G_4)V_2 = 0$$

$$(G_5 + G_6 + G_7)V_2 - G_7 V_{out} = EG_5$$

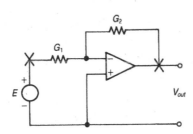

FIGURE 5.7.6   Analyze the network in Fig. 5.7.2a without Thévenin transformation.

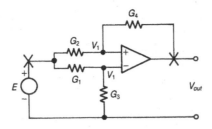

FIGURE 5.7.7   Analyze the network in Fig. 5.7.7a without Thévenin transformation.

In matrix form:

$$\begin{bmatrix} G_1+G_2+G_3 & -G_2 & -G_3 \\ -G_2 & G_2+G_4 & 0 \\ 0 & G_5+G_6+G_7 & -G_7 \end{bmatrix} \begin{bmatrix} V_1 \\ V_2 \\ V_{out} \end{bmatrix} = \begin{bmatrix} EG_1 \\ 0 \\ EG_5 \end{bmatrix}.$$

Inserting numerical values:

$$\begin{bmatrix} 4 & -1 & -2 \\ -1 & 2 & 0 \\ 0 & 6 & -1 \end{bmatrix} \begin{bmatrix} V_1 \\ V_2 \\ V_{out} \end{bmatrix} = \begin{bmatrix} E \\ 0 \\ 2E \end{bmatrix}.$$

The solution is $V_1 = \dfrac{6}{5} = 1.2V$, $V_2 = \dfrac{3}{5} = 0.6V$, $V_{out} = \dfrac{8}{5} = 1.6V$.

We will end this chapter with some recommendations.

5.8.   RECOMMENDATIONS

The various methods we discussed so far give us the possibility of analyzing any network by either the nodal or the mesh formulation. In some cases this involves quite a bit of redrawing and preprocessing before we are able to write the equations. The purpose of these steps is to reduce the number of equations to a minimum because the solution process is the most time consuming.

In our explanations we treated both the nodal and mesh formulations equally, but they are not equally important. As long as the networks are small, like our examples, it does not matter much which one we select. If we go to larger networks, then the nodal method is much more practical. Its results are voltages which are usually easier to measure than currents. Even more importantly, the nodal method is general while the mesh method applies to planar networks only, that is to networks which can be drawn on paper without any element crossing some other element. It is sometimes quite difficult to establish that a large network is planar.

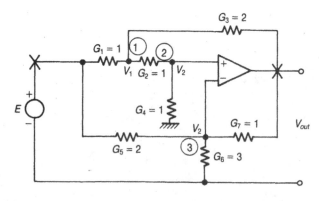

FIGURE 5.7.8   Analyze without Thévenin transformation.

We will remember the following rule of thumb for our studies and for all hand analyses:

1. If all sources in a planar network are voltage sources then it is probably easiest to use the mesh formulation.

2. If the sources are mixed, it is generally better to transform the voltage sources into current sources and apply nodal formulation. In such cases use conductances, G, to handle resistors.

The preprocessing which we must go through and the redrawing of figures which is sometimes necessary are reasonable when we are learning the tricks of the trade. In addition, the transformations sometimes reduce the size of the system, a point of utmost importance in hand calculations. Unfortunately, the transformations are also sources of possible errors and programs for analysis never use them. Another method, called modified nodal, is preferred; it will be explained later.

PROBLEMS CHAPTER 5

P.5.1 Transform the voltage source into a current source and calculate the nodal voltage V_1.

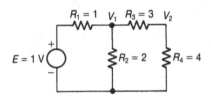

P.5.2 Transform the voltage source into a current source and calculate the nodal voltage V_1.

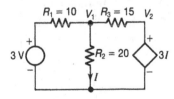

P.5.3 Transform the voltage into a current source, calculate V and I.

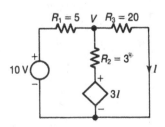

P.5.4 Add the two voltages on the left, transform all voltage sources into current sources and obtain V_1.

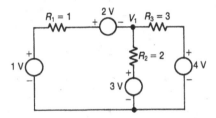

P.5.5 Transform the voltage sources into current sources and calculate V_A.

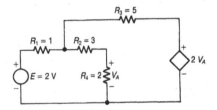

P.5.6 Transform the current source into a voltage source and calculate the voltages V_1 and V_2.

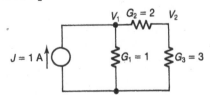

P.5.7   Transform the current sources into voltage sources and calculate I_A.

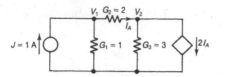
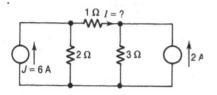

P.5.8   Transform the current sources into voltage sources and calculate the current I.

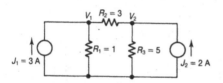

P.5.9   Transform the current sources into voltage sources and calculate V_1 and V_2.

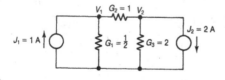

P.5.10   Transform the current sources into voltage sources. Calculate the voltage across G_2.

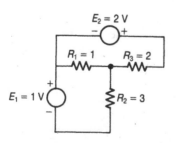

P.5.11   Use voltage splitting of the voltage source on the left and calculate the voltage across R_3.

P.5.12   Use source splitting and then transform the voltage sources into current sources. Apply nodal formulation to get the voltage across R_3.

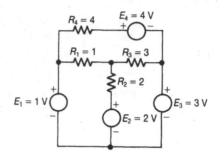

P.5.13   Use voltage source splitting, transform the sources into current sources and calculate the voltage across R_2.

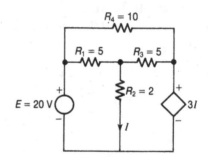

P.5.14   Apply current source splitting to J_2 and transform the current sources into voltage sources and use mesh formulation. Calculate the three nodal voltages. Note that the network is directly suitable for nodal formulation, but we wish to practice something else.

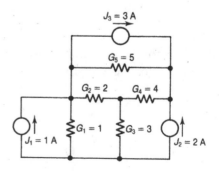

P.5.15 Calculate the Thevenin and Norton equivalents. The auxiliary source is indicated by dotted connection. Write the equations with both sources simultaneously but in the solution always use only one of them.

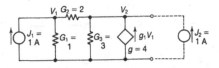

P.5.16 Do the same as in P.5.15.

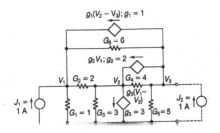

P.5.17 Do the same as in P.5.15.

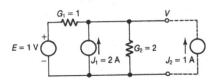

P.5.18 Do the same as in P.5.15

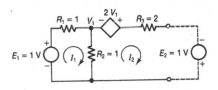

P.5.19 Do the same as in P.5.15.

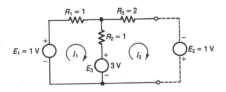

P.5.20 Do the same as in P.5.15.

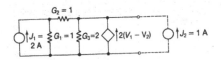

P.5.21 Express the current I in terms of E, V_1 and G_E, transform the voltage source into a current source and use nodal formulation to calculate the nodal voltages V_1 and V_2. Keep α as a variable.

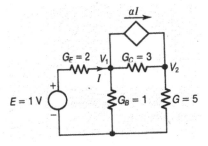

P.5.22 Use voltage splitting to obtain the voltage V. Also find the Thevenin equivalent for the indicated output.

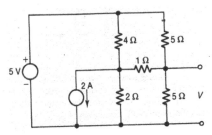

P.5.23 Find the Thevenin and Norton equivalent.

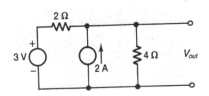

P.5.24   Find the output voltage by first transforming the voltage source into a current source.

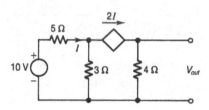

P.5.25   Do the same as in P.5.24.

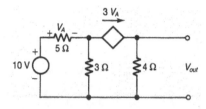

P.5.26   Find the output voltage by transforming first the voltage source into a current source and applying nodal formulation.

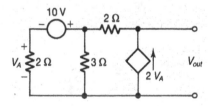

P.5.27   Do the same as in P.5.26.

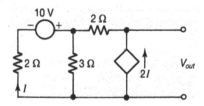

P.5.28   Find the Thevenin and Norton equivalents.

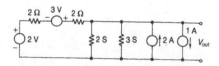

P.5.29   Find the Thevenin and Norton equivalents.

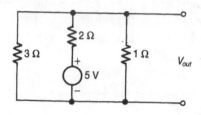

P.5.30   Use voltage splitting and nodal formulation to calculate the voltages V_1 and V_2. Keep G_3 as a variable.

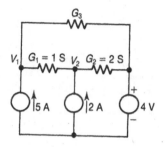

P.5.31   Solve the problem P.5.1 by keeping all resistors as variables.

P.5.32   Solve problems P.5.2 through P.5.5 by keeping the constants of the dependent sources as variables.

P.5.33   In problem P.5.7 keep the conductances and the α of the CC as variables.

P.5.34   In the problem P.5.13 keep the constant r of the CV as a variable and take the output across the dependent source. Select such value of r that the output voltage will be 10 V.

P.5.35 Transform the voltage source and use nodal formulation to get the nodal voltages V_1 and V_2.

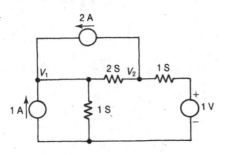

P.5.36 Transform the voltage source, use nodal formulation and obtain V_{out}.

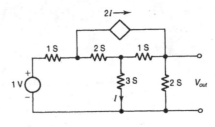

6 CAPACITORS AND INDUCTORS

INTRODUCTION

This chapter introduces two elements of fundamental importance: the capacitors and inductors. The relationships of their currents and voltages are more complicated than those for the resistors, and are expressed by differentiation and integration. We also define two new variables, the charge of the capacitor and the flux of the inductor. For the sake of completeness we give the basic relationships for nonlinear elements but our discussions will consider linear elements only.

The use of capacitors and inductors leads to differential and algebraic equations. Unfortunately, writing the necessary differential equations, and also solving them, is a fairly complicated process. In the case of linear networks it is much more convenient to use the Laplace transform. It can unify writing of the equations for time domain as well as frequency domain and is much easier to apply. All the necessary steps will be covered in the following chapters. In this chapter we restrict the explanations to the definitions and to connections of capacitors and inductors. The theoretical steps may even be skipped, but it is important to explain series and parallel connections of capacitor or inductors

6.1. CAPACITORS

A capacitor is an element composed of two conducting materials, separated by an insulator. We will first say a few words about its use and how it can be manufactured.

Technological advances constantly change the way capacitors are produced, but the student can make a reasonable capacitor quite easily by using the simplest materials available to him. The insulating material can be paper. If it is waxed, even better. The conducting material can be tin foil from a chocolate bar. Take a narrow strip of the paper, place somewhat narrower tin foil strips on each side and put another strip of paper on top. Connect a wire to each of the foils and roll together making sure that the two wires or foils do not touch. The result is a useful capacitor, similar to capacitors produced commercially for many years. If we connect the wires to the terminals of a car battery, a certain amount of electrons will be transferred to the capacitor. If we disconnect it from the battery, the electrons will stay and we will be able to measure a voltage on the two leads. If the insulating material is a good one, the electrons will stay for a long time; thus the capacitor can store energy. For an initial understanding of its usefulness let us say that an ideal capacitor stops dc current but lets changing currents pass through, making it easier for signals which

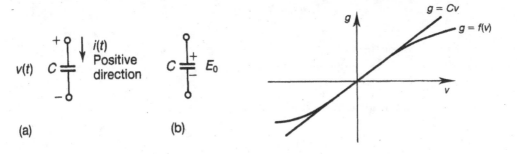

FIGURE 6.1.1   Symbol for a capacitor: (a) Voltage and current orientation.(b) Initial voltage E_0.

FIGURE 6.1.2 q–v characteristic of capacitors.

are fast. The property is somewhat similar to the resistance, but a resistance which is small for fast signals and infinite for dc.

If electrons are stored on the capacitor, we say that there is charge on it. The charge is denoted by q or $q(t)$ (if we wish to indicate dependence on time) and is measured in coulombs, C. The symbol for the capacitor is in Fig. 6.1.1a and the same letter, C, is used for it.

In a general case, the charge is a function of the voltage across the capacitor,

$$q(t) = f[v(t)] \tag{6.1.1}$$

and we can plot q as function of v, as shown in Fig. 6.1.2. The current through the capacitor is the derivative of the charge with respect to time,

$$i(t) = \frac{d\,q(t)}{d\,t} \tag{6.1.2}$$

We can also express the charge as function of the current,

$$q(t) = \int_{-\infty}^{t} i(\tau)d\tau \tag{6.1.3}$$

but it is inconvenient to work with an integral having its lower bound at $-\infty$. We prefer to split (6.1.3) into two integrals

$$q(t) = \int_{-\infty}^{0} i(\tau)d\tau + \int_{0}^{t} i(\tau)d\tau, \tag{6.1.4}$$

denote the first integral as initial charge at time $t = 0$ and use the symbol q_0 or $q(0)$,

$$q(t) = q_0 + \int_{0}^{t} i(\tau)d\tau \tag{6.1.5}$$

These expressions are valid for any capacitor.

A special case is the linear capacitor for which the charge is directly proportional to the voltage,

$$q(t) = C v(t). \tag{6.1.6}$$

and the proportionality constant is called the *capacitance*, C. It is measured in farads, F. The linear relationship of the voltage and charge is indicated in Fig. 6.1.2. by the straight line.

The current through a linear capacitor is given by

$$i(t) = C \frac{d v(t)}{d t} \tag{6.1.7}$$

Positive direction of current is related to the voltage across the capacitor by the same rules as before: If the current flows from + to −, then it is taken as positive. These conventions are also illustrated in Fig. 6.1.1a.

If we invert equation (6.1.7), we get an expression for the voltage on a linear capacitor

$$v(t) = \frac{1}{C} \int_{-\infty}^{t} i(\tau) d\tau \tag{6.1.8}$$

Similarly as above, we split this integral into two,

$$v(t) = \frac{1}{C} \int_{-\infty}^{0} i(\tau) d\tau + \frac{1}{C} \int_{0}^{t} i(\tau) d\tau$$

denote the first integral by E_0 and call it the *initial* voltage on the capacitor,

$$v(t) = E_0 + \frac{1}{C} \int_{0}^{t} i(\tau) d\tau \tag{6.1.9}$$

We are using the letter E because, as we will see later, this initial voltage acts, in some sense, like an independent source. The way we will indicate the initial voltage in networks is shown in Fig. 6.1.1b.

The power relationship of the voltage across the capacitor and the current flowing through it is given by

$$p(t) = v(t) \, i(t). \tag{6.1.10}$$

The energy stored in a capacitor is given by the integral of (6.1.10),

$$w_C(t) = \int_{-\infty}^{t} p(\tau) d\tau = C \int_{-\infty}^{t} v(\tau) \frac{d v(\tau)}{d\tau} d\tau = C \int_{-\infty}^{t} v(\tau) d v(\tau) = \frac{1}{2} C v^2(t) \tag{6.1.11}$$

6.2. CONNECTIONS OF CAPACITORS

Capacitors can be connected in parallel or in series and we now derive the rules how to obtain the resulting capacitor values.

Consider first the parallel connection shown in Fig. 6.2.1. It is some kind of a current divider because all capacitors have the same voltage but different currents can flow through them. Let the current source change with time which we indicate by writing $j(t)$. The current flowing through each capacitor is given by equation (6.1.7):

$$i_1(t) = C_1 \frac{dv(t)}{dt}$$

$$i_2(t) = C_2 \frac{dv(t)}{dt} \qquad (6.2.1)$$

$$i_3(t) = C_3 \frac{dv(t)}{dt}$$

The respective currents add and give the overall current flowing into the combination,

$$j(t) = i_1(t) + i_2(t) + i_3(t)$$

$$= (C_1 + C_2 + C_3) \frac{dv(t)}{dt} \qquad (6.2.2)$$

We can denote

$$C_{parallel} = C_1 + C_2 + C_3 \qquad (6.2.3)$$

and obtain the total capacitance for a parallel connection. It is equal to the sum of the individual capacitances. The formula can be extended to any number of capacitors connected in parallel.

Series connection of capacitors is shown in Fig. 6.2.2. The combination resembles a voltage divider because the same current flows through all elements but individual voltages can differ. For each capacitor we can write

$$v_1(t) = \frac{1}{C_1} \int_0^t i(\tau) d\tau$$

$$v_2(t) = \frac{1}{C_2} \int_0^t i(\tau) d\tau \qquad (6.2.4)$$

$$v_3(t) = \frac{1}{C_3} \int_0^t i(\tau) d\tau$$

The overall voltage is equal to

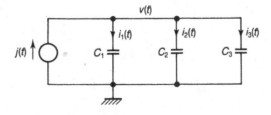

FIGURE 6.2.1 Parallel connection of capacitors.

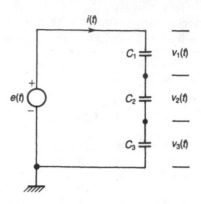

FIGURE 6.2.2   Series connection of capacitors.

$$e(t) = v_1(t) + v_2(t) + v_3(t) \tag{6.2.5}$$

Inserting from (6.2.4) we get

$$e(t) = (\frac{1}{C_1} + \frac{1}{C_2} + \frac{1}{C_3}) \int_0^t i(\tau) \, d\tau$$

and the overall capacitance for series connection is

$$\frac{1}{C_{series}} = \frac{1}{C_1} + \frac{1}{C_2} + \frac{1}{C_3} \tag{6.2.6}$$

The formula is clearly valid for any number of capacitors connected in series.

If we compare these networks with similar resistive networks we realize that the capacitor is in some sense similar to the conductance, because the values add for parallel connections. For series connections we must add inverted values of capacitances and then take the reciprocal of this sum to get the equivalent capacitance.

6.3.   INDUCTORS

An inductor is usually manufactured as a coil of insulated wire, often with an iron core or with a material which has properties similar to iron. The symbol for an inductor is in Fig. 6.3.1a.

Similarly as in the case of the capacitor, we give first an informal and very simplified explanation of its application. If we wish to have a good inductor, we must use a wire which has low resistance, for instance silver. Suppose that we manage to produce an inductor with a wire that has no resistance. Then the inductor will have properties which are opposite to those of a capacitor. It lets the dc current pass through, but presents increasingly larger 'resistance' to rapidly changing signals. If a current flows through the inductor, it creates around it magnetic field which stores energy and which can have influence on other inductors placed nearby. This property is used in the construction of transformers but we postpone such details. The word we normally use is *flux*, we denote it by Φ and measure it in units called *webers*, Wb.

In a general case, the flux is a function of the current flowing through the inductor,

$$\Phi(t) = f[i(t)] \tag{6.3.1}$$

We can plot this relationship, as shown in Fig. 6.3.2. The voltage across the inductor is defined as the derivative of the flux with respect to time,

$$v(t) = \frac{d\Phi(t)}{dt} \tag{6.3.2}$$

We can also invert this relationship and get

$$\Phi(t) = \int_{-\infty}^{t} v(\tau)d\tau \tag{6.3.3}$$

It is not convenient to work with the infinite bounds. We usually split this integral into two

$$\Phi(t) = \int_{-\infty}^{0} v(\tau)d\tau + \int_{0}^{t} v(\tau)d\tau \tag{6.3.4}$$

and denote the first integral as an initial flux, $\Phi(0)$ or Φ_0,

$$\Phi(t) = \Phi_0 + \int_{0}^{t} v(\tau)d\tau \tag{6.3.5}$$

These expressions are valid for any inductor.

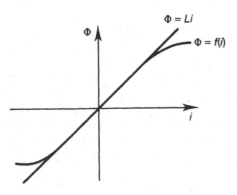

(a)   **(b)**

FIGURE 6.3.1   Symbol for an inductor: (a) Voltage and current orientations. (b) Initial current J_0.

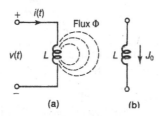

FIGURE 6.3.2 Φ–i characteristic of inductors.

A special case is the linear inductor. In this case the flux is directly proportional to the current,

$$\Phi(t) = L i(t).$$
(6.3.6)

and the proportionality constant is called the *inductance*, L. It is measured in henries, H. This relationship is represented by the straight line in Fig. 6.3.2.

The voltage across a linear inductor is given by

$$v(t) = L \frac{d i(t)}{d t}$$
(6.3.7)

and is related to the positive direction of the current by the same rules as always: the current is positive if it flows from $+$ to $-$. The convention is also illustrated in Fig. 6.3.1a.

If we invert equation (6.3.7), we get an expression for the current through a linear inductor

$$i(t) = \frac{1}{L} \int_{-\infty}^{t} v(\tau) d\tau$$

Similarly as above, we split this integral into two integrals,

$$i(t) = \frac{1}{L} \int_{-\infty}^{0} v(\tau) d\tau + \frac{1}{L} \int_{0}^{t} v(\tau) d\tau$$

denote the first integral by J_0 and call it the *initial* current through the inductor,

$$i(t) = J_0 + \frac{1}{L} \int_{0}^{t} v(\tau) d\tau.$$
(6.3.9)

We are using the letter J because this initial current acts, in some way, similarly as an independent current source. We will mark the initial current through the inductor as shown in Fig. 6.3.1b.

The power relationship of the voltage across the inductor and the current flowing through it is

$$p(t) = v(t) i(t)$$
(6.3.10)

The inductor is an energy storing device and the energy stored in it is given by the integral of (6.3.10),

$$w_L(t) = \int_{-\infty}^{t} p(\tau) d\tau = L \int_{-\infty}^{t} i(\tau) \frac{d i(\tau)}{d\tau} d\tau = L \int_{-\infty}^{t} i(\tau) d i(\tau) = \frac{1}{2} L i^2(t)$$
(6.3.11)

The reader may have noticed that the material of this section is remarkably similar to the material of the first section of this chapter. This is no coincidence: the capacitor and the inductor are indeed very similar in the sense that all we do is interchange voltages by currents and vice versa.

6.4. CONNECTIONS OF INDUCTORS

Inductors can be connected in series or in parallel and we will derive the rules how to obtain the resulting inductance values.

We first consider the series connection, shown in Fig. 6.4.1. It resembles the voltage divider: the same current flows through the individual elements but the voltages across them differ. The sum of individual voltages is equal to the voltage of the source, $e(t)$. Using equation (6.3.7) we write the voltage across each inductor

$$v_1(t) = L_1 \frac{di(t)}{dt}$$

$$v_2(t) = L_2 \frac{di(t)}{dt} \qquad (6.4.1)$$

$$v_3(t) = L_3 \frac{di(t)}{dt}$$

The respective voltages add and give the overall voltage across the combination

$$e(t) = v_1(t) + v_2(t) + v_3(t) \qquad (6.4.2)$$

Inserting from (6.4.1)

$$e(t) = (L_1 + L_2 + L_3) \frac{di(t)}{dt}$$

We can denote

$$L_{series} = L_1 + L_2 + L_3 \qquad (6.4.3)$$

and obtain the total inductance for the series connection. It is equal to the sum of the individual inductances. The formula can be extended to any number of inductors connected in series.

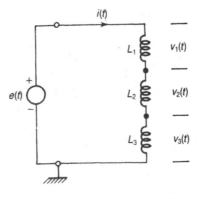

FIGURE 6.4.1   Series connection of inductors.

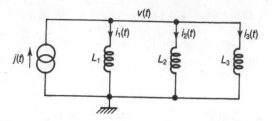

FIGURE 6.4.2   Parallel connection of inductors.

Parallel connection of inductors is shown in Fig. 6.4.2. The combination resembles a current divider where the voltage is common to all elements but the currents through the individual elements differ. For each inductor we can write

$$i_1(t) = \frac{1}{L_1} \int_0^t v(\tau)\, d\tau$$

$$i_2(t) = \frac{1}{L_2} \int_0^t v(\tau)\, d\tau \tag{6.4.4}$$

$$i_3(t) = \frac{1}{L_3} \int_0^t v(\tau)\, d\tau$$

The total current is equal to

$$j(t) = i_1(t) + i_2(t) + i_3(t) \tag{6.4.5}$$

Inserting from (6.4.4) we get

$$j(t) = \left(\frac{1}{L_1} + \frac{1}{L_2} + \frac{1}{L_3}\right) \int_0^t v(\tau)\, d\tau$$

and the overall inductance for parallel connection is

$$\frac{1}{L_{parallel}} = \frac{1}{L_1} + \frac{1}{L_2} + \frac{1}{L_3} \tag{6.4.6}$$

The formula is clearly valid for any number of inductors connected in parallel.

If we compare the connections of inductors with similar connections of resistors, we see that inductors are in some sense similar to resistors, because their inductances are added in series connections. In parallel connections we must add the inverted values of individual inductances. We thus again observe the dual nature of capacitors and inductors.

PROBLEMS CHAPTER 6

P.6.1(a)–(c) Find the equivalent capacitances of the networks.

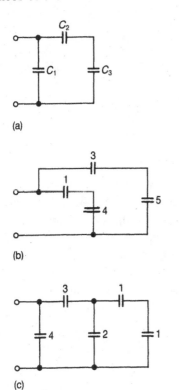

(a)

(b)

(c)

P.6.2 Find the equivalent capacitance of the combination

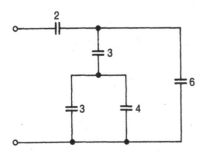

P.6.3(a)–(c) Find equivalent inductances of the networks.

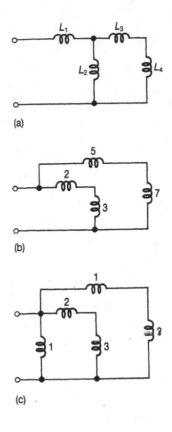

(a)

(b)

(c)

P.6.4 Find equivalent inductance for the combination.

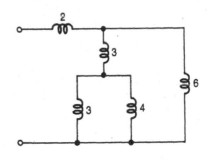

7 NETWORKS WITH CAPACITORS AND INDUCTORS

INTRODUCTION

This chapter covers the first steps in the analysis of linear networks in the Laplace transform. The concepts of impedance and admittance are defined and it is shown that analysis of networks which have resistors, dependent sources, inductors and capacitors, proceeds along exactly the same lines as were covered in the previous chapters. The only difference is that instead of only R or G we use other expressions. Examples show that in the Laplace transform the solutions of networks are expressed as ratios of two polynomials in the variable s. The only purpose of this chapter is to introduce to the student the necessary mechanics how to analyze arbitrary networks. Detailed explanations and the theories are covered in subsequent chapters.

7.1. IMPEDANCES AND ADMITTANCES

Laplace transform is a method used for solving linear differential equations and all linear networks can be solved by it. Its main feature is that it replaces differentiation by a multiplication with a special variable, s. It also replaces integration by the division by s and thus transforms transcendental operations into algebraic operations. For instance, the expression (6.1.7), repeated here for convenience

$$i = C\frac{dv}{dt} \tag{7.1.1}$$

is replaced by

$$I = sCV \tag{7.1.2}$$

and Eq. (6.3.7),

$$v = L\frac{di}{dt} \tag{7.1.3}$$

is replaced by

$$V = sLI \tag{7.1.4}$$

If we take the integral expressions, then (6.1.8), repeated here

$$v = \frac{1}{C} \int_0^t i(\tau)d\tau \qquad (7.1.5)$$

is replaced by

$$V = \frac{1}{sC}I \qquad (7.1.6)$$

and (6.3.8), also repeated,

$$i = \frac{1}{L} \int_0^t v(\tau)d\tau \qquad (7.1.7)$$

is replaced by

$$I = \frac{1}{sL}V \qquad (7.1.8)$$

The reader may have noticed that the integral forms (7.1.5) and (7.1.7) were slightly modified and start from zero. In all the above expressions we have neglected the possibility that an initial charge can be stored on the capacitor or that an initial flux may be created around the inductor. This we do intentionally, to make the initial explanations as simple as possible. We will return to the full possibilities later.

At this moment we do not say what s exactly means and we will see later that it can be given different meanings, depending on whether we are interested in time domain or frequency domain responses. Until now, the source was always a dc source; this may not be so in the future. From now on, E or J will be *symbols* for the two independent sources, essentially indicating that such terms should get to the right side of equations. Their exact meaning and/or values will differ from application to application and different substitutions will be needed.

Looking at the above equations, we realize that they all have the form of the Ohm's law. In fact, we can collect the terms and write

$$Y = \begin{cases} G \\ sC \\ \dfrac{1}{sL} \end{cases} \qquad (7.1.9)$$

The symbol Y is called the *admittance* and applies to all three forms. We can also collect the other set and write

$$Z = \begin{cases} R \\ \dfrac{1}{sC} \\ sL \end{cases} \qquad (7.1.10)$$

The symbol Z applies for all three terms and is called the *impedance*. The set (7.1.9) is used whenever we write the nodal equations, the set (7.1.10) whenever we write the mesh equations. What is most important, all the analysis steps we covered in previous chapters are directly applicable to networks with capacitors and inductors. All we do is substitute the proper expression for the element.

7.2. NETWORKS WITH CAPACITORS

If the network contains only resistors and capacitors, then it is natural to use nodal formulation and transform all sources (dependent or independent) to current sources. As before, the resistors are expressed in their conductance equivalents.

The first network which we will analyze is shown in Fig. 7.2.1. The currents are and $I_G = GV$, $I_C = sCV$ the nodal equation is

$$(G + sC)V = J$$

The unknown voltage V is

$$V = \frac{1}{G + sC} J$$

We stress again that, depending on application, s as well as J may be later replaced by something else.

Suppose that instead of the current source we have a voltage source and the network in Fig. 7.2.2. In this case it is easy to use the mesh method and write

$$(R + \frac{1}{sC})I = E$$

The current is

$$I = \frac{sC}{sCR + 1} E$$

and the output voltage is given by

$$V_{out} = Z_C I = \frac{1}{sCR + 1} E$$

The network could also be looked upon as a voltage divider. Using impedances we have

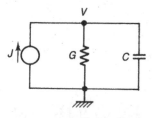

FIGURE 7.2.1   GC network with a current source.

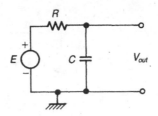

FIGURE 7.2.2 RC network with a voltage source.

$$V_{out} = \frac{Z_c}{Z_R + Z_c} E = \frac{1}{sCR + 1} E$$

Another network is in Fig. 7.2.3a, with a voltage source, two capacitors and two resistors. The voltage source is first transformed into a current source as shown in Fig. 7.2.3b. The two nodal equations are

$$(G_1 + sC_1 + G_2)V_1 - G_2V_2 = EG_1$$
$$-G_2V_1 + (G_2 + sC_2)V_2 = 0$$

The equations can be put into matrix form

$$\begin{bmatrix} (G_1 + G_2 + sC_1) & -G_2 \\ -G_2 & G_2 + sC_2 \end{bmatrix} \begin{bmatrix} V_1 \\ V_2 \end{bmatrix} = \begin{bmatrix} EG_1 \\ 0 \end{bmatrix}$$

Inserting numerical values

$$\begin{bmatrix} 2s + 7 & -4 \\ -4 & 3s + 4 \end{bmatrix} \begin{bmatrix} V_1 \\ V_2 \end{bmatrix} = \begin{bmatrix} 3E \\ 0 \end{bmatrix}$$

Solution proceeds exactly the same way as we have been doing before, only the steps are now somewhat more complicated because we must keep the s variable as a letter. The determinant of the system matrix is

$$D = \begin{vmatrix} 2s + 7 & -4 \\ -4 & 3s + 4 \end{vmatrix} = (2s + 7).(3s + 4) - 16 = 6s^2 + 29s + 12$$

The numerator for V_1 is

$$N_1 = \begin{bmatrix} 3E & -4 \\ 0 & 3s + 4 \end{bmatrix} = (9s + 12)E$$

and the numerator for V_2 is

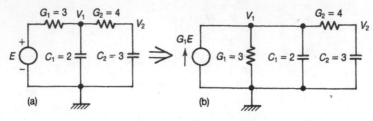

FIGURE 7.2.3 Network with two capacitors: (a) Original. (b) Transformed for nodal formulation.

$$N_2 = \begin{bmatrix} 2s+7 & 3E \\ -4 & 0 \end{bmatrix} = 12E$$

The solution in the Laplace domain is

$$V_1 = \frac{9s+12}{6s^2+29s+12}E$$

$$V_2 = \frac{12}{6s^2+29s+12}E$$

Example 1.

The network in Fig. 7.2.4 has one capacitor and a voltage controlled current source. Find the two nodal voltages.

The nodal equations are

$$(G_1+sC)V_1 - sCV_2 = J$$
$$-sCV_1 + (G_2+sC)V_2 - g(V_1-V_2) = 0$$

In matrix form

$$\begin{bmatrix} G_1+sC & -sC \\ -sC-g & G_2+sC+g \end{bmatrix}\begin{bmatrix} V_1 \\ V_2 \end{bmatrix} = \begin{bmatrix} J \\ 0 \end{bmatrix}$$

Inserting numerical values

$$\begin{bmatrix} 2s+2 & -2s \\ -2s-3 & 2s+6 \end{bmatrix}\begin{bmatrix} V_1 \\ V_2 \end{bmatrix} = \begin{bmatrix} J \\ 0 \end{bmatrix}$$

The denominator is $D = 10s + 12$, the numerators are $N_1 = (2s+6)J$ and $N_2 = (2s+3)J$. The voltages are

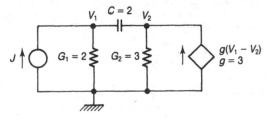

FIGURE 7.2.4   Network with a *VC*.

$$V_1 = \frac{2s+6}{10s+12}J$$

$$V_2 = \frac{2s+3}{10s+12}J$$

Example 2.

The network in Fig. 7.2.5a has a current controlled current source and we wish to obtain the voltages V_1 and V_2. Following the steps explained in Section 5.2 we first express the current I by using the original figure: $I = sC_1(E - V_1)$. As can be seen, nothing has changed as far as the analysis steps are concerned. Once this current has been determined and written to the dependent source, we can apply the source transformations, again exactly the same way as before. The transformed network is in Fig. 7.2.5b where we have written the controlling current to the controlled current source. The nodal equations are

$$(G_1 + sC_1 + sC_2)V_1 - sC_2V_2 = EsC_1$$
$$-sC_2V_1 + (sC_2 + sC_3 + G_2)V_2 - \alpha sC_1(E - V_1) = 0$$

Inserting numerical values and transferring the terms with E to the right hand side we get the system

$$\begin{bmatrix} 4s+3 & -2s \\ 2s & 3s+4 \end{bmatrix}\begin{bmatrix} V_1 \\ V_2 \end{bmatrix} = \begin{bmatrix} 2sE \\ 4sE \end{bmatrix}$$

The solutions are: $D = 16s^2 + 25s + 12$, $N_1 = (14s^2 + 8s)E$ and $N_2 = (12s^2 + 12s)E$.

The nodal voltages are:

$$V_1 = \frac{14s^2 + 8s}{16s^2 + 25s + 12}E$$

$$V_2 = \frac{12s^2 + 12s}{16s^2 + 25s + 12}E$$

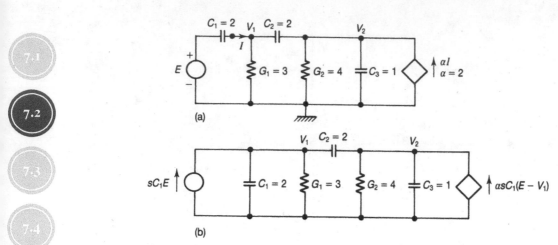

FIGURE 7.2.5   Network with a CC. (a) Original. (b) Transformed for nodal formulation.

Example 3.

The network in Fig. 7.2.6a has only voltage sources and thus can be immediately analyzed by the mesh method. It can also be transformed to make it suitable for nodal formulation, Fig. 7.2.6b. Use both methods to obtain the output voltage. In the mesh method we first find that the controlling current is $I = I_1 - I_2$. The mesh equations are

$$(\frac{1}{2s} + 3)I_1 - 3I_2 = E$$

$$-3I_1 + (3 + \frac{1}{s})I_2 + 2(I_1 - I_2) = 0$$

Collecting terms multiplied by the same current we get

$$\begin{bmatrix} \dfrac{6s+1}{2s} & -3 \\ -1 & \dfrac{s+1}{s} \end{bmatrix} \begin{bmatrix} I_1 \\ I_2 \end{bmatrix} = \begin{bmatrix} E \\ 0 \end{bmatrix}$$

The determinant is

$$D = \frac{6s^2 + 7s + 1}{2s^2} - 3 = \frac{7s+1}{2s^2}$$

The numerators are $N_1 = \dfrac{s+1}{s} E$ and $N_2 = E$, the currents are

$$I_1 = \frac{2s^2 + 2s}{7s+1} E$$

$$I_2 = \frac{2s^2}{7s+1}$$

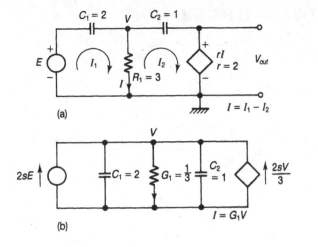

FIGURE 7.2.6   Network with a *CC*. (a) Original. (b) Transformed for nodal formulation.

The output voltage is $V_{out} = rI = 2(I_1 - I_2)$ and the result is

$$V_{out} = \frac{4s}{7s+1}E$$

The problem simplifies considerably if we transform the voltage sources into current sources and use nodal formulation. The network is redrawn in Fig. 7.2.6b, the controlling current is now $I = GV = \frac{1}{3}V$. Transferring this information to the dependent source we solve just one equation for the unknown V,

$$(3s + \frac{1}{3})V - \frac{2s}{3}V = 2sE,$$

with the result

$$V = \frac{6s}{7s+1}E$$

This voltage is not yet the desired output voltage, it is the voltage at the node in the middle of Fig. 7.2.6a. However, since the output voltage is $V_{out} = rI = 2I$ and the current was determined as $I = GV$, all we have to do is collect the partial results and get

$$V_{out} = rGV = \frac{4s}{7s+1}E$$

This is the same result as above.

7.3. NETWORKS WITH INDUCTORS

If the network contains only resistors, inductors and voltage sources, then it is natural to use the mesh formulation. In this case the resistors are expressed in terms of their resistances.

The first network we consider is in Fig. 7.3.1 and the mesh equation is

$$(R + sL)I = E$$

with the solution

$$I = \frac{E}{sL + R}$$

The voltage across the inductor is

$$V_L = sLI = \frac{sL}{sL + R}E$$

We could have solved this network also as a voltage divider by using

$$V_L = \frac{Z_L}{Z_R + Z_L}E$$

and get the result immediately.

A parallel combination of a current source and the same elements is in Fig. 7.3.2. In this case we apply nodal formulation:

$$(G + \frac{1}{sL})V = J$$

and the solution is

$$V = \frac{sL}{sLG + 1}J$$

A network with two meshes is in Fig. 7.3.3 and here it also pays to use mesh analysis. The equations are

$$(R_1 + sL_1)I_1 - sL_1I_2 = E$$
$$-sL_1I_1 + (sL_1 + sL_2 + R_2)I_2 = 0$$

Inserting numerical values

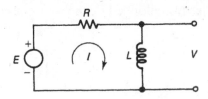

FIGURE 7.3.1   RL Network with a voltage source.

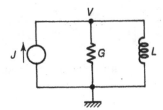

FIGURE 7.3.2   GL Network with a current source.

$$\begin{bmatrix} 5+s & -s \\ -s & 4s+10 \end{bmatrix}\begin{bmatrix} I_1 \\ I_2 \end{bmatrix}\begin{bmatrix} E \\ 0 \end{bmatrix}$$

The determinant is

$$D = (5+s)(4s+10) - s^2 = 3s^2 + 30s + 50$$

and $N_2 = sE$. The current of the second mesh is

$$I_2 = \frac{s}{3s^2 + 30s + 50}E$$

and the output voltage is

$$V_{out} = sL_2I_2 = \frac{3s^2}{3s^2 + 30s + 50}E.$$

Additional examples show more complicated situations.

Example 1.

Find the output voltage of the network in Fig. 7.3.4.

The network is suitable for mesh equations. The current controlling the CC is $I = I_2$. The mesh equations are:

$$(3+2s)I_1 - 2sI_2 + 2I = E$$
$$-2sI_1 - 2I + (7s+6)I_2 = 0$$

Inserting for I and collecting terms multiplied by the same variable

$$\begin{bmatrix} 3+2s & 2-2s \\ -2s & 7s+4 \end{bmatrix}\begin{bmatrix} I_1 \\ I_2 \end{bmatrix} = \begin{bmatrix} E \\ 0 \end{bmatrix}$$

The determinant is

$$D = (3+2s)(7s+4)-(-2s)(2-2s) = 10s^2 + 33s + 12$$

The numerator $N_2 = 2sE$ and the current of the second mesh is

$$I_2 = \frac{2s}{10s^2 + 33s + 12}E.$$

The output voltage is

$$V_{out} = sL_2 I_2 = \frac{10s^2}{10s^2 + 33s + 12}E.$$

Example 2.

The network in Fig. 7.3.5a is suitable for nodal formulation and we use it first. Next we apply transformations of the sources and use mesh formulation.

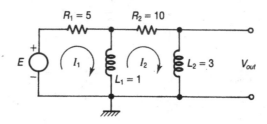

FIGURE 7.3.3   Network with two inductors.

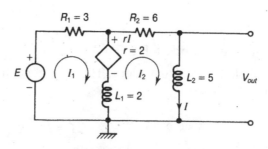

FIGURE 7.3.4   Network with a CC.

The nodal equations are

$$(\frac{1}{3s}+\frac{1}{4})V_1 - \frac{1}{4}V_2 = J$$

$$-\frac{1}{4}V_1 + (\frac{1}{4}+\frac{1}{2s})V_2 - 2(V_1 - V_2) = 0$$

Collecting terms we get

$$\begin{bmatrix} \dfrac{3s+4}{12s} & \dfrac{-1}{4} \\ \dfrac{-9}{4} & \dfrac{9s+2}{4s} \end{bmatrix} \begin{bmatrix} V_1 \\ V_2 \end{bmatrix} = \begin{bmatrix} J \\ 0 \end{bmatrix}$$

The determinant is

$$D = \frac{3s+4}{12s} \cdot \frac{9s+2}{4s} - \frac{9}{16} = \frac{21s+4}{24s^2}$$

The numerator $N_2 = \frac{9}{4}J$ and the output voltage is $V_{out} = \frac{54s^2}{21s+4}J.$

Transformation of the sources reduces the network to only one mesh, shown in Fig. 7.3.5b. The controlling voltage is now $V_1 - V_2 = RI = 4I$ and the expression for the controlled source is written into the figure. The single mesh equation is

$$(5s+4)I + 16sI = 3sJ$$

and the solution is $I = \frac{3s}{21s+4}J$. The output voltage, V_2, is marked in the figure and is obtained as

$$V_{out} = 16sI + sL_2I = 18sI = \frac{54s^2}{21s+4}J,$$

the same result as above.

Example 3.

The network in Fig. 7.3.6a has one voltage controlled voltage source and one current controlled voltage source. Find the output voltage using the mesh method.

In the first step we realize that the controlling variables are: $I = I_2$ and $V_A = R_1(I_1 - I_2) = 3(I_1 - I_2)$. The values are transferred into Fig. 7.3.6b and the mesh equations are:

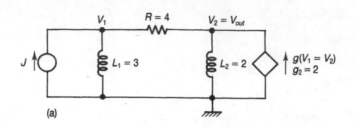

(a)

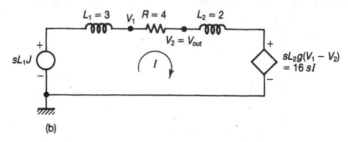

(b)

FIGURE 7.3.5　Network with a VC: (a) Original. (b) Transformed for mesh formulation.

$$(2s+3)I_1 - 3I_2 + 4I_2 = E$$
$$-3I_1 + (7+s)I_2 - 4I_2 - 9I_1 + 9I_2 = 0$$

Collecting terms

$$\begin{bmatrix} 2s+3 & 1 \\ -12 & 12+s \end{bmatrix}\begin{bmatrix} I_1 \\ I_2 \end{bmatrix} = \begin{bmatrix} E \\ 0 \end{bmatrix}$$

The determinant is

$$D = (2s+3)(12+s) + 12 = 2s^2 + 27s + 48$$

The numerators are $N_1 = (s+12)E$ and $N_2 = 12E$. The currents are

$$I_1 = \frac{s+12}{2s^2 + 27s + 48}E$$

$$I_2 = \frac{12}{2s^2 + 27s + 48}E$$

The output can be calculated, for instance, as the voltage of the independent voltage source on the left, less the voltage across L_1, less the voltage across R_2. Thus $V_{out} = E - sL_1 I_1 - R_2 I_2$. The result is

$$V_{out} = \frac{3s}{2s^2 + 27s + 48}E$$

7.4. NETWORKS WITH CAPACITORS AND INDUCTORS

If the network contains both inductors and capacitors, we cannot avoid the unpleasant fractions, at least not with the methods we have learned so far. Theoretically, it does not matter which method we select, provided the network is planar. In general, it is advantageous to favor the nodal method.

The first network we will consider is a parallel combination of R, L and C, as shown in Fig. 7.4.1. This network has a special name, it is called the *parallel tuned circuit*. We will analyze it by the nodal method and by using G for the resistor.

$$(G + sC + \frac{1}{sL})V = J$$

and the voltage is

$$V = \frac{sL}{s^2 LC + sLG + 1}J.$$

If we are interested in the currents through the elements, we obtain them by using $I_G = GV$, $I_C = sCV$ and $I_L = \frac{1}{sL}V$,

Another frequently used network is in Fig. 7.4.2, a series combination of the same elements. The network is called *series tuned circuit* and mesh formulation is clearly advantageous:

$$(R + sL + \frac{1}{sC})I = E$$

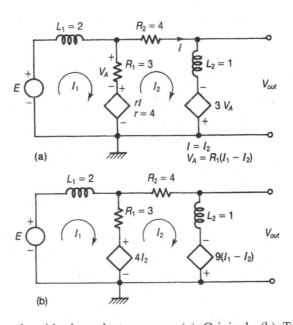

(a)

(b)

FIGURE 7.3.6 Network with dependent sources: (a) Original. (b) Transformed for mesh formulation.

7·4

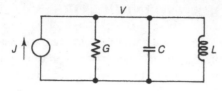

Fig. 7.4.1

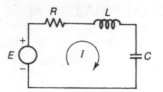

Fig. 7.4.2

FIGURE 7.4.1　Parrallel tuned circuit.

FIGURE 7.4.2　Series tuned circuit.

and the current is

$$I = \frac{sC}{s^2LC + sRC + 1}E.$$

If we wish to get the voltages across the elements, then they are obtained by:

$$V_R = RI,\ V_L = sLI,\ \text{and}\ V_C - \frac{1}{sC}I$$

Example 1.

Analyze the network in Fig. 7.4.3a and find the output voltage.

It is advantageous to use nodal formulation, but first we must determine the controlling voltage on the original network. It is $V_A = E - V_1$. This is transferred into the modified network in Fig. 7.4.3b and we can proceed with the equations:

$$(2 + 3s + \frac{1}{s})V_1 - \frac{1}{s} = 3sE$$

$$-\frac{1}{s}V_1 + \frac{1}{s}V_2 - 2E + 2V_1 = 0$$

Transferring the term with E to the right we arrive at

$$\begin{bmatrix} \dfrac{3s^2 + 2s + 1}{s} & -\dfrac{1}{s} \\ \dfrac{2s - 1}{s} & \dfrac{1}{s} \end{bmatrix} \begin{bmatrix} V_1 \\ V_2 \end{bmatrix} = \begin{bmatrix} 3sE \\ 2E \end{bmatrix}$$

The determinant is

$$D = \frac{3s^2 + 2s + 1}{s^2} + \frac{2s - 1}{s^2} = \frac{3s + 4}{s}$$

For the numerator we must evaluate

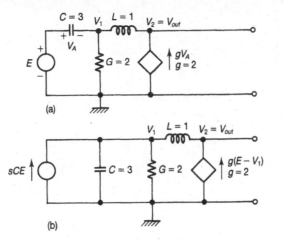

FIGURE 7.4 3 GLC network with a *VC*: (a) Original. (b) Transformed for nodal formulation.

$$N_2 = \begin{vmatrix} \dfrac{3s^2 + 2s + 1}{s} & 3sE \\[2mm] \dfrac{2s-1}{s} & 2E \end{vmatrix} = \dfrac{7s+2}{s} E$$

and the output voltage is

$$V_2 = \dfrac{7s+2}{3s+4} E.$$

Example 2.

Analyze the network in Fig. 7.4.4a by the nodal method.

In the first step we find the controlling current of the *CC*: it is $I = \dfrac{E - V_1}{sL}$. Writing this value to the dependent source and transforming the independent voltage source into a current source we arrive at the network Fig. 7.4.4b for which we can write

$$(\dfrac{1}{3s} + 1 + 2s)V_1 - 2sV_2 = \dfrac{E}{3s}$$

$$-2sV_1 + (2s+2)V_2 - \dfrac{2}{3s}(E - V_1) = 0$$

Transferring the term with E to the right and collecting terms

$$\begin{bmatrix} \dfrac{6s^2 + 3s + 1}{3s} & -2s \\[3mm] \dfrac{2 - 6s^2}{3s} & 2s + 2 \end{bmatrix} \begin{bmatrix} V_1 \\[2mm] V_2 \end{bmatrix} = \begin{bmatrix} \dfrac{E}{3s} \\[3mm] \dfrac{2E}{3s} \end{bmatrix}$$

The determinant is

$$D = \frac{18s^2 + 12s + 2}{3s}$$

For the numerator we must find the determinant

$$N_2 = \begin{vmatrix} \dfrac{6s^2 + 3s + 1}{3s} & \dfrac{E}{3s} \\ \dfrac{2 - 6s^2}{3s} & \dfrac{2E}{3s} \end{vmatrix} = \frac{6s + 2}{3s} E.$$

The result is

$$V_2 = \frac{6s + 2}{18s^2 + 12s + 2} E.$$

Example 3.

For the network in Fig. 7.4.5 we wish to derive the output voltage as a function of the capacitor C. This means that we must keep the letter C in all equations and solutions. The network is in a form suitable for nodal formulation and the equations are:

$$(sC + 1 + \frac{1}{2s})V_1 - \frac{1}{2s}V_2 = J$$

$$-\frac{1}{2s}V_1 + (\frac{1}{2s} + 2)V_2 - gV_1 = 0$$

In matrix form this leads to

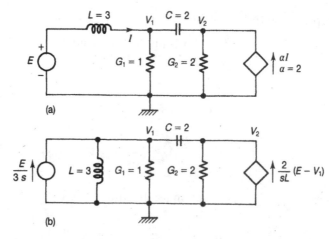

(a)

(b)

FIGURE 7.4.4   RLC network with a CC: (a) Original. (b) Transformed for nodal formulation.

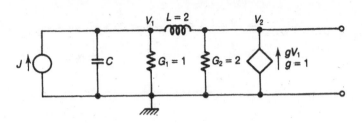

FIGURE 7.4.5 RLC network with a VC.

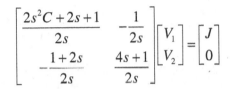

$$\begin{bmatrix} \dfrac{2s^2C+2s+1}{2s} & -\dfrac{1}{2s} \\ -\dfrac{1+2s}{2s} & \dfrac{4s+1}{2s} \end{bmatrix}\begin{bmatrix} V_1 \\ V_2 \end{bmatrix} = \begin{bmatrix} J \\ 0 \end{bmatrix}$$

The determinant is

$$D = \frac{2s^2C+2s+1}{2s}\cdot\frac{4s+1}{2s} - \frac{1}{2s}\cdot\frac{1+2s}{2s} = \frac{8Cs^2+s(8+2C)+4}{4s}$$

For the numerator we find $N_2 = \dfrac{2s+1}{2s}J$ and the ratio gives the output voltage

$$V_2 = \frac{2s+1}{4Cs^2+s(4+C)+2}J.$$

Example 4.

Find the output voltage for the network in Fig. 7.4.6 by keeping the inductance L as a variable element. Use numerical values for all other elements.

The network is suitable for nodal formulation and we write directly

$$\begin{bmatrix} 1+s+\dfrac{1}{sL} & -\dfrac{1}{sL} \\ -\dfrac{1}{sL} & 1+s+\dfrac{1}{sL} \end{bmatrix}\begin{bmatrix} V_1 \\ V_2 \end{bmatrix} = \begin{bmatrix} J \\ 0 \end{bmatrix}$$

The determinant, after a few intermediate steps, is

$$D = \frac{s^3L+2s^2L+s(L+2)+2}{sL}$$

and the numerator for the output is $N_2 = \dfrac{J}{sL}$. The output voltage is

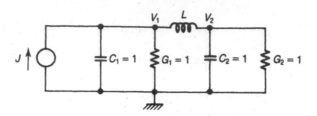

FIGURE 7.4.6   RLC network.

$$V_2 = \frac{1}{s^3 L + 2s_2 L + s(L+2) + 2} J.$$

All examples we solved lead to the ratio of two polynomials. Some had the highest power of s equal to 1, and such networks would be called *first order* networks. If the highest power of s of any of the polynomials is 2, we speak about a *second order* network and so on. Important is that analysis of all networks composed of the elements we introduced in Chapter 1 will *always* lead to the ratio of two polynomials in s, and all coefficients of these polynomials will be real.

PROBLEMS CHAPTER 7

P.7.1(a)–(d)   Use nodal analysis for all problems; in (c) first transform the source into a current source. For each capacitor substitute $Y = sC$, for conductances use $Y = G$. Find V_{out}.

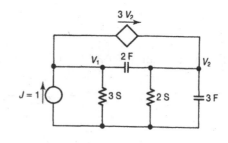

(d)

P.7.2   Dependent sources are handled exactly the same way as in resistive networks. Find the voltages V_1 and V_2 by using $Y = sC$ for the capacitors.

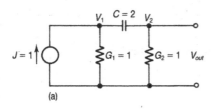

(a)

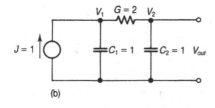

(b)

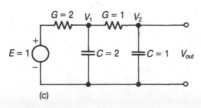

(c)

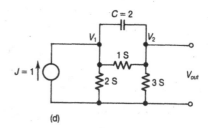

P.7.3 Find the output voltage $V_2 = V_{out}$.

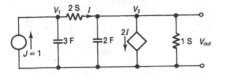

P.7.4 Find the output voltage.

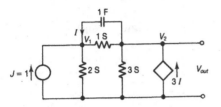

P.7.5 Mesh analysis is applied the same way as before, with $Z = \dfrac{1}{sC}$ for capacitors and $Z = R$ for resistors. Find the output voltage.

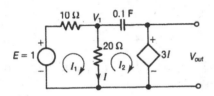

P.7.6 Use nodal formulation and find V_{out}

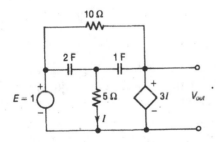

P.7.7 Thevenin or Norton equivalents are calculated the same way as before, only now the independent sources may become functions of s and the equivalent resistance or admittance is changed into an equivalent impedance or admittance. Find the Thevenin and Norton equivalents for this circuit.

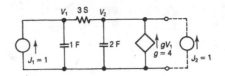

P.7.8 Find the Thevenin and Norton equivalents.

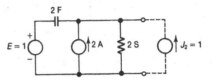

P.7.9 Find the Thevenin and Norton equivalents.

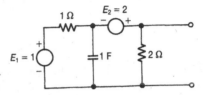

P.7.10 Express the controlling current in terms of the input voltage and V_1, substitute for I in the dependent source and then transform the voltage source into a current source. Use nodal formulation to obtain $V_{out} = V_2$.

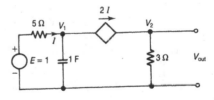

P.7.11 In nodal formulation use $Y = \dfrac{1}{sL}$ for inductors and $Y = G$. Find V_1 and V_2.

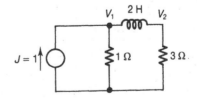

P.7.12   In mesh formulation use $Z = sL$ for inductors and $Z = R$ for resistors. Find the output voltage.

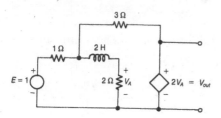

P.7.13   Calculate the voltage V_{out} using mesh formulation.

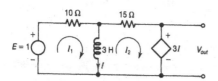

P.7.14   Find the voltae V. This can be done in three ways: conveniently by finding first I_1 and then calculating $V = 10 - sLI_1$. It can also be done by calculating $I = I_2$ and then $V = sLI$. Finally, less conveniently, it can be obtained by calculating both I_1 and I_2 and then $V = 3I_2 + 5(I_1 - I_2)$. In all cases you must get the same result. Perform all three steps.

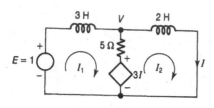

P.7.15   Find the output voltage.

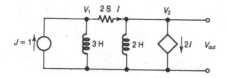

P.7.16   Find the output voltage.

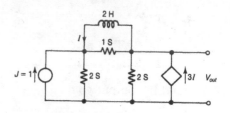

P.7.17   Calculate V_1 and V_2.

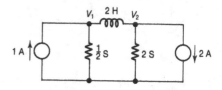

P.7.18   Find both nodal voltages.

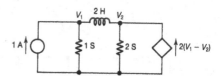

P.7.19   Find the Thevenin and Norton equivalents.

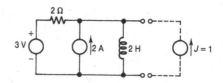

P.7.20   Calculate V_{out} using mesh formulation. Note that you must first express the current I in terms of the source voltage and V_1. Next you must substitute this current into the dependent source and only then you can transform the voltage source into a current source. The current source will be a function of s.

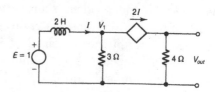

P.7.21 Calculate V_{out} by first transforming the voltage source into a current source. Note that you must use $Y = sC$ for the capacitor, $Y = \dfrac{1}{sL}$ for the inductor and $Y = G$ for resistances.

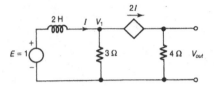

P.7.22 Find the output voltage by using nodal formulation.

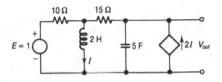

P.7.23 Find the output voltage by using nodal formulation.

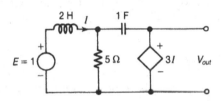

P.7.24 Find the voltage V. This is a similar network as in P.7.14, only now we have an inductor and a capacitor. Find the voltage V the same three ways as in P.7.14.

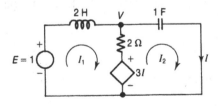

P.7.25 Find the output voltage V_{out} using nodal formulation.

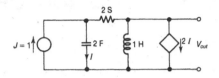

P.7.26 Find the output voltage V_{out} using nodal formulation.

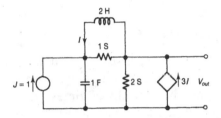

P.7.27 Calculate the Thevenin and Norton equivalents.

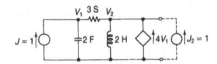

P.7.28 Find the output voltage.

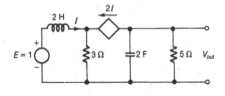

8   FREQUENCY DOMAIN

INTRODUCTION

One of the typical signals used in electrical engineering is the sinusoidal signal. The electric power, distributed in North America, has sinusoidal form and the current changes its direction periodically 60 times per second. Europe uses alternating current distribution as well, but the frequency is 50 Hz. Unless there is power failure, the system is connected all the time. If you turn on the light or connect an electric motor, you disturb the system a little but in general we can say that the system is not changing and is, as we say, in a steady state. Occasionally, there may be power failure and then other processes take place; we call them transients.

This chapter will study methods for the analysis of systems with sinusoidal signals in steady state.

8.1.   SINUSOIDAL SIGNALS AND PHASORS

A sinusoidal signal belongs to a class of *periodic* functions which have the property that

$$v(t + nT) = v(t) \qquad\qquad n = ..., -2, -1, 0, 1, 2, ... \qquad (8.1.1)$$

Here T is called the *period* of the signal, measured in seconds, s. The signal repeats itself after the period and we cannot discover which value of n applies to the instant we consider. The simplest of all periodic functions are the cosine and sine, defined by the equations

$$v(t) = V \cos(\omega t + \phi)$$
$$v(t) = V \sin(\omega t + \phi) \qquad (8.1.2)$$

The letter v is used for voltages, i is used for currents. In (8.1.2) V is called the *amplitude*, t is time in seconds, s, ω is angular frequency measured in radians per second, rad/s, and ϕ is an angle. The angular frequency, ω, is related to the period, T, by the expression

$$\omega = \frac{2\pi}{T} \qquad (8.1.3)$$

and the inverted value of the period

$$f = \frac{1}{T} \qquad (8.1.4)$$

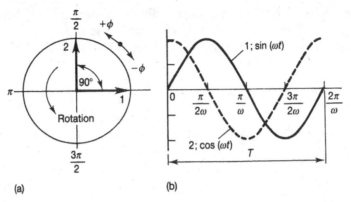

FIGURE 8.1 1   (a) Rotating vectors. (b) Equivalent sinusoids.

is called the *frequency* and is measured in hertz, Hz. The angle, ϕ, can be measured either in degrees or in radians, related by the correspondence of 2π and 360°. Thus π corresponds to 180°, $\frac{\pi}{2}$ corresponds to 90° and so on. In calculations we must always know which of these two units is being considered.

Fig. 8.1.1a shows a circle with two vectors in it. The vectors rotate counterclockwise with constant and equal angular speed, ω, but we now stop the motion. The relative position of the two vectors does not change with rotation and both vectors stay 90° apart all the time. We can stop the rotation at any instant and get a picture similar to the one in Fig. 8.1.1a, only rotated by some angle.

Consider first the vector at position 1 and assume that you are a viewer positioned to the right of the line which is an extension of the vector. For the instant shown you will not see the vector, just one point. If the vector starts rotating, you will see a growing line until the vector reaches position 2, when you will see the longest line. Afterwards the line will start decreasing, will become zero, will start growing on the other side, reach its maximum value and start decreasing again. If we also keep track of the times needed to reach the particular projected point and draw these times on the horizontal axis, we will get a sinusoid, shown in the Fig. 8.1.1b. A similar consideration for vector 2 will result in a cosine, as shown. The figure also shows that adding an angle ϕ results in a counterclockwise move by the given amount. Subtraction of ϕ moves the vector clockwise.

Rotate vector 1 counterclockwise by adding to it the angle $\phi = 90°$. It will reach the position 2 and will become $v(t) = V \sin(\omega t + 90°)$. However, as we have already stated above, this vector represents the cosine, $v(t) = V \cos \omega t$. It must follow that

$$\sin(\omega t + 90°) = \cos \omega t,$$

a well known formula. We can consider similarly the vector 2 as the original one, shift it clockwise by $\phi = -90°$ and thus place it in the position of the original vector 1. This leads to another well known formula

$$\cos(\omega t + 90°) = \sin \omega t.$$

Fig. 8.1.2a shows three vectors. Their corresponding sinusoids are in b. Vector 1 is $v_1(t) = V \sin \omega\, t$, vector 2 is $v_2(t) = V \sin(\omega\, t + 45°)$, and vector 3 is $v_3(t) = -V \cos \omega\, t$. If we shift vector 1 backwards (clockwise) by 90°, we will get $V \sin(\omega\, t - 90°)$, but this is equal to $v_3(t)$. Thus another formula is obtained

$$\sin(\omega\, t - 90°) = -\cos \omega\, t.$$

Similarly, if we start with vector 3 and shift it by + 90°, it will become $v_1(t)$ and the resulting formula is

$$-\cos(\omega\, t + 90°) = \sin \omega\, t.$$

The examples show that sines and cosines are actually a single function looked upon from different reference points. A summary of various trigonometric equations is in Table 8.1.1.

The words *leading* or *lagging* are often used when speaking about the various sinusoids or their equivalent vectors. Place yourself in Fig. 8.1.1a on the circle at the point marked by π and start the counterclockwise rotation of both vectors. Vector 2 will reach your place first, and we would say that vector 2 is leading vector 1 by 90°. By the same token we could say that vector 1 is lagging vector 2 by 90°. These statements are correct but they depend on the point of observation. If you place your observation point between the two vectors, to the position marked by $\dfrac{\pi}{4}$, and start the counterclockwise rotation, then vector 1 will reach you first, and will thus be the leading vector. It will lead vector 2 by 270°.

The correspondence of variously shifted sine waves and their vectors has one big advantage: instead of drawing shifted sinusoids, which is time consuming and confusing, we can consider vectors like in Fig. 8.1.1a or 8.1.2a. Such vectors are also given a special name, the *phasors*. What we still need is a method which can describe the vector sizes and orientations mathematically. This is done by introducing the *plane of complex numbers*.

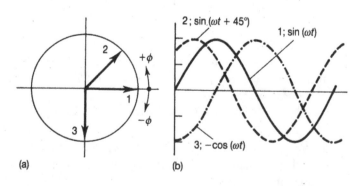

(a)   (b)

FIGURE 8.1.2   (a) Rotating vectors. (b) Equivalent sinusoids.

Table 8.1.1.

Trigonometric equivalences

$\sin(A + B) = \sin A \cos B + \cos A \sin B$	$\sin A \sin B = \dfrac{1}{2}[\cos(A - B) - \cos(A + B)]$
$\sin(A - B = \sin A \cos B - \cos A \sin B$	$\cos A \cos B = \dfrac{1}{2}[\cos(A - B) + \cos(A + B)]$
$\cos(A + B) = \cos A \cos B - \sin A \sin B$	$\sin A \cos B = \dfrac{1}{2}[\sin(A - B) + \sin(A + B)]$
$\cos(A - B) = \cos A \cos B + \sin A \sin B$	$\sin A + \sin B = 2 \sin \dfrac{A + B}{2} \cos \dfrac{A - B}{2}$
$\sin(-A) = -\sin A$	$\sin A - \sin B = 2 \cos \dfrac{A + B}{2} \sin \dfrac{A - B}{2}$
$\cos(-A) = \cos A$	$\cos A + \cos B = 2 \cos \dfrac{A + B}{2} \cos \dfrac{A - B}{2}$
$\sin(A + 90^0) = \cos A$	$\cos A - \cos B = -2 \sin \dfrac{A + B}{2} \sin \dfrac{A - B}{2}$
$\cos(A - 90^0) = \sin A$	

Consider a plane in which we mark the origin and two perpendicular axes. Usually the variable x is assigned to the horizontal axis and variable y to the vertical axis. This is not suitable for our situation because we must also take into account the fact that a vector which has rotated by 360° has reached its original position and not a new one. The problem is resolved by using complex numbers. In the plane of complex numbers, the so called $s - plane$, distances measured on the horizontal axis are marked by real numbers. Distances measured on the vertical axis are multiplied by the imaginary number

$$j = \sqrt{-1} \tag{8.1.5}$$

The notation differs from mathematics where we always use the letter i. Since in electrical engineering i usually denotes the current, it is accepted practice to use j for the imaginary unit. The properties of (8.1.5) are such that

$$j^2 = -1$$

$$j^3 = -j = -\sqrt{-1} \tag{8.1.6}$$

$$j^4 = 1$$

and the values will start repeating with further multiplications by j. Every multiplication by j represents a rotation by 90° counterclockwise. In Fig. 8.1.3 the vector v_1 is pointing to the right from the origin. If we multiply this vector by $j = \sqrt{-1}$, then in the complex plane it will rotate by 90° counterclockwise and will point upwards as a vector $v_2 = jv_1$. If we multiply by another j, we will get $v_3 = j^2 v_1 = -v_1$. Multiplication by another j will get us to $v_4 = j^3 v_1 = -jv_1$. Still another multiplication by j will return us to the original position of the vector v_1.

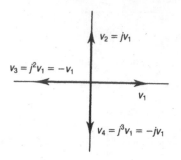

FIGURE 8.1.3   Rotations by 90.⁰

All other vectors in the plane will have real and imaginary parts. Fig. 8.1.4 shows four vectors, all starting at the origin and all having real and imaginary parts. Since all vectors always start at the same origin, we can alternatively talk about points in the complex plane, by tacitly understanding that these complex points represent the endpoints of the vectors. Thus the points in Fig. 8.1.4 are $s_1 = 3 + j2$, $s_2 = -1 + j3$ and so on. We thus arrive at the concept of the general complex number

$$s = a + jb. \tag{8.1.7}$$

The real an imaginary parts are given special notation

$$\begin{aligned} a &= \operatorname{Re}(s) \\ b &= \operatorname{Im}(s) \end{aligned} \tag{8.1.8}$$

Here Im(s) itself is a real number representing the distance in vertical direction. Thus a complex number can also be expressed by

$$s = \operatorname{Re}(s) + j \operatorname{Im}(s). \tag{8.1.9}$$

If the imaginary part of the complex number changes its sign, then we say that this new number is *complex conjugate* to the previous one and we mark it by a bar,

$$\bar{s} = Re(s) - j\operatorname{Im}(s). \tag{8.1.10}$$

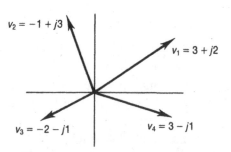

FIGURE 8.1.4   General vectors in complex plane.

Complex numbers can be added, subtracted, multiplied or divided. Addition and subtraction are performed on the real and imaginary parts independently. For instance, if we have

$$s_1 = a_1 + jb_1$$

and

$$s_2 = a_2 + jb_2$$

then their sum is

$$s = s_1 + s_2 = (a_1 + a_2) + j(b_1 + b_2) \tag{8.1.11}$$

and their difference is

$$s = s_1 - s_2 = (a_1 - a_2) + j(b_1 - b_2) \tag{8.1.12}$$

For the vectors s and $\bar{s}$ we thus get

$$\begin{aligned} s + \bar{s} &= a + jb + a - jb = 2a = 2\operatorname{Re}(s) \\ s - \bar{s} &= a + jb - a + jb = j2b = j2\operatorname{Im}(s) \end{aligned} \tag{8.1.13}$$

Complex numbers can also be expressed by their distance from the origin, A, and by the angle, ϕ, shown in Fig. 8.1.5. The two descriptions are equivalent; the first one is in *Cartesian* or *rectangular* coordinates, the second in *polar* coordinates. It follows from the figure that

$$A = \left[a^2 + b^2 \right]^{1/2} \tag{8.1.14}$$

$$\tan\phi = \frac{b}{a}$$

or

$$\phi = \arctan\frac{b}{a} \tag{8.1.15}$$

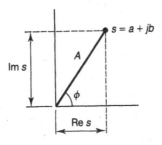

FIGURE 8.1.5   Rectangular and polar coordinates.

The distance, A, is also called the absolute value and we use the notation

$$A = |s|$$

Example 1.

Fig. 8.1.6a shows two vectors, $s_1 = 1 + j2$ and $s_2 = 3 + j1$. Their sum is $s = s_1 + s_2 = 4 + 3j$. Note the way this addition is accomplished by drawing the dotted lines which are parallel to the two vectors.

Fig. 8.1.6b shows two vectors $s_1 = 3 + 2j$ and $s_2 = 1 + 2j$. Their difference is $s = s_1 - s_2 = 2 + j0$. Note how this can be achieved graphically by first copying the vector to be subtracted in opposite direction and then using this new vector and the parallel lines with the first vector like in the addition.

Another set of useful formulas can be found by considering the unit circle in Fig. 8.1.7. Since the radius is equal to one, then the real part is $a = \cos \phi$ and the imaginary part is $b = \sin \phi$. The point on the unit circle is thus $s = \cos \phi + j \sin \phi$. If we increase the amplitude to A, we can reach any point in the complex plane and describe it by

$$s = A(\cos \phi + j \sin \phi) \tag{8.1.16}$$

There exists still another, more general expression for the term in the bracket:

$$e^{j\phi} = \cos \phi + j \sin \phi \tag{8.1.17}$$

Using this formula we can express an arbitrary point in the complex plane by

$$s = Ae^{j\phi} \tag{8.1.18}$$

and its complex conjugate by

$$\bar{s} = A(\cos \phi - j \sin \phi) = Ae^{-j\phi} \tag{8.1.19}$$

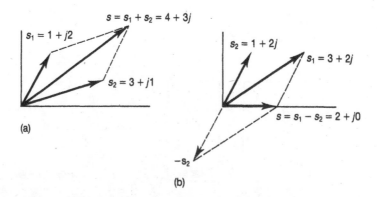

FIGURE 8.1.6 Addition and subtraction of vectors: (a) Addition. (b) subtraction.

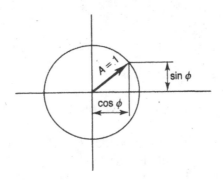

FIGURE. 8.1.7 Unit circle relation to sine and cosine.

Summarizing the development, we have three equivalent expressions for a point in the complex plane s:

$$s = a + jb$$
$$= Ae^{j\phi} \qquad (8.1.20)$$
$$= A(\cos \phi + j \sin \phi)$$

Addition and multiplication are always performed in Cartesian coordinates, the way we showed in (8.1.11) and (8.1.12). For multiplication and division we have two choices. In Cartesian coordinates the product is

$$s_1 \cdot s_2 = (a_1 + jb_1)(a_2 + jb_2) = (a_1a_2 - b_1b_2) + j(a_1b_2 + a_2b_1) \qquad (8.1.21)$$

In polar coordinates,

$$s_1 \cdot s_2 = R_1 e^{j\phi_1} R_2 e^{j\phi_2} = R_1 R_2 e^{j(\phi_1 + \phi_2)} \qquad (8.1.22)$$

Division in Cartesian coordinates is expressed by

$$\frac{s_1}{s_2} = \frac{a_1 + jb_1}{a_2 + jb_2} = \frac{(a_1 + jb_1)(a_2 - jb_2)}{(a_2 + jb_2)(a_2 - jb_2)}$$
$$= \frac{a_1a_2 + b_1b_2 + j(b_1a_2 - a_1b_2)}{a_2^2 + b_2^2} \qquad (8.1.23)$$

and in polar coordinates by

$$\frac{s_1}{s_2} = \frac{R_1 e^{j\phi_1}}{R_2 e^{j\phi_2}} = \frac{R_1}{R_2} e^{j(\phi_1 - \phi_2)} \qquad (8.1.24)$$

Since today even modestly priced calculators have some kind of conversion between the Cartesian and polar coordinates, it does not matter much which of the two definitions is used. If a calculator is not available, then it is easier to work with the Cartesian formulas.

Example 2.

Find the product and ratio of the complex numbers $s_1 = 3 + j4$ and $s_2 = 2 - j3$.

(a) The product in Cartesian coordinates is

$$s = s_1 \times s_2 = (3 + 4j)(2 - 3j) = 6 + 8j - 9j + 12 = 18 - j1$$

In polar coordinates $s_1 = 5e^{j53.13}$, $s_2 = 3.60555e^{-j56.3099}$ and the result is $s = 18.0278e^{-j3.18}$. The angle is given in degrees with decimal fractions.

(b) The ratio in Cartesian coordinates is

$$s = \frac{s_1}{s_2} = \frac{(3+4j)}{(2-3j)} = \frac{(3+4j)(2+3j)}{(2-3j)(2+3j)}$$

$$= \frac{6+9j+8j-12}{4+9} = \frac{-6+17j}{13} = -0.4615 + j1.3077$$

In polar coordinates the result is the ratio of the two polar representations, $s = 1.387e^{j109.44}$, the angle given again in degrees with decimal fractions. Note that it does not matter whether we write $3j$ or $j3$.

Example 3.

Convert the following vectors in rectangular coordinates to polar coordinates. The angles are given in degrees with decimal fractions.

$2 + j3 = 3.6055e^{j56.31}$

$5 - 2j = 5.385e^{-j21.801}$

$-3 - 4j = 5e^{-j126.87}$

$-2 + 1j = 2.236e^{j153.4}$

Convert the following vectors in polar form into rectangular form. The angles are given degrees.

$5e^{j30} = 4.33 + j2.5$

$4e^{-j25} = 3.625 - j1.690$

$2e^{-j135} = -1.414 - j1.414$

$4e^{j160} = -3.759 + j1.368$

The exponential expressions are also valid for functions of time,

$$Ae^{j(\omega t+\phi)} = A\big[\cos(\omega t + \phi) + j\sin(\omega t + \phi)\big] \tag{8.1.25}$$

from which we also have

$$A\cos(\omega t + \phi) = \text{Re } Ae^{j(\omega t + \phi)}$$
$$A\sin(\omega t + \phi) = \text{Im } Ae^{j(\omega t + \phi)} \qquad (8.1.26)$$

The formulas enable us to use vectors when dealing with sinusoids. The left part of (8.1.25) can be rewritten in the form

$$Ae^{j(\omega t + \phi)} = \left[Ae^{j\phi} \right] e^{j\omega t} \qquad (8.1.27)$$

where the term in the square brackets represents a vector with amplitude A, rotated by the angle ϕ. Permanent rotation in counterclockwise direction is expressed by the term $e^{j\phi t}$. To describe a vector we often use bold letters and thus

$$\mathbf{A} = Ae^{j\phi} \qquad (8.1.28)$$

The name *phasor* is used for (8.1.28).

Suppose that we are asked to perform the addition

$$v(t) = 4\cos(\omega t - 20^0) + 3\cos(\omega t + 30^0) \qquad (8.1.29)$$

Applying (8.1.26) and (8.1.27) we express the functions in exponential form

$$v(t) = \text{Re}\left\{ \left[4e^{-j20} + 3e^{j30} \right] e^{j\omega t} \right\} \qquad (8.1.30)$$

All we have to do is consider the terms in the square brackets, $4e^{-j20} + 3e^{j30}$. Direct addition is not possible but we can convert to rectangular form, perform the addition and then convert back to polar form:

$$4e^{-j20} + 3e^{j30} = (3.7588 - j1.36808) + (2.598 + j1.5) = 6.3568 + j0.13192$$
$$= 6.3582e^{j1.18885}$$

Using this result we go back to (8.1.27), replace the term in the square brackets

$$v(t) = \text{Re}\{[6.3582e^{j1.18885}]e^{j\omega t}\}.$$

and return to formula (8.1.26) and to the cosine function

$$v(t) = 6.3582\cos(\omega t + 1.18885^0)$$

Example 4.

Find the equivalent cosine for

$$v(t) = 5\cos(\omega t + 30°) - 4\sin(\omega t - 20°)$$

The second function must be first converted to a cosine:

$$v(t) = 5 \cos(\omega t + 30°) - 4 \cos(\omega t - 20° - 90°) = 5 \cos(\omega t + 30°) - 4 \cos(\omega t - 110°)$$

$$= \text{Re}\{[5e^{j30} - 4^{-j110}]e^{j\omega t}\}$$

We now deal with the part in the square brackets

$$5e^{j30} - 4e^{-j110} = (4.33 + j2.5) - (-1.368 - j3.7588) = 5.698 + j6.2588$$
$$= 8.464e^{j47.684}$$

Returning back to cosine we have the result

$$v(t) = 8.464 \cos(\omega t + 47.684°)$$

Example 5.

Find the sum of

$$v(t) = 2 \cos(\omega t + 30°) - 3 \cos(\omega t - 50°) + 4 \cos(\omega t + 130°)$$

The vectors to be added are

$$2e^{j30} - 3e^{-j50} + 4e^{j130} = (1.7321 + j) - (1.9284 - j2.2981) + (-2.5712 + j3.0642)$$
$$= -2.7675 + j6.3623 = 6.9381e^{j113.508}$$

and

$$v(t) = 6.9381 \cos(\omega t + 113.508°)$$

The above developments were done by using cosine functions, but the same applies to sine functions, with the only difference that the symbol Re is replaced by the symbol Im. Important is that we must select one type and use it throughout. Operations can be summarized in the following steps:

1. All functions on which we operate *must* have the same ω.
2. Convert all functions to the form $v(t) = V \cos(\omega t + \phi)$.
3. Find the vector representation $\mathbf{V} = Ve^{j\phi}$ and, if necessary, convert to rectangular form $\mathbf{V} = a + jb$.
4. Perform the necessary operations and convert the result, $\mathbf{R} = c + jd$, into polar form, $\mathbf{R} = R^{j\psi}$.
5. The resulting time function is $R \cos(\omega t + \psi)$.

The points remain valid if the word cosine is replaced everywhere by the word sine. The operations are fundamental to frequency domain analysis of networks.

8.2. FREQUENCY DOMAIN ANALYSIS

In frequency domain analysis we are interested in the reaction of a linear network to a permanently acting cosine signal. By permanently acting we mean that any switching or changes of the signal have happened long ago, theoretically an infinitely long time ago. To proceed with the theory we introduce a few definitions based on the results of the previous section. Let the independent voltage source be represented by the signal and its corresponding phasor

$$e(t) = E \cos(\omega t + \phi)$$
$$\mathbf{E} = Ee^{j\phi}$$

(8.2.1)

For the independent current source we have

$$j(t) = J \cos(\omega t + \phi)$$
$$\mathbf{J} = Je^{j\phi}$$

(8.2.2)

Network voltages and currents are similarly defined

$$v(t) = V \cos(\omega t + \phi)$$
$$\mathbf{V} = Ve^{j\phi}$$

(8.2.3)

and

$$i(t) = I \cos(\omega t + \phi)$$
$$\mathbf{I} = Ie^{j\phi}$$

(8.2.4)

The bold letters formally denote that we are dealing with vectors.

In Chapter 7 we introduced the impedances and admittances for capacitors and inductors. We also stated that every independent source is represented by a letter E or J, to be replaced by appropriate substitution. In frequency domain analysis we do the following:

1. For any given angular frequency we replace s by $j\omega$.
2. Every independent source is replaced by its phasor.
3. The resulting voltages and currents are also phasors.
4. Return to the functions of time by using the above equivalences.

We will start our frequency domain studies by performing the operations on impedances and admittances. The resistor voltage and current are coupled by the Ohm's law,

$$V_R = RI_R.$$

(8.2.5)

Let the independent variable, in this case the current, be the function

$$i(t) = I \cos \omega t.$$

(8.2.6)

The phasor of the current is

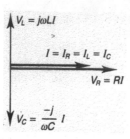

FIGURE. 8.2.1   Voltage phasors for given current.

$$\mathbf{I} = Ie^{j0} = I.$$ (8.2.7)

The voltage phasor, $\mathbf{V}$, will be R times the current phasor and will point in the same direction as the current phasor. It is an accepted practice to draw the vector of the independent variable horizontally, pointing to the right. Thus for a resistor both the current and voltage phasors will point in the same direction, as shown in Fig. 8.2.1, where the phasors are drawn as two parallel lines for clarity. We say that the voltage across the resistor and the current through it are *in phase*. If we return to the original time function, we will have

$$v_R(t) = RI \cos \omega t$$ (8.2.8)

For an inductor we defined in (7.1.4)

$$V_L = sLI_L.$$ (8.2.9)

After the substitution $s = j\omega$,

$$V_L = j\omega LI_L$$ (8.2.10)

Let I_L represent a vector pointing horizontally to the right. The voltage across the inductor will be a phasor whose length will be ωL times the length of the current phasor, but will point upwards, because of the multiplication by j. In agreement with our discussions in section 1 we will say that the voltage across the inductor *leads* the current through the inductor by 90°. Returning to functions of time, this means that

$$v_L(t) = \omega L \cos(\omega t + 90°).$$ (8.2.11)

The phasor V_L is plotted in Fig. 8.2.1.

Consider next the capacitor by using (7.1.6)

$$V_C = \frac{1}{j\omega C} I_C = -j\frac{1}{\omega C} I_C.$$ (8.2.12)

We conclude that the voltage phasor of the capacitor will be $\dfrac{1}{\omega C}$ times the length of the current phasor but will point down and will *lag* the current by 90°. In time domain, the voltage across the capacitor will be

$$v_C(t) = \frac{1}{\omega C}\cos(\omega t - 90°) \tag{8.2.13}$$

The phasor of the voltage, V_C, is plotted in Fig. 8.2.1.

Work with admittances is similar. For the resistor we have the other form of the Ohm's law

$$I_R = GV_R \tag{8.2.14}$$

The independent variable is now the voltage, which we draw as a horizontal vector pointing to the right. The current will be G times the length of the voltage phasor and will point in the same direction, as shown in Fig. 8.2.2. For the capacitor we have, after the substitution,

$$I_C = j\omega CV_C \tag{8.2.15}$$

which means that the current phasor is ωC times the length of the voltage phasor and points upwards, due to the multiplication by j. We say that the capacitive current leads the voltage by 90°. The phasor is shown in Fig. 8.2.2. If we consider the inductor, then

$$I_L = \frac{1}{j\omega L}V_L = -j\frac{1}{\omega L}V_L \tag{8.2.16}$$

The current phasor will point downwards, as shown in Fig. 8.2.2, and its length will be $\dfrac{1}{\omega L}$ times the length of the voltage phasor.

If the network is more complicated, we use the same network analysis methods as in previous chapters. Let us demonstrate it on the network in Fig. 8.2.3. The source is $j(t) = 1\cos \omega t$, its phasor is $\mathbf{J} = 1e^{j0} = 1$. We set up the equations exactly the same way as we have been doing until now by using admittances. We will keep the elements as variables and the symbol J for the source. Eventually, the phasor will replace the symbol. The matrix equation is

$$\begin{bmatrix} sC_1 + G & -G \\ -G & sC_2 + G \end{bmatrix}\begin{bmatrix} V_1 \\ V_2 \end{bmatrix} = \begin{bmatrix} J \\ 0 \end{bmatrix}$$

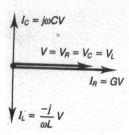

FIGURE 8.2.2   Current phasors for given voltage.

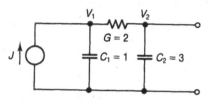

FIGURE 8.2.3   Network with two capacitors.

If we are asking for the output voltage, $V_2(s) = V_2$, and keep the variables throughout our solutions, then the determinant is

$$D = s^2 C_1 C_2 + sG(C_1 + C_2),$$

the numerator is

$$N_2 = GJ,$$

and the output voltage is

$$V_2(s) = \frac{G}{s^2 C_1 C_2 + sG(C_1 + C_2)} J.$$

Inserting element values

$$V_2(s) = \frac{2}{3s^2 + 8s} J$$

Substituting $s = j\omega$ and $\mathbf{J} = J = 1$, as derived above,

$$V_2(j\omega) = \frac{2}{-3\omega^2 + 8j\omega}$$

This expression gives us the possibility to evaluate the response for various ω. If we select, for instance, $s = j\omega = j1$, then

$$V_2(j1) = \frac{2}{-3+8j} = 0.2341e^{-j110.556}$$

This indicates that for an input signal $j(t) = 1 \cos(1\ t)$ we get the output voltage $v_2(t) = 0.2341 \cos(1\ t - 110.556°)$.

We can equally well substitute $s = j1$ and $\mathbf{J} = J = 1$ at the very beginning of our work and get the system

$$\begin{bmatrix} 2+j & -2 \\ -2 & 2+3j \end{bmatrix} \begin{bmatrix} V_1 \\ V_2 \end{bmatrix} = \begin{bmatrix} 1 \\ 0 \end{bmatrix},$$

obtain the determinant

$$D = (2+j)(2+3j) - 4 = -3 + 8j,$$

calculate the numerator

$$N_1 = \begin{vmatrix} 2+j & 1 \\ -2 & 0 \end{vmatrix} = 2,$$

and get

$$V_2(j1) = \frac{2}{-3+8j},$$

the same result as above. As we can see, it is only a matter of convenience when we substitute $s = j\omega$ and the phasor value for the independent source.

Example 1.

The signal applied to the network in Fig. 8.2.4 is $e(t) = 3 \cos(2t + 30°)$. Thus the phasor for the source is $\mathbf{E} = 3e^{j30} = 2.598 + j1.5$ and $s = j2$. Calculate I_2 and V_{out} in two ways:

(a) Insert element values but keep s as a variable and use the symbol E for the source. Obtain the result and then substitute for s and E. (b) Insert all numerical values at the beginning.

(a) The mesh equations are

$$(1 + 2s)I_1 - 2sI_2 = E$$

$$-2sI_1 + (2 + 4s)I_2 = 0$$

In matrix form

$$\begin{bmatrix} 1+2s & -2s \\ -2s & 2+4s \end{bmatrix} \begin{bmatrix} I_1 \\ I_2 \end{bmatrix} = \begin{bmatrix} E \\ 0 \end{bmatrix}$$

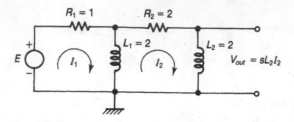

FIGURE 8.2.4   Network with two inductors.

The denominator is $D = 4s^2 + 8s + 2$, the numerator is $N_2 = 2sE$, the current is

$$I_2 = \frac{2sE}{4s^2 + 8s + 2}$$

and the output voltage is

$$V_{out}(s) = sL_2 I_2 = \frac{4s^2 E}{4s^2 + 8s + 2}.$$

Now we can substitute $s = j2$ and $E = 2.598 + j1.5$. The output voltage will be

$$V_{out}(j2) = \frac{-16(2.598 + j1.5)}{-14 + j16} = 0.438 + j2.2148.$$

Conversion to polar form provides $V_{out}(j2) = 2.2577e^{j78.8148}$ and the output voltage in time domain is $v_{out}(t) = 2.2577 \cos(2t + 78.8148)$.

(b) Inserting all numerical values immediately leads to

$$\begin{bmatrix} 1 + j4 & -j4 \\ -j4 & 2 + 8j \end{bmatrix} \begin{bmatrix} I_1 \\ I_2 \end{bmatrix} = \begin{bmatrix} 2.598 + j1.5 \\ 0 \end{bmatrix}$$

The denominator is $D = -14 + j16$, the numerator is $N_2 = j4E = -6 + j10.392$. The current $I_2 = 0.5537 - j0.1095$ and the output voltage is $j4I_2 = 0.438 + j2.2148$, the same result as above.

We may occasionally encounter a problem in which the network has several sources, *all* with the same ω but with different amplitudes and phases. In such case we can proceed in several ways, as demonstrated on the network of Fig. 8.2.5. The first signal and its phasor are

$$j_1(t) = 2 \cos(2t + 45°)$$

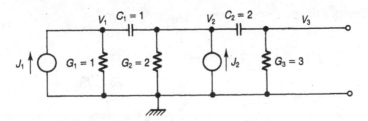

FIGURE 8.2.5   Network with two current sources.

$$J_1 = 2e^{j45^0} = 2\left(\cos 45^0 = j\sin 45^0\right) = \sqrt{2} + j\sqrt{2}$$

The second signal and its phasor are

$$j2(t) = 3\cos(2t + 90°)$$

$$\mathbf{J}_2 = 3e^{j90°} = 3(\cos 90° + j\sin 90°) = 3(0 + j1) = 3j$$

In both cases $\omega = 2$. Using nodal formulation, the system equation is

$$\begin{bmatrix} 1+s & -s & 0 \\ -s & 2+3s & -2s \\ 0 & -2s & 3+2s \end{bmatrix} \begin{bmatrix} V_1 \\ V_2 \\ V_3 \end{bmatrix} = \begin{bmatrix} J_1 \\ J_2 \\ 0 \end{bmatrix}$$

If we do the calculations by hand, then we can

1. Keep the variables s, J_1 and J_2 and solve. Substitute into the result $s = j2$ and the above numerical values for J_1 and J_2.
2. Substitute $s = j2$ into the matrix, use the form $\sqrt{2} + j\sqrt{2}$ and $j3$ for the sources and do all calculations in complex algebra.

We will show both cases. In the first case we evaluate the determinant

$$D(s) = 12s^2 + 19s + 6$$

and by using the Cramer's rule find

$$N(s) = 2s^2 J_1 + (2s + 2s^2)J_2$$

The output voltage is

$$V_3(s) = \frac{2s^2 J_1 + (2s + 2s^2)J_2}{12s^2 + 19s + 6}.$$

If we now substitute the phasors and set $s = j2$, we obtain

$$V_3(j2) = \frac{-8(\sqrt{2}+j\sqrt{2})+(4j-8)3j}{-42+j38} = \frac{-23.3137 - j35.3137}{-42+j38} = 0.7471e^{j98.705°}$$

The steady state output is $v_{out}(t) = 0.7471 \cos(2t + 98.705°)$. If we start with the substitutions immediately, then we must solve

$$\begin{bmatrix} 1+j2 & -j2 & 0 \\ -j2 & 2+j6 & -j4 \\ 0 & -4j & 3+j4 \end{bmatrix}\begin{bmatrix} V_1 \\ V_2 \\ V_3 \end{bmatrix} = \begin{bmatrix} \sqrt{2}+j\sqrt{2} \\ j3 \\ 0 \end{bmatrix}$$

The result is the same but operations with complex numbers are often more difficult.

We should mention that there exists still another possibility: invoking the law of superposition. In such case we find the determinant of the system and then solve twice, first by using J_1 only and next time using J_2 only. We will get two independent solutions and their sum will be equal to the complete steady-state frequency domain response. Try this possibility yourself.

Example 2.

The network in Fig. 8.2.6 has two independent sources and one current controlled current source. Find the two nodal voltages.

The phasors are $\mathbf{J}_1 = 2$, $\mathbf{J}_2 = 3$ and the controlling current, expressed in terms of the nodal voltage, is $I = G_1 V_1 = 0.5V_1$. The controlled current is $\alpha I = 2I = 2 \times 0.5V_1 = V_1$. The two nodal equations, with J denoting the symbol for a current source, are

$$(s + 0.5)V_1 + \alpha I = J_1$$

$$(0.2 + 2s)V_2 - \alpha I = J_2$$

Substituting for $\alpha I = V_1$ we get

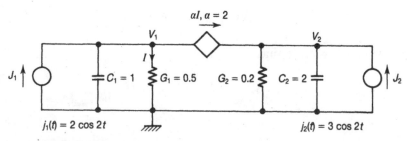

FIGURE. 8.2.6.   Network with two current sources and a *CC*.

$$\begin{bmatrix} s+1.5 & 0 \\ -1 & 2s+0.2 \end{bmatrix}\begin{bmatrix} V_1 \\ V_2 \end{bmatrix} = \begin{bmatrix} J_1 \\ J_2 \end{bmatrix}$$

The determinant is $D = 2s^2 + 3.2s + 0.3$,

$$N_1 = \begin{vmatrix} J_1 & 0 \\ J_2 & 2s+0.2 \end{vmatrix} = (2s+0.2)J_1$$

$$N_2 = \begin{vmatrix} s+1.5 & J_1 \\ -1 & J_2 \end{vmatrix} = (s+1.5)J_2 + J_1$$

Substituting numerical values for the phasors and $s = j2$ we obtain

$$V_1 = \frac{0.4 + j8}{-7.7 + j6.4} = 0.48 - j0.64 = 0.8e^{-j53.13}$$

$$V_1 = \frac{6.5 + j6}{-7.7 + j6.4} = -0.11621 - j0.8758 = 0.883e^{-j97.558}$$

The resulting time functions are

$$v1(t) = 0.8 \cos(2t - 53.13°)$$

$$v2(t) = 0.98835 \cos(2t - 97.558°)$$

In the past, phasors were used extensively because electrical engineering dealt mostly with power distribution and thus with only one frequency. In present applications we deal usually with ranges of frequencies and this requires repeated evaluations for various $s = j\omega$.

Let us summarize the developments of this section.

1. Frequency domain analysis can proceed only if all independent sources have the same frequency.

2. All independent sources must be given as cosines or transformed into cosines.

3. Express all sources in terms of their phasors.

4. Replace every s by $j\omega$.

5. Use phasors for the independent sources.

6. Obtain the desired output as a voltage, V, or current, I. These results are phasors.

7. Return to the original cosine functions using (8.2.3) and (8.2.4).

The above steps remain valid if we replace everywhere the word cosine by sine.

8.3. RC AND RL NETWORKS

Frequency domain analysis may ask for the solution of one of the two typical cases:

1. What is the steady state response of the network to a given signal or several given signals, all with the same frequency but with different amplitudes and phases.

2. What is the steady state response of the network to one input signal, always with the same amplitude and phase, but with various frequencies.

The first case was covered in Section 2, the second will be the subject of this section. The signal is assumed to be the function $\cos \omega t$, the phasor is either $\mathbf{E} = E = 1$ or $\mathbf{J} = J = 1$ and s is replaced by $j\omega$ with a sequence of values of ω. For each frequency, the result of the analysis is a complex number, $R = a + jb$, as we have seen in section 2. For each such number we can calculate the absolute value and the phase. Two plots can be prepared, both having on the horizontal axis either the frequency f in Hz, or ω in rad/s. If we plot on the vertical axis the absolute value, $|R|$, then we get the *amplitude response* of the network. If we plot the phase, then the curve is called the *phase response* of the network. In many cases the absolute value is not a convenient unit and we may decide to plot on the vertical axis the amplitude response in *decibels*, dB, which is $20 \log_{10} |R|$.

We will introduce the concept of frequency domain responses by taking four simple RC and RL voltage dividers. The first network is in Fig. 8.3.1. Its output voltage is

$$V_{out}(s) = \frac{\dfrac{1}{sC}}{\dfrac{1}{sC} + R} = \frac{1}{sCR + 1} E \qquad (8.3.1)$$

Substituting $E = 1$ and $s = j\omega$

$$V_{out}(j\omega) = \frac{1}{j\omega CR + 1}$$

The absolute value is

$$|V_{out}| = \frac{1}{\sqrt{\omega_2 C_2 R_2 + 1}}$$

and the phase angle is

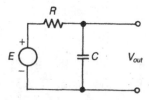

FIGURE 8.3.1   RC network.

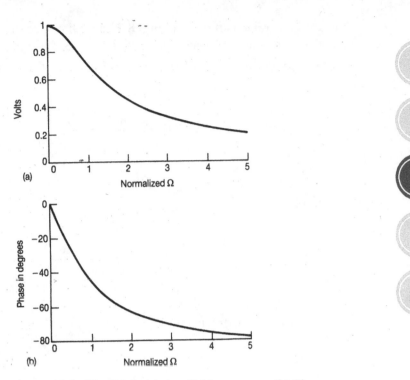

FIGURE 8.3.2 Respones of network in Fig. 8.3.1: (a) Amplitide response. (b) Phase response.

$$\phi = - \, arctan \; \omega CR$$

We could now insert various values of C and R, select frequencies and evaluate the absolute value and phase, but for this simple network it is possible to introduce *normalization* and get a result which is valid for all such networks. Define the substitution

$$\Omega = \omega CR \qquad (8.3.2)$$

The expressions simplify to

$$|V_{out}| = \frac{1}{\sqrt{\Omega^2 + 1}} \qquad (8.3.3)$$

and

$$\phi = - \, arctan \; \Omega \qquad (8.3.4)$$

The table given below gives results for a few values of Ω.

Ω	$|V_{out}|$	ϕ^0
0	1	0
0.5	0.8944	-26.57
1	0.7071	-45.00
2	0.4470	-63.44
3	0.3162	-71.57
4	0.2425	-75.96
100	0.010	-89.43

The absolute value is plotted in Fig. 8.3.2a with respect to the *normalized* frequency Ω; this is the amplitude response of the network. The phase response is in Fig. 8.3.2b.

The second network we will study is in Fig. 8.3.3. Its output voltage is

$$V_{out}(s) = \frac{R}{\dfrac{1}{sC} + R} = \frac{sCR}{sCR + 1}E \tag{8.3.5}$$

Substituting $E = 1$ and $s = j\omega$

$$V_{out}(j\omega) = \frac{j\omega CR}{j\omega CR + 1}$$

We can now introduce normalization (8.3.2)

$$V_{out}(j\Omega) = \frac{j\Omega}{j\Omega + 1}$$

and get

$$\left|V_{out}\right| = \frac{\Omega}{\sqrt{\Omega^2 + 1}} \tag{8.3.6}$$

and

$$\phi = 90° - arctan\ \Omega \tag{8.3.7}$$

The table below gives results for a few values of Ω.

Ω	$\left|V_{out}\right|$	ϕ^0
0	0	90
0.5	0.4472	63.43
1	0.7071	45.00
2	0.8944	26.57
3	0.9487	18.44
4	0.9701	14.04
100	0.9999	0.57

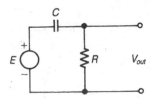

FIGURE 8.3.3 CR network.

The amplitude response is plotted in Fig. 8.3.4a, the phase response in Fig. 8.3.4b.

Before discussing the results, we will consider two additional cases. The first one is in Fig. 8.3.5. The output is

$$V_{out}(s) = \frac{R}{sL + R} E = \frac{1}{s\frac{L}{R} + 1} E \tag{8.3.8}$$

Substituting $E = 1$ and $s = j\omega$ and defining

$$\Omega = \omega \frac{L}{R} \tag{8.3.9}$$

we get

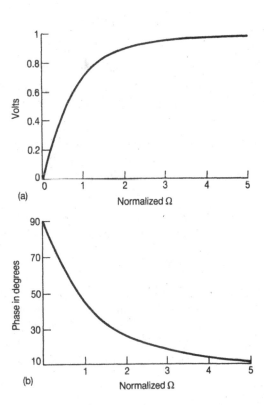

FIGURE 8.3.4 Responses of network in Fig. 8.3.3: (a) Amplitude response. (b) Phase response.

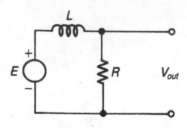

FIGURE 8.3.5   LR network.

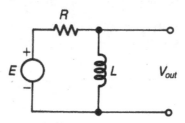

FIGURE 8.3.6   RL network.

$$|V_{out}| = \frac{1}{\sqrt{\Omega^2 + 1}}$$

and

$$\phi = -\arctan \Omega$$

These equations have *exactly* the same form as the equations describing the properties of the network in Fig. 8.3.1 and the curves in Fig. 8.3.2 are valid for the network in Fig. 8.3.5 as well, only the normalization variable is now (8.3.9) instead of (8.3.2).

The last network is in Fig. 8.3.6. Its output voltage is

$$V_{out}(s) = \frac{sL}{sL + R} = \frac{s\dfrac{L}{R}}{s\dfrac{L}{R} + 1}E \qquad (8.3.10)$$

Proceeding similarly as above we discover that properties of this network are *exactly* equivalent to the properties of the network in Fig. 8.3.3 and that the plots in Fig. 8.3.4 are valid for this network as well, only the normalization variable is now given by (8.3.9) instead of (8.3.2).

Let us now discuss the curves and how they can be used for similar networks with arbitrary component values. Consider, for instance, the network in Fig. 8.3.1 and let $R = 10^3\Omega$ and $C = 10^{-6}F$. We wish to know at what frequency will the response drop to 0.7071 of its value at dc. Reading from the curve (or from the table) we find that for this case $\Omega = \omega RC = 1$. Substituting the element values we have $\omega \times 10^3 \times 10^{-6} = 1$, from which $\omega = 10^3$ or $f = \dfrac{1000}{2\pi} = 159 Hz$

RC or RL networks may be more complicated that the above cases. Dependent sources may also be present and we will give two simple examples.

Example 1.

Consider the network in Fig. 8.3.7 and find the output as a function of the s variable. Let the phasor for the signal be $\mathbf{J} = 1$.

The nodal equations are

$$(1 + s)V_1 - sV_2 = 1$$

$$-sV_1 + (2 + s)V_2 - V_1 + V_2 = 0$$

As a matrix equation

$$\begin{bmatrix} 1+s & -s \\ -1-s & 3+s \end{bmatrix} \begin{bmatrix} V_1 \\ V_2 \end{bmatrix} = \begin{bmatrix} 1 \\ 0 \end{bmatrix}$$

The determinant is $D = 3s + 3$, the numerator is $N_2 = s + 1$. The output voltage. $V_2 = \dfrac{s+1}{3s+3} = \dfrac{1}{3}$ In this case it so happens that the output is independent of frequency.

Example 2.

The network in Fig. 8.3.8 has a current controlled current source. Find V_2 as a function of s and plot the amplitude response in absolute units. The source phasor is $\mathbf{J} = 1$.

The controlling current is $I = G_1 V_1$ and thus the controlled current is $\alpha I = 2G_1 V_1 = 4V_1$. The nodal equations are

$$\left(2 + \frac{1}{2s}\right)v_1 - \frac{1}{2s}v_2 = 1$$

$$-\frac{1}{2s}V_1 + \left(3 + \frac{1}{2s}\right)V_2 - 4V_1 = 0$$

In matrix form

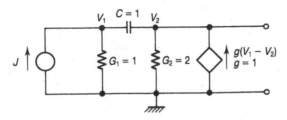

FIGURE 8.3.7   Network with one *VC*.

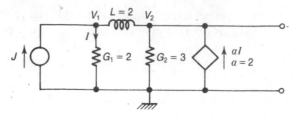

FIGURE 8.3.8   Network with one *CC*.

$$
\begin{bmatrix}
\dfrac{4s+1}{2s} & -\dfrac{1}{2s} \\[3mm]
\dfrac{-8s-1}{2s} & \dfrac{6s+1}{2s}
\end{bmatrix}
\begin{bmatrix} V_1 \\ V_2 \end{bmatrix}
=
\begin{bmatrix} 1 \\ 0 \end{bmatrix}
$$

The determinant is $D = \dfrac{12s+1}{2s}$, the numerator for the output is $N_2 = \dfrac{8s+1}{2s}$ and the

output voltage is $V_2 = \dfrac{8s+1}{12s+1}$ Inserting various values of $s = j\omega$ we can evaluate and plot the amplitude or phase response. Since plotting without a computer is always a time consuming task, it is useful to make an estimate of the frequency response at a few selected frequencies. The easiest is to find how the network will behave at $s = 0$ and at s approaching infinity. In the above example this will give $V_2(0) = 1$ and $V_2(\infty) = \dfrac{8}{12}$.

8.4.   RLC NETWORKS

Networks consisting of one capacitor, one inductor and one or more resistors have important frequency domain properties which we will study in this section. The simplest ones, with just one resistor, are also called tuned circuits.

The series tuned circuit is shown in Fig. 8.4.1. Using mesh analysis and the phasor value $\mathbf{E} = E$ (corresponding to the input signal $E \cos \omega t$)

$$
I = \frac{sC}{s^2 LC + sCR + 1} E = \frac{1}{L} \frac{s}{s^2 + s\dfrac{R}{L} + \dfrac{1}{LC}} E
\tag{8.4.1}
$$

Consider first the special case when $R = 0$; the current becomes

$$
I = \frac{1}{L} \frac{s}{s^2 + \dfrac{1}{LC}} E
$$

If we now select $s = j\omega = \dfrac{j}{\sqrt{LC}}$, the denominator becomes zero and the current will be infinitely large. The voltage across the capacitor,

$$V_C = \frac{1}{sC} I,$$

also becomes infinitely large for this frequency. The same will be true for the voltage across the inductor, because

$$V_L = sLI$$

Clearly, such frequency (in radians) represents a special property of the tuned circuit and is given a special name, the *resonant* frequency. It is commonly denoted by ω_0 and is given by

$$\omega_o = \frac{1}{\sqrt{LC}} \qquad (8.4.2)$$

If $R \neq 0$, then at the resonant frequency

$$I_{res} = \frac{1}{R} E$$

The current has the same phase as the input voltage (real value) and grows if R is made smaller. The voltage across the resistor will be

$$V_{R,res} = R \frac{E}{R} = E,$$

exactly equal to the amplitude of the source signal. The voltage across the capacitor will become

$$V_{C,res} = \frac{1}{j\omega_0 C} \frac{E}{R} = -j \frac{1}{R} \left[\frac{L}{C} \right]^{1/2} E$$

and the voltage across the inductor will be

$$V_{L,res} = j\omega_0 L \frac{E}{R} = j \frac{1}{R} \left[\frac{L}{C} \right]^{1/2} E$$

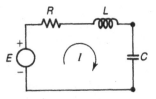

FIGURE 8.4.1 Series tuned circuit.

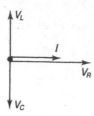

FIGURE 8.4.2   Phasors at resonance.

Since the voltage source is expressed as a phasor, these voltages are also phasors and can be plotted in a diagram, Fig. 8.4.2. The voltage across the resistor is in phase with the current, the voltage across the capacitor is lagging behind the resistor voltage by 90^0 and the voltage across the inductor is leading the resistor voltage by $90°$. The voltages across the capacitor and inductor have at resonance exactly opposite phases, have equal value and become larger if the resistor becomes smaller. Thus there is some kind of "quality" of the series tuned circuit; the smaller the resistor, the better is the quality of the tuned circuit. The concept of a *quality factor*, Q, is introduced,

$$Q = \frac{1}{R}\left[\frac{L}{C}\right]^{1/2} \tag{8.4.3}$$

We have now introduced two special names, the resonant frequency, ω_0, and the quality factor, Q. Their ratio

$$\frac{\omega_0}{Q} = \left[\frac{1}{LC}\right]^{1/2} R\left[\frac{C}{L}\right]^{1/2} = \frac{R}{L} \tag{8.4.4}$$

turns out to be exactly the middle coefficient of the denominator in equation (8.4.1). We can introduce both these new definitions into the second order polynomial in (8.4.1) and obtain

$$I = \frac{1}{L}\frac{s}{s^2 + s\dfrac{\omega_0}{Q} + \omega_0^2}E. \tag{8.4.5}$$

We can also express the resonant voltages in terms of Q. The voltage across the capacitor becomes

$$V_{C,res} = -jQE \tag{8.4.6}$$

and across the inductor

$$V_{L,res} = jQE \tag{8.4.7}$$

It probably comes as a surprise that these voltages can be much larger than the voltage actually supplied by the independent voltage source.

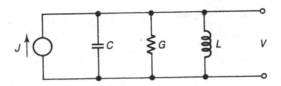

FIGURE 8.4.3 Parallel tuned circuit.

Another possibility is a parallel tuned circuit shown in Fig. 8.4.3. If the input current is $j(t) = J \cos \omega t$, the phasor for the current is $\mathbf{J} = J$ and the voltage across the combination is

$$V = \frac{1}{C} \frac{s}{s^2 + s\dfrac{G}{C} + \dfrac{1}{LC}} J \tag{8.4.8}$$

Using our generalized definitions, we can identify

$$\omega_0^2 = \frac{1}{LC} \tag{8.4.9}$$

and

$$\frac{\omega_0}{Q} = \frac{G}{C}$$

from where we get

$$Q = \frac{\omega_0 C}{G} = \omega_0 CR = R \left[\frac{C}{L} \right]^{1/2} \tag{8.4.10}$$

Note that this expression for the Q of a parallel tuned circuit differs from that of the series tuned circuit. Inserting (8.4.9) and (8.4.10) into (8.4.8)

$$V = \frac{1}{C} \frac{s}{s^2 + s\dfrac{\omega_0}{Q} + \omega_0^2} J \tag{8.4.11}$$

At the resonant frequency the voltage

$$V_{res} = \frac{Q}{\omega_0 C} J = \frac{J}{G} = RJ \tag{8.4.12}$$

It is in phase with the current of the source and grows with the increase of R. If the resistor becomes very large, $R \to \infty$, then $Q \to \infty$ and the voltage becomes infinitely large.

It is advantageous to introduce normalized expressions because they can be used for any values of L and C. Divide the numerator and denominator of Eq. (8.4.11) by ω_0^2 and set J to a unit phasor

$$V = \frac{1}{\omega_0 C} \frac{\dfrac{s}{\omega_0}}{\dfrac{s^2}{\omega_0^2} + \dfrac{1}{\omega_0}\dfrac{1}{Q} + 1} \tag{8.4.13}$$

Substitute $s = j\omega$ and define *normalized* Ω,

$$\Omega = \frac{\omega}{\omega_0} \qquad (8.4.14)$$

This modifies (8.4.13) to

$$V = \frac{1}{\omega_0 C} \frac{j\Omega}{(1-\Omega^2) + j\Omega \frac{1}{Q}} \qquad (8.4.15)$$

The product $\omega_0 C$ is a constant for any given network and the frequency dependent properties are expressed by the second fraction only

$$S = \frac{j\Omega}{(1-\Omega^2) + j\Omega \frac{1}{Q}} \qquad (8.4.16)$$

The absolute value of S is

$$|S| = \frac{\Omega}{\left[(1-\Omega^2)^2 + \frac{\Omega^2}{Q^2}\right]^{1/2}} \qquad (8.4.17)$$

and the phase is

$$\phi = arctan \frac{Q(1-\Omega^2)}{\Omega} \qquad (8.4.18)$$

Expression (8.4.17), sometimes called the *selectivity* curve, was evaluated for several Q and the plot is shown in Fig. 8.4.4. These normalized curves are valid for a parallel tuned circuit having the same Q and any values of L and C. They can be used to find suppression of signals away from resonance.

The concept of Q is extremely useful and we will derive a method how it can be determined from measurements. While the derivations may be skipped at first reading, the conclusions and the way how to measure Q should be remembered. Consider (8.4.11), insert $s = j\omega$ and take absolute value

$$|V(j\omega)| = \frac{1}{C} \frac{\omega}{\left[(\omega_0^2 - \omega^2)^2 + \frac{\omega^2 \omega_0^2}{Q^2}\right]^{1/2}} \qquad (8.4.19)$$

Also obtain the value at resonance

$$|V(j\omega_0)| = \frac{Q}{\omega_0 C} \qquad (8.4.20)$$

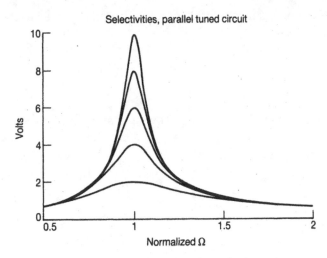

FIGURE 8.4.4 Selectivities for various Q.

Form the ratio and make it equal to $\sqrt{2}$

$$\frac{\left|V(j\omega_0)\right|}{\left|V(j\omega)\right|} = \sqrt{2} \tag{8.4.21}$$

This choice has the advantage that the derivations below become simple. Inserting from above and performing a few algebraic operations we arrive at

$$(\omega_0^2 - \omega^2)^2 = \frac{\omega^2 \omega_0^2}{Q^2} \tag{8.4.22}$$

Take the root

$$\omega_0^2 - \omega^2 = \pm \frac{\omega\omega_0}{Q}$$

and write

$$\omega^2 \pm \omega \frac{\omega_0}{Q} - \omega_0^2 = 0 \tag{8.4.23}$$

This represents two quadratic equations which must be considered separately. Taking first the positive sign in (8.4.23) we get two roots

$$\omega_{1,2} = -\frac{\omega_0}{2Q} \pm \omega_0 \left[\frac{1}{4Q^2} + 1\right]^{1/2}$$

and only the positive one is meaningful (negative frequency does not exist)

$$\omega_A = -\frac{\omega_0}{2Q} + \omega_0 \left[\frac{1}{4Q^2} + 1 \right]^{1/2}$$

Taking next the negative sign in Eq. (8.4.23) we get another two roots

$$\omega_{3,4} = -\frac{\omega_0}{2Q} \pm \omega_0 \left[\frac{1}{4Q^2} + 1 \right]^{1/2}$$

and only the positive root is meaningful,

$$\omega_B = \frac{\omega_0}{2Q} + \omega_0 \left[\frac{1}{4Q^2} + 1 \right]^{1/2}$$

Form the difference

$$\Delta\omega = \omega_B - \omega_A = \frac{\omega_0}{Q}$$

and rewrite as

$$Q = \frac{\omega_0}{\Delta\omega} = \frac{f_0}{\Delta f} \tag{8.4.24}$$

The result is interpreted as follows: Measure the voltage across the tuned circuit and find such frequency f_0 of the signal generator at which the voltage is maximum, V_{max}. Change the signal frequency below f_0 until the voltage across the tuned circuit drops to $\frac{V_{max}}{\sqrt{2}}$; this is f_A. Next change the signal frequency above f_0 until the voltage is again $\frac{V_{max}}{\sqrt{2}}$; this frequency is f_B. Insert into (8.4.24) to get the $Q = \frac{f_0}{f_B - f_A}$ of the tuned circuit. Application of the above is indicated schematically in Fig. 8.4.5.

The concept of Q and ω_0 is widely used in the literature and can be extended to any network function with a second order polynomial. We will demonstrate this generalization on the network shown in Fig. 8.4.6. Losses of inductors are usually modeled by a resistor in series with the inductance. Losses of capacitors are modeled by a resistor in parallel to the capacitance. Thus the network takes into account losses in both elements. We are interested in the output voltage. The analysis is simple, because the network is essentially a voltage divider in which the first impedance is

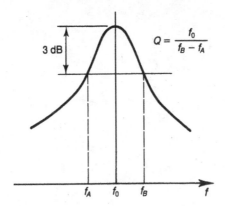

FIGURE 8.4.5 Determining Q from measurements.

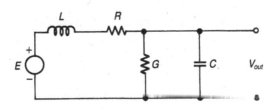

FIGURE 8.4.6 Tuned circuit with losses.

$$Z_1 = R + sL$$

and the second one is

$$Z_2 = \frac{1}{sC + G}$$

The transfer function is

$$V_{out} = \frac{Z_2}{Z_1 + Z_2} E = \frac{1}{s^2 LC + s(CR + LG) + (1 + GR)} E. \qquad (8.4.25)$$

In order to get the form (8.4.5) we factor out the product LC

$$V_{out} = \frac{1}{LC} \frac{1}{s^2 + s\dfrac{CR + LG}{LC} + \dfrac{1 + GR}{LC}} E.$$

Now we can identify

$$\omega_0^2 = \frac{1 + GR}{LC} \qquad (8.4.26)$$

and

$$\frac{\omega_0}{Q} = \frac{CR+LG}{LC}$$

from which we get

$$Q = \frac{\omega_0 LC}{CR+LG} = \frac{\sqrt{LC(1+RG)}}{CR+LG} \tag{8.4.27}$$

We will use similar generalized properties of the Q and ω_0 when we get to the analysis of active networks. We will also derive some additional formulas there. At this moment we remember the following steps:

1. Factor out the coefficient at s^2 so that the second-order polynomial is reduced to the form $s^2 + as + b$.

2. Then $\omega_0^2 = b$ and $\dfrac{\omega_0}{Q} = a$

Example 1.

Analyze the network in Fig. 8.4.7 and find its ω_0 and Q. The nodal equations are

$$(s + 1)V_1 - sV_2 = J$$

$$-sV_1 + (s + \frac{1}{2s} + 1)V_2 + 2.5V_1 = 0$$

In matrix form

$$\begin{bmatrix} s+1 & -s \\ 2.5-s & \dfrac{2s^2+2s+1}{2s} \end{bmatrix} \begin{bmatrix} V_1 \\ V_2 \end{bmatrix} = \begin{bmatrix} J \\ 0 \end{bmatrix}$$

The denominator is $D = \dfrac{9s^2+3s+1}{2s}$, $N_2 = (s-2.5)J$ and the output voltage is

$$V_2 = \frac{(s-2.5)2s}{9s^2+3s+1}J = \frac{1}{9}\frac{2s^2-5s}{s^2+\dfrac{1}{3}s+\dfrac{1}{9}}J$$

Thus the resonant frequency is

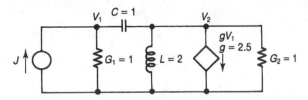

FIGURE 8.4.7 RLC network with a *VC*.

$$\omega_0 = \frac{1}{3}$$

and the Q is obtained by comparing the coefficient at the first power of s

$$\frac{\omega_0}{Q} = \frac{1}{3}$$

from which we obtain $Q = 1$.

8.5. PHASORS AND POWER

We introduced the concept of power in chapter one and we stated, in equation (1.5.2), that it is equal to the product of the voltage and current,

$$p(t) = v(t)i(t). \tag{8.5.1}$$

The equation is valid for any signal at any instant of time and in this section we will apply it to the situations when the voltage and current are cosine functions. For generality we will take

$$v(t) = V\cos(\omega t + \alpha) \tag{8.5.2}$$
$$i(t) = I\cos(\omega t + \beta)$$

where V and I are *amplitudes* of the signals. The situation is sketched in Fig. 8.5.1 in the form of two phasors, their rotation stopped at one instant. Inserting (8.5.2) into (8.5.1) we get

$$p(t) = VI\cos(\omega t + \alpha)\cos(\omega t + \beta) \tag{8.5.3}$$

We will use several substitutions in order to get an easier understandable result. Using the formula

$$\cos A \cos B = \frac{1}{2}\left[\cos(A - B) + \cos(A + B)\right]$$

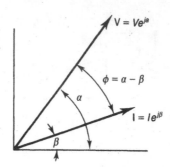

FIGURE 8.5.1   Phasors for voltage and current.

we rewrite (8.5.3) in the form

$$p(t) = \frac{VI}{2}\left[\cos(\omega t + \alpha - \omega t - \beta) + \cos(\omega t + \alpha + \omega t + \beta)\right]$$
$$= \frac{VI}{2}\left[\cos(\alpha - \beta) + \frac{VI}{2}\cos(2\omega t + \alpha + \beta)\right]$$

If we introduce the concept of phase difference of the two phasors, see Fig. 8.5.1,

$$\phi = \alpha - \beta \tag{8.5.4}$$

then the above can be further simplified to

$$p(t) = \frac{VI}{2}\cos\phi + \frac{VI}{2}\cos(2\omega t + 2\alpha - \phi)$$

For the complicated expression in the second cosine we can further apply the formula

$$\cos(A - B) = \cos A \cos B + \sin A \sin B$$

and obtain

$$p(t) = \frac{VI}{2}\cos\phi + \frac{VI}{2}\cos\phi\cos(2\omega t + 2\alpha) + \frac{VI}{2}\sin\phi\sin(2\omega t + 2\alpha)$$

Finally, since we know that the angle α is the result of our stopping the rotation, we can neglect it and get the final expression

$$p(t) = \frac{VI}{2}\cos\phi + \frac{VI}{2}\cos\phi\cos(2\omega t) + \frac{VI}{2}\sin\phi\sin(2\omega t) \tag{8.5.5}$$

The first term represents a dc component of the power, the other two components are changing with twice the angular speed, are shifted 90^0 with respect to each other and have zero average value.

Let us now consider some special situations, the first being that the voltage and current are in phase, as they would be when $i(t)$ flows through a resistor and creates a voltage $v(t)$ across it. In such case $\phi = 0$ and (8.5.5) simplifies to

$$p(t) = \frac{VI}{2} + \frac{VI}{2}\cos(2\omega t).$$

The first component is dc power, consumed in the resistor, the second term is oscillating with zero average and thus the average power consumed by the resistor is equal to the dc component. The same power would be consumed by the same resistor if we applied a dc voltage having an *effective* value

$$V_{eff} = \frac{V}{\sqrt{2}} \qquad (8.5.6)$$

In such case the current would have an effective value

$$I_{eff} = \frac{I}{\sqrt{2}}. \qquad (8.5.7)$$

Effective values are often used in equations but we will keep the absolute values.

If the element is a capacitor or inductor, then we know that the phase difference between the voltage and the current is $\pm 90^0$ and (8.5.5) simplifies to

$$p(t) = \pm \frac{VI}{2}\sin(2\omega t).$$

This is an oscillating power with zero average value and thus no average power is lost in any of these two elements. If the element is composed of several simpler components, the angle between the voltage and current will differ from 0^0 and $\pm 90^0$ and all three components of (8.5.5) will be nonzero.

Since the phasors were such a help in the analysis, it would be nice to use them for power calculations as well. This is indeed possible. The phasors for the signals (8.5.2) are

$$\mathbf{V} = Ve^{j\alpha}$$
$$\mathbf{I} = Ie^{j\beta} \qquad (8.5.8)$$

Form the product of the voltage phasor, the *complex conjugate* of the current phasor, and divide by 2

$$S = \frac{1}{2}V\bar{I} = \frac{VI}{2}e^{j(\alpha-\beta)} = \frac{VI}{2}e^{j\phi}$$

$$S = \frac{VI}{2}\cos\phi + j\frac{VI}{2}\sin\phi = P + jQ \qquad (8.5.9)$$

This complex power determines the amplitudes of all three terms in (8.5.5) and is widely used. The first component expresses the real power and is measured in watts, W, the second component is measured in reactive voltamperes, VAR, and their sum, **S**, is measured in volt-amperes, VA. Equation (8.5.9) gives us the possibility to use impedances and admittances to evaluate not only voltages and currents but also the components of the power, as we will show on several examples

Example 1.

Consider a system with $f = 60\ Hz$, which means that $\omega = 2\pi60 = 377\ rad/s$. The voltage $v(t) = 20\cos\omega t$ is applied to a parallel combination of a conductance $G = 2$ and $C = \frac{3}{377}$, as sketched in Fig. 8.5.2. Calculate the power and reactive power delivered into this combination.

The phasor of the voltage is $\mathbf{V} = 20e^{j0} = 20$. Admittance of the element combination is, at the given frequency, $Y = G + j\omega C = 2 + j3$. The current is

$$\mathbf{I} = \mathbf{V}Y = 20(2 + j3) = 40 + j60$$

Inserting into (8.5.9) we get

$$S = \frac{1}{2}\mathbf{V}\bar{\mathbf{I}} = \frac{1}{2}20(40 - j60) = 400 - j600$$

and thus

$$P = 400\ W$$

$$Q = -600\ VAR$$

Example 2.

Apply the same voltage as in example 1 to the series combination in Fig. 8.5.3 and obtain the components of the complex power.

For the given frequency the impedance of this combination is

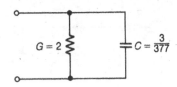

FIGURE 8.5.2 Simple GC network.

$$Z = 12 + j\omega L = 12 + j10$$

The current is

$$I = \frac{V}{Z} = \frac{20}{12 + j10} = 0.9836 - j0.8196$$

The complex power is

$$S = \frac{1}{2}V\bar{I} = \frac{1}{2}20(0.9836 + j0.8196) = 9.836 + j8.196$$

and thus

$$P = 9.836 \ W$$

$$Q = 8.196 \ VAR$$

Example 3.

Calculate the complex power components for the network and signal in Fig. 8.5.4.

The phasor for the signal $2 \cos 3t$ is $\mathbf{V} = 2$, with $\omega = 3$. Using standard mesh formulation and substituting $s = j3$ we get

$$\begin{bmatrix} 5+j6 & -j6 \\ -j6 & 3+j6 \end{bmatrix}\begin{bmatrix} I_1 \\ I_2 \end{bmatrix} = \begin{bmatrix} 2 \\ 0 \end{bmatrix}$$

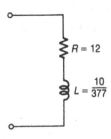

FIGURE 8.5.3 Simple RL network.

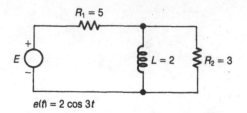

$e(t) = 2 \cos 3t$

FIGURE 8.5.4   Network with one inductor.

The determinant is $D = 15 + j48$, $N_1 = 6 + j12$ and the current

$$I_1 = \frac{6 + j12}{15 + j48}$$

To get a complex conjugate of I_1 change all signs of the imaginary parts. Thus the complex power is

$$S = \frac{1}{2}2\frac{6 - j12}{15 - j48} = 0.2633 + j0.0427$$

The components are

$$P = 0.2633 \ W$$

$$Q = 0.0427 \ VAR$$

Example 4.

Find the power consumed by the network in Fig. 8.5.5. The signal phasor is $\mathbf{J} = 3e^{j0} = 3$ and $s = j5$. Nodal formulation provides

$$\begin{bmatrix} 1 + 3s & -2s \\ -2s & 2s + 3 \end{bmatrix}\begin{bmatrix} V_1 \\ V_2 \end{bmatrix} = \begin{bmatrix} 3 \\ 0 \end{bmatrix}$$

Inserting $s = j5$ we must solve

$$\begin{bmatrix} 1 + j15 & -j10 \\ -j10 & 3 + j10 \end{bmatrix}\begin{bmatrix} V_1 \\ V_2 \end{bmatrix} = \begin{bmatrix} 3 \\ 0 \end{bmatrix}$$

The determinant is $D = -47 + j55$, the numerator is $N_1 = 9 + j30$, and the voltage

$$V_1 = \frac{9 + j30}{-47 + j10}.$$

Since the phasor for the current does not have an imaginary component, the complex power is

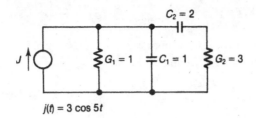

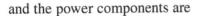

FIGURE 8.5.5   Network with two capacitors.

$$S = \frac{1}{2} \frac{9 + j30}{-47 + j55} 3 = 0.3516 - j0.5459$$

and the power components are

$$P = 0.3516 \ W$$

$$Q = -0.5459 \ VAR$$

Example 5.

Suppose that we measured the voltage across a device and the current through it

$$v(t) = 10 \cos(60t + 70°)$$

$$i(t) = 5 \cos(60t - 50°)$$

Find the complex power components.
The phasors are

$$\mathbf{V} = 10e^{j70}$$

$$\mathbf{I} = 5e^{-j50}$$

and the complex power is

$$S = \frac{1}{2} 10e^{j70} 5e^{j50} = 25e^{j120} = -12.5 + j21.65.$$

Since the real component is negative, the device actually delivers power 12.5 W.

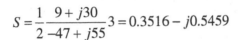

PROBLEMS CHAPTER 8

P.8.1   Given are three complex numbers:
$v_1 = 3 + j2$, $v_2 = -1\ j4$, $v_3 = -2 - j3$ and $v_4 = 4 - j2$. Perform the following operations

(a) $v_1 + v_2$

(b) $v_1 - v_2 + v_3$

(c) $v_1 + v_3 - v_2 - v_4$

(d) $v_1 v_2$

(e) $\dfrac{v_1 v_3}{v_2}$

(f) $\dfrac{(v_1 - v_2)v_4}{v_3}$

(g) $v_1\ v_2\ v_3\ v_4$

(h) $\dfrac{v_1 + v_2}{v_3 + v_4}$

P.8.2   Convert the following vectors in rectangular form into polar form:

(a) $v_1 = 3 + j5$

(b) $v_2 = -1 + j2$

(c) $v_3 = -2 - j4$

(d) $v_4 = 4 - j1$

(e) $v_5 = -4 - j3$

(f) $v_6 = 0 + j3$

(g) $v_7 = -1 - j0$

(h) $v_8 = 0 - j5$

P.8.3   Convert the following polar coordinates into rectangular form: The angles are given in degrees.

(a) $v_1 = 5e^{j60}$

(b) $v_2 = 2e^{j120}$

(c) $v_3 = 4e^{-j150}$

(d) $v_4 = 3e^{-j20}$

(e) $v_5 = 5e^{j150}$

(f) $v_6 = 4e^{j90}$

(g) $v_7 = 5e^{-j180}$

(h) $v_s = 4e^{-j90}$

P.8.4   Write the phasor representations for the following functions:

(a) $3 \cos(2t + 40^0)$

(b) $2 \cos(3t + 150^0)$

(c) $4 \cos(4t - 30^0)$

(d) $5 \sin \omega t$

(e) $4 \sin(\omega t + 20^0)$

(f) $2 \sin(377t + 50^0)$

P.8.5 Using phasors, find the resulting single cosine (or sine) functions.

(a) $4 \cos (2t + 40^0) + 3 \cos(2t - 50^0)$

(b) $\cos \omega t + \sin \omega t$

(c) $3\cos(\omega t + 20^0) + 2 \cos(\omega t + 40^0)$

(d) $2 \cos(\omega t - 120^0) - 3 \cos(\omega t + 140^0)$

(e) $2 \cos \omega t + 4 \cos(\omega t + 45^0)$

(f) $\cos(\omega t + 20^0) + \cos(\omega t + 40^0) + \cos(\omega t + 60^0)$

(g) $2 \cos(\omega t + 180^0) + 4 \cos(\omega t + 90^0)$

(h) $\sin(\omega t + 30^0) - 2 \sin(\omega t - 30^0)$

(i) $3 \cos(\omega t + 50^0) + 2 \cos(\omega t + 150^0) + \cos(\omega t - 30^0)$

(j) $2 \cos \omega t + 3 \cos(\omega t + 90^0) + \cos(\omega t + 180^0) + 4 \cos(\omega t - 90^0)$

P.8.6 Find the impedance and admittance for the series RL combination. In each case evaluate at $\omega = 1, 2, 3$.

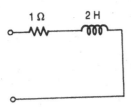

P.8.7 Find the impedance and admittance for the series CR combination. In each case evaluate at $\omega = 1, 2, 3$.

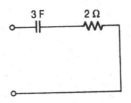

P.8.8 Find the impedance and admittance for the parallel CR combination. In each case evaluate at $\omega = 1, 2, 3$.

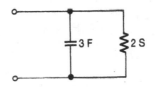

P.8.9 Find the impedance and admittance for the parallel LR combination. In each case evaluate at $\omega = 1, 2, 3$.

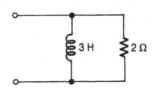

P.8.10 Find the impedance of the series RCL combination at $\omega = 2$.

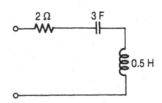

P.8.11   Find the input admittance of the CGRL combination at $\omega = 5$.

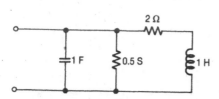

P.8.12   The input signal is $e(t) = 2 \cos(2t + 30^0)$. Find the phasor of the current flowing into the network. Also express the result in terms of the input signal.

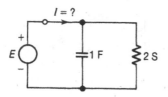

P.8.13   Find the output voltage for the current source being the function $j(t) = 3 \cos(3t + 25^0)$. (a) Keep the variable s, get the result and then insert the phasor of the signal. (b) Substitute for $s = j3$ from the start and do the same.

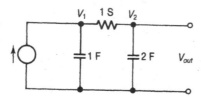

P.8.14   Find the output voltage if the input signal is $j(t) = \cos \omega t$ and get the results for $\omega = 1, 2, 3$. Note that in this case it is to your advantage to keep the variable s, since you can re-use the solution step.

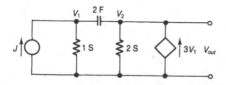

P.8.15   Do the same as in P.8.14.

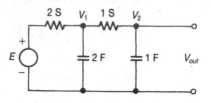

P.8.16   Find the output voltage V_2 for $j(t) = \cos \omega t$ with $\omega = 1, 2$

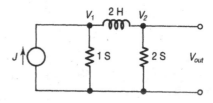

P.8.17   Using mesh formulation, obtain the output voltage and evaluate for $e(t) = \cos \omega t$ with $\omega = 1, 2$.

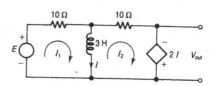

P.8.18   Find the output current for $e(t) = \cos \omega t$ with $\omega = 1$. Obtain the result by (a) keeping the variable s and (b) inserting $s = j1$ from the start.

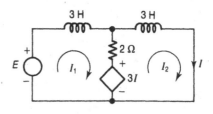

P.8.19   Find the output voltage for $e(t) = \cos \omega t$ with $\omega = 1, 2$.

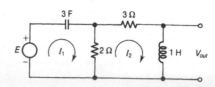

P.8.20 Calculate the complex power components for problem P.8.6 for (a) input voltage $e(t) = \cos\omega t$ and (b) input current $j(t) = \sin \omega t$, using $\omega = 1, 2$.

P.8.21 Do the same as in P.8.20 for the combination in P.8.9.

P.8.22 Do the same as in P.8.20 for the combination in P.8.10.

P.8.23 Do the same as in P.8.20 for the combination in P.8.11.

P.8.24 Find the complex power components consumed in the R,L part of the network in P.8.11. Let the source be a current source $j(t) = 2 \cos \cdot 2t$.

P.8.25 In problem P.8.14 calculate the complex power components delivered by the source. The input signal is $2 \cos 2t$.

P.8.26 Do the same as in P.8.25 for the network in P.8.15.

9   LAPLACE TRANSFORMATION

INTRODUCTION

Laplace transformation is the most efficient method for solving linear networks. It was invented by Pierre Simon de Laplace in the eighteenth century, but was widely accepted only in the twentieth century.

The purpose of a mathematical transformation is to introduce definitions which considerably simplify some complicated steps. In the case of the Laplace transform, differentiation is replaced by multiplication by the transform variable s. Integration is made even simpler, because it is replaced by division by s.

We will not give a fully comprehensive treatment of the Laplace transform theory and only those subjects which have direct relationship to linear networks will be selected. A detailed study is left to specialized mathematical courses.

9.1. DEFINITION OF THE LAPLACE TRANSFORMATION

The Laplace transform works with the complex variable

$$s = \sigma + j\,\omega \tag{9.1.1}$$

The real part, σ, does not have any direct practical explanation but the imaginary part does. If we set $\sigma = 0$ and thus

$$s = j\,\omega = j2\pi f \tag{9.1.2}$$

then we can speak about actual frequency, f, of a steady sinusoidal signal applied to the network. This was already explained in detail in previous Chapters.

Let us have a function $f(t)$ of time t such that $f(t) = 0$ for $t < 0$. Then the (one-sided) Laplace transform is defined by

$$L[f(t)] = F(s) = \int_{0^-}^{\infty} f(t)e^{-st}\,dt \tag{9.1.3}$$

It is common practice to use capital letters for the transformed variables (functions of s) and lower case letters for the functions of time, t.

The inverse transformation is defined by the integral

$$L^{-1}[f(t)] = \frac{1}{2\pi j} \int_{c-j\infty}^{c+j\infty} F(s)e^{+st} ds \qquad (9.1.4)$$

Note that the inverse transformation has the exponent with + sign and the original transformation has − sign. Also note the symbolic notation with L and L^{-1}. The integration in (9.1.4) is along a line parallel with the imaginary axis.

Evaluations of (9.1.3) are relatively simple to apply and require straightforward integration in the real variable, t. Evaluations of the inverse transform require considerably more mathematical background than can be expected from the student at this time. Fortunately, it is not necessary to go via (9.1.4). It is possible to get the transformations for a certain number of practical functions and prepare tables. The inverse transformations are then done by consulting the tables. This will be our way of handling the inverses.

Two fundamental rules of the Laplace transform will be needed in the following derivations. For a function $f(t)$, having the transform $F(s)$, we have

$$L[af(t)] = aF(s) \qquad (9.1.5)$$

This follows immediately from (9.1.3). The second rule considers addition. If we have two functions, $f_1(t)$ and $f_2(t)$ and their transforms are known to be $F_1(s)$ and $F_2(s)$, then

$$L[f_1(t) + f_2(t)] = \int_{0^-}^{\infty}(f_1(t) + f_2(t))e^{-st} dt = \int_{0^-}^{\infty} f_1(t)e^{-st} dt +$$

$$\int_{0^-}^{\infty} f_2(t)e^{-st} dt = F_1(s) + F_2(s) \qquad (9.1.6)$$

In other words, if an overall function is formed by the sum of simpler functions, then the same summation applies to the Laplace inversion.

9.2. SIMPLE FUNCTIONS AND THEIR TRANSFORMS

Finding the Laplace transform of more complicated functions may sometimes be achieved by applying various Laplace transform rules. We will introduce some of these rules later but here we will obtain all the necessary transforms by integrating (9.1.3) with selected functions of time.

The simplest and often used function is the unit step, see Fig. 9.2.1a. It is defined by

$$u(t) = 1 \text{ if } t \geq 0$$
$$= 0 \text{ if } t < 0 \qquad (9.2.1)$$

Inserting into (9.1.3) we must integrate

$$F_u(s) = \int_{0^-}^{\infty} 1 \cdot e^{-st} dt = \frac{-e^{-st}}{s} \Big|_{0^-}^{\infty} = \frac{1}{s}$$

or

$$L[u(t)] = F_u(s) = \frac{1}{s} \qquad (9.2.2)$$

A rectangular pulse $p(t)$ having duration T and height $1/T$ is shown in Fig. 9.2.1b. Its transform is

$$F_p(s) = \frac{1}{T}\int_{0^-}^{T} e^{-st}\,dt = \frac{e^{-st}}{-sT}\Bigg|_{0^-}^{T} = \frac{1-e^{-sT}}{sT} \qquad (9.2.3)$$

The definition is such that the area of the function is equal to one, no matter what T we select. This function is used sometimes but we will use it to define the Dirac impulse.

A Dirac impulse, $\delta(t)$, is the limit of the pulse $p(t)$ for $T \to 0$. This strange function, sketched in Fig. 9.2.1c, is simultaneously infinitely high and infinitely narrow but still having an area equal to 1. Its Laplace transform is found by applying the L'Hospital limiting rule to (9.2.3)

$$L[\delta(t)] = F_\delta(s) = \lim_{T\to 0}\frac{1-e^{-sT}}{sT} = 1 \qquad (9.2.4)$$

The function is extensively used in theoretical studies. As will be shown later, it is the derivative of $u(t)$.

Another commonly occurring function is the exponential,

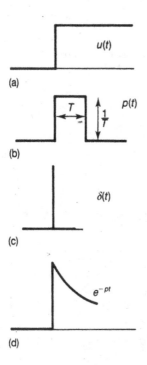

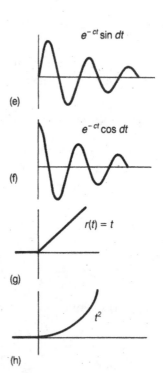

FIGURE 9.2.1   Typical time functions.

$$f(t) = e^{-pt} \tag{9.2.5}$$

It is sketched in Fig. 9.2.1d for real p, although p can also be complex. The transform is

$$L[e^{-pt}] = \int_{0^-}^{\infty} e^{-pt} e^{-st} dt = \frac{e^{-(s+p)t}}{-(s+p)}\bigg|_{0^-}^{\infty} = \frac{1}{s+p} \tag{9.2.6}$$

A sinusoid multiplied by an exponential is another important function

$$f(t) = e^{-ct} \sin d\,t \tag{9.2.7}$$

It is sketched in Fig. 9.2.1e for positive real c and d. The integral to be solved is

$$F(s) = \int_{0^-}^{\infty} e^{-ct} \sin d\,t e^{-st} dt = \int_{0^-}^{\infty} e^{-(s+c)t} \sin d\,t\, dt$$

Integration is done by parts using the formula $\int u' v = uv - \int uv'$. Define $u' = e^{-(s+c)t}$ and $v = \sin d\,t$. Then $u = \dfrac{-e^{-(s+c)t}}{s+c}$, $v' = d\cos d\,t$, and

$$F(s) = \frac{-e^{-(s+c)t} \sin d\,t}{s+c}\bigg|_{0^-}^{\infty} + \frac{d}{s+c}\int_{0^-}^{\infty} e^{-(s+c)t} \cos d\,t\, dt$$

Inserting limits we get zero for the first term. The second term is integrated by parts using the same formula, selecting u and u' as above but setting $v = \cos d\,t$ and $v' = -d\sin d\,t$. This leads to

$$F(s) = \frac{d}{s+c}\left[\frac{-e^{-(s+c)t} \cos d\,t}{s+c}\bigg|_{0^-}^{\infty} - \frac{d}{s+c}\int_{0^-}^{\infty} e^{-(s+c)t} \sin d\,t\, dt\right]$$

Inserting the limits again and realizing that the integral on the right side is nothing but $F(s)$ we obtain

$$F(s) = \frac{d}{s+c}\left\{\frac{1}{s+c} - \frac{d}{s+c}F(s)\right\}$$

Transferring terms with $F(s)$ to the left side we get

$$L[e^{-ct} \sin d\,t] = \frac{d}{(s+c)^2 + d^2} \tag{9.2.8}$$

Proceeding similarly we can also find

$$L[e^{-ct} \cos d\,t] = \frac{s+c}{(s+c)^2 + d^2} \tag{9.2.9}$$

This function is sketched in Fig. 9.2.1f.

If the function does not have the exponential component, we set $c = 0$ in the function as well as in the Laplace transform and obtain

$$L[\sin d\,t] = \frac{d}{s^2 + d^2} \qquad (9.2.10)$$

$$L[\cos d\,t] = \frac{s}{s^2 + d^2} \qquad (9.2.11)$$

Finally, realizing that $\sin(d\,t + \phi) = \sin d\,t \cos\phi + \cos d\,t \sin\phi$ and taking into account the fact that the Laplace transform is a linear operation, see (9.1.6),

$$L[e^{-ct}\sin(d\,t + \phi)] = \frac{d\cos\phi + (s+c)\sin\phi}{(s+c)^2 + d^2} \qquad (9.2.12)$$

Should $c = 0$, then

$$L[\sin(d\,t + \phi)] = \frac{d\cos\phi + s\sin\phi}{s^2 + d^2} \qquad (9.2.13)$$

Another function which can have some practical application is the ramp function, $r(t) = kt$ where k is the slope. Its sketch is in Fig. 9.2.1g. If $k = 1$, the ramp function is clearly the integral of the unit step. Its transform is

$$F_r(s) = \int_{0^-}^{\infty} t\,e^{-st}\,dt$$

Setting $u' = e^{-st}$ and $v = t$ we find for $k = 1$

$$L[t] = \frac{1}{s^2} \qquad (9.2.14)$$

We can go further and consider higher powers; t^2 is sketched in Fig. 9.2.1h. Inserting into (9.1.3) and integrating by parts we get after the first integration

$$L[t^2] = -\frac{t^2 e^{-st}}{s}\bigg|_{0^-}^{\infty} + \frac{2}{s}\int_{0^-}^{\infty} t\,e^{-st}\,dt$$

The first term is zero after inserting the limits and the integral has just been solved in (9.2.14). Using this result we get

$$L[t^2] = \frac{2}{s^3} \qquad (9.2.15)$$

We can continue similarly to higher powers and obtain, by induction,

$$L[t^n] = \frac{n!}{s^{n+1}} \qquad (9.2.16)$$

The transformations are summarized in Table 9.2.1 and we will be referring to it frequently. Most of the terms are multiplied by $u(t)$ to indicate, in a formally correct way, that the functions all start at $t = 0$. The table contains all the transformations which can reasonably appear in our studies.

<table>
<tr><td colspan="4" align="center">Table 9.2.1.

Laplace transform pairs</td></tr>
</table>

No	$F(s)$	$f(t)$	Remark
1	1	$\delta(t)$	Dirac impulse
2	$\dfrac{1}{s}$	$u(t)$	unit step
3	$\dfrac{1}{s^2}$	$t\,u(t)$	ramp function
4	$\dfrac{1}{s^n}$	$\dfrac{t^{n-1}}{(n-1)!}u(t)$	$0!=1$
5	$\dfrac{1}{s+p}$	$e^{-pt}u(t)$	p real or complex
6	$\dfrac{1}{(s+p)^n}$	$\dfrac{t^{n-1}e^{-pt}}{(n-1)!}u(t)$	p real or complex
7	$\dfrac{d}{s^2+d^2}$	$\sin d\,t\,u(t)$	d real
8	$\dfrac{s}{s^2+d^2}$	$\cos d\,t\,u(t)$	d real
9	$\dfrac{d\cos\phi+s\sin\phi}{s^2+d^2}$	$\sin(d\,t+\phi)\,u(t)$	d,ϕ real
10	$\dfrac{d}{(s+c)^2+d^2}$	$e^{-ct}\sin d\,t\,u(t)$	c,d real
11	$\dfrac{s+c}{(s+c)^2+d^2}$	$e^{-ct}\cos dt\,u(t)$	c,d real
12	$\dfrac{As+B}{(s+c)^2+d^2}$	$e^{-ct}[A\cos d\,t+\dfrac{B-Ac}{d}\sin d\,t]\,u(t)$	A,B,c,d real
13	$\dfrac{A+jB}{s+c+jd}+\dfrac{A-jB}{s+c-jd}$	$2e^{-ct}[A\cos d\,t+B\sin d\,t]\,u(t)$	A,B,c,d real

9.3. RATIONAL FUNCTIONS WITH SIMPLE POLES

Laplace domain functions which describe properties of linear time independent networks are rational functions in the s variable. We have already established this in some of our previous examples. Such functions can be written in the following forms:

$$F(s)=\frac{N(s)}{D(s)}=\frac{\displaystyle\sum_{0}^{M}a_i s^i}{\displaystyle\sum_{i=0}^{N}b_i s^i}=K_0\frac{\displaystyle\prod_{i=1}^{M}(s-z_i)}{\displaystyle\prod_{1}^{N}(s-p_i)} \tag{9.3.1}$$

where the coefficient in K_0 is equal to

$$K_0=\frac{a_M}{b_N}$$

In the above equation, $N(s)$ and $D(s)$ are polynomials. The roots of the numerator, $N(s)$, are called the zeros, z_i, the roots of the denominator, $D(s)$, are called poles, p_i. For inversion of the Laplace transform we *must* know the poles. Knowledge of the zeros is not a necessity but is useful in most cases. This section will cover cases where all poles are *distinct*, which means that there are no cases of the type $(s - p_i)^m$ where m is larger than one. Problems with multiple poles, $m > 1$, will be handled in the next section.

We start our explanations with two examples:

$$F_1(s) = \frac{3s^2 + 27}{2s^2 + 4s + 4} = \frac{3(s^2 + 9)}{2(s^2 + 2s + 2)} = \frac{3}{2}\frac{(s + j3)(s - j3)}{(s + 1 + j)(s + 1 - j)} \tag{9.3.2}$$

and

$$F_2(s) = \frac{3s^2 + 27}{2s^3 + 6s^2 + 8s + 4} = \frac{3}{2}\frac{s^2 + 9}{s^3 + 3s^2 + 4s + 2}$$
$$= \frac{3}{2}\frac{(s + j3)(s - j3)}{(s + 1)(s + 1 + j)(s + 1 - j)} \tag{9.3.3}$$

In equation (9.3.2), it was easy to find both the poles and the zeros by a simple and well known formula for quadratic polynomials. In most situations we are not so lucky.

Very often, students do not clearly understand what is involved in the step from the polynomial to its roots. Let us state here that finding the roots is simple for a polynomial of degree two, but not much can be done without a computer for polynomials of higher degrees. Theoretically, formulas exist to find roots of polynomials of third and fourth degree, but such formulas are difficult to apply. Should the polynomials be in symbolic form (with letters describing the types of elements having yet unknown values), then it will be difficult to find the roots of polynomials of degree as low as three. Our suggestion is not even to try in the case of a symbolic function. In numerical cases use the computer or a suitable calculator.

Recall the functions in Section 2 or look at Table 9.2.1. The Laplace transforms there have, in the denominator, either a term of the form $s + p$, or a polynomial of the type $(s + c)^2 + d^2$, which has complex conjugate roots. If we can break a more complicated rational function into a sum of such simple rational functions, we can use Table 9.2.1 for each of them and find the inverse transformation as a sum of simple time functions. The steps are called *partial fraction expansion* and will be the method we will use.

Decomposition can be done by several methods which we will explain. We will also remember the following rule: If the numerator has the same or higher degree than the denominator, then we must first reduce the numerator degree by division and make it one less than the denominator. For instance, should we have the function

$$F(s) = \frac{s^3 + 4s^2 + 5s + 1}{s^2 + s + 1}$$

then we first apply long division to get

$$F(s) = s + 3 + \frac{s - 2}{s^2 + s + 1}$$

and only then do we look for a decomposition of the remaining rational function. The numerator is usually not higher than the denominator but we will have cases when the degrees are equal. In such case we apply one step of such division and obtain a constant. Should its value be k then the inverse transformation will have a term $k\,\delta(t)$, as can be seen from the first pair in Table 9.2.1.

9.3.1 Decomposition Using Real Arithmetic

We will first introduce a method which operates with real numbers. It is based on the decomposition $\dfrac{K_i}{s+p_i}$ for each real pole and on the decomposition $\dfrac{A_i s + B_i}{(s+c_i)^2 + d_i^2}$ for each complex conjugate pair.

Before we proceed we derive a useful formula for the inversion. Consider the following sequence of equivalent operations

$$\frac{As+B}{(s+c)^2+d^2} = \frac{(As+Ac)+(B-Ac)}{(s+c)^2+d^2} = A\frac{s+c}{(s+c)^2+d^2} + \frac{B-Ac}{d}\frac{d}{(s+c)^2+d^2}$$

Comparing the last two terms with formulas 10 and 11 in Table 9.2.1 we see that

$$L^{-1}[\frac{As+B}{(s+c)^2+d^2}] = e^{-ct}[A\cos dt + \frac{B-Ac}{d}\sin dt]u(t) \qquad (9.3.4)$$

This is formula 12 in Table 9.2.1.

The function (9.3.2) has the same degree in the numerator and denominator. The degree of the numerator is first lowered by division

$$F_1(s) = \frac{3}{2}[\frac{s^2+9}{s^2+2s+2}] = \frac{3}{2}[1+\frac{-2s+7}{(s+1)^2+1}].$$

Comparing with the above formula we identify $A = -2$, $B = 7$, $c = 1$ and $d = 1$ and the inversion is

$$f_1(t) = \frac{3}{2}[\delta(t) - 2e^{-1}\cos t + 9e^{-1}\sin t]u(t) \qquad (9.3.5)$$

The second example, equation (9.3.3), already has the degree of the numerator less than the degree of the denominator and no division is needed. We use the decomposition

$$F_2(s) = \frac{3}{2}[\frac{s^2+9}{(s+1)(s^2+2s+2)}] = \frac{3}{2}[\frac{K}{s+1} + \frac{As+B}{s^2+2s+2}]$$

Cross-multiply the terms on the right to get a single fraction and compare the resulting numerator with the numerator of the left side. This leads to

$$s^2 + 9 = Ks^2 + 2Ks + 2K + As^2 + As + Bs + B$$

The equation can be satisfied only if coefficients at equal powers are equal. This leads to the system

$$K + A = 1$$
$$2K + A + B = 0$$
$$2K + B = 9.$$

(9.3.6)

Its solution is $K = 10$, $A = -9$ and $B = -11$ and

$$F_2(s) = \frac{3}{2}[\frac{10}{s+1} + \frac{-9s-11}{(s+1)^2 + 1}]$$

(9.3.7)

Identify $A = -9$, $B = -11$, $c = 1$, $d = 1$ and $\dfrac{B - Ac}{d} = -2$. Using the table, the inversion is

$$f_2(t) = \frac{3}{2}[10e^{-t} - 9e^{-t}\cos t - 2e^{-t}\sin t]u(t)$$
$$= \frac{3}{2}e^{-t}[10 - 9\cos t - 2\sin t]u(t)$$

(9.3.8)

Summarizing the method, assume the decomposition $\dfrac{K_i}{s + p_i}$ for each real simple pole and the decomposition $\dfrac{A_i s + B_i}{(s + c_i)^2 + d_i^2}$ for each pair of complex conjugate poles. Express the sum of fractions as one fraction by cross-multiplying the terms and compare coefficients with the numerator of the given function. This provides a set of linear equations which are solved for the unknown constants. Then consult Table 9.2.1 for the inverse functions.

Example 1.

Find the time domain response of the function

$$F(s) = \frac{s^2 + 4}{(s^2 + 9)(s^2 + 4s + 5)}$$

The function has complex conjugate poles $p_{1,2} = \pm j3$ and $p_{3,4} = -2 \pm j1$. We need the decomposition in the form

$$\frac{s^2 + 4}{(s^2 + 9)(s^2 + 4s + 5)} = \frac{A_1 s + B_1}{s^2 + 9} + \frac{A_2 s + B_2}{s^2 + 4s + 5}$$

Cross-multiplying on the right side we get the equation for the numerators

$$s^2 + 4 = A_1 s^3 + A_1 4s^2 + A_1 5s + B_1 s^2 + B_1 4s$$
$$+ B_1 5 + A_2 s^3 + A_2 9s + B_2 s^2 + B_2 9$$

and equating coefficients at the same powers of s we get the system

$$A_1 + A_2 = 0$$
$$4A_1 + B_1 + B_2 = 1$$
$$5A_1 + 4B_1 + 9A_2 = 0$$
$$5B_1 + 9B_2 = 4$$

Since we see that $A_2 = -A_1$, we can reduce the size of the problem by substituting into the last three equations. This leads to

$$\begin{bmatrix} 4 & 1 & 1 \\ -4 & 4 & 0 \\ 0 & 5 & 9 \end{bmatrix} \begin{bmatrix} A_1 \\ B_1 \\ B_2 \end{bmatrix} = \begin{bmatrix} 1 \\ 0 \\ 4 \end{bmatrix}.$$

The solution is $D = 160$, $A_1 = \dfrac{1}{8}$, $B_1 = \dfrac{1}{8}$, $A_2 = \dfrac{1}{8}$ and $B_2 = \dfrac{3}{8}$. The problem is decomposed into

$$\frac{s^2 + 4}{(s^2 + 9)(s^2 + 4s + 5)} = \frac{\dfrac{1}{8}s + \dfrac{1}{8}}{s^2 + 9} + \frac{-\dfrac{1}{8}s + \dfrac{3}{8}}{(s+2)^2 + 1}$$

At this point we can use formula 12 in Table 9.2.1, identify $A_1 = \dfrac{1}{8}$, $B_1 = \dfrac{1}{8}$, $c_1 = 0$, $d_1 = 3$

and calculate $\dfrac{B_1 - A_1 c_1}{d_1} = \dfrac{1}{24}$. For the second term identify $A_2 = -\dfrac{1}{8}$, $B_2 = \dfrac{3}{8}$, $c_2 = 2$, $d_2 = 1$

and calculate $\dfrac{B_2 - A_2 c_2}{d_2} = \dfrac{5}{8}$. The inverted function is

$$f(t) = \{\frac{1}{8}\cos 3t + \frac{1}{24}\sin 3t + e^{-2t}[-\frac{1}{8}\cos t + \frac{5}{8}\sin t]\}u(t)$$

9.3.2 Decomposition Using Complex Arithmetic

Instead of using second order polynomials as above, we can break the denominator polynomial into products $(s + c + jd)(s + c - jd)$ and treat each term as a simple pole using (9.2.6). Since for any complex pole there must be another complex conjugate pole, it will be *always* true that the decomposition constants will also be complex conjugate.

Similarly as above we will first derive a useful formula. Consider a complex conjugate pair with complex conjugate numerator constants and use (9.2.6) for inversion

$$L^{-1}[\frac{A + jB}{s + c + jd} + \frac{A - jB}{s + c - jd}] = (A + jB)e^{(-c-jd)t} + (A - jB)e^{(-c+jd)t}$$

$$= e^{-ct}[2A\frac{e^{jdt} + e^{-jdt}}{2} + 2B\frac{e^{jdt} - e^{-jdt}}{2j}]$$

This leads to the formula

$$L^{-1}[\frac{A+jB}{s+c+jd}+\frac{A-jB}{s+c-jd}]=2e^{-ct}[A\cos dt + B\sin dt] \qquad (9.3.9)$$

which is also in Table 9.2.1, formula 13.

For instance, in equation (9.3.2) we could use the decomposition

$$\frac{3}{2}\frac{s^2+9}{s^2+2s+2}=\frac{3}{2}[1+\frac{A+jB}{s+1+j}+\frac{A-jB}{s+1-j}]$$

Proceeding similarly as above, we cross multiply the terms on the right to get the equation

$$s^2 + 9 = s^2 + 2s + 2 + (A + jB)(s + 1 - j) + (A - jB)(s + 1 + j)$$

After the multiplications we obtain

$$s^2 + 9 = s^2 + s(2 + 2A) + 2 + 2A + 2B$$

Comparing terms at equal powers of s we get the system

$$2 + 2A = 0$$

$$2 + 2A + 2B = 9$$

with the solution $A = -1$, $B = \frac{9}{2}$. The decomposition is

$$F_1(s) = \frac{3}{2}[1+\frac{-1-j9/2}{s+1-j}+\frac{-1+j9/2}{s+1+j}] \qquad (9.3.10)$$

Consulting Table 9.2.1, formula 13 we get the final result

$$f_1(t) = \frac{3}{2}\delta(t) + 3e^{-t}(-\cos t + \frac{9}{2}\sin t)u(t) \qquad (9.3.11)$$

which is the same as obtained in (9.3.5).

9.3.3   Residues of Simple Poles

In this section, we cover the best method for evaluating the constants of the partial fractions under the condition that *all poles are simple*. The method is suitable for programming as well as hand calculations, even if the poles are complex.

Let us have a rational function in which we know the poles. The numerator may (but need not) be in factored form but the denominator must be. Thus we assume that the function has already the one of the forms

$$F(s) = \frac{\sum\limits_0^M a_i s^i}{\prod\limits_1^N (s - p_i)} = K_0 \frac{\prod\limits_{i=1}^M (s - z_i)}{\prod\limits_1^N (s - p_i)} \tag{9.3.12}$$

In the first expression we assume that all numerator coefficients have been divided by a possible term $b_N \neq 1$. In the second expression

$$K_0 = \frac{a_M}{b_N}$$

Since we assumed that the poles are simple, we know from the previous discussion that we can decompose this function into two forms. If $M < N$, then

$$F(s) = \sum_{i=1}^N \frac{K_i}{s - p_i} \tag{9.3.13}$$

if $M = N$, then

$$F(s) = K_0 + \sum_{i=1}^N \frac{K_i}{s - p_i} \tag{9.3.14}$$

The numbers K_i, $i = 1, 2, \ldots N$, are called the *residues* of the function. The steps are easiest to explain by taking an example. Consider

$$F(s) = \frac{(s+2)(s+4)}{(s+1)(s+3)(s+5)} = \frac{K_1}{s+1} + \frac{K_2}{s+3} + \frac{K_3}{s+5} \tag{9.3.15}$$

and let us be interested in K_1. Multiply the equation by the term $s + 1$. This results in

$$(s+1)F(s) = \frac{(s+2)(s+4)}{(s+3)(s+5)} = K_1 + K_2 \frac{s+1}{s+3} + K_3 \frac{s+1}{s+5}$$

Note that in the middle expression the multiplication *removed* the term $s + 1$. If we now substitute $s = -1$, we will get

$$[(s+1)F(s)]_{s=-1} = \frac{(-1+2)(-1+4)}{(-1+3)(-1+5)} = K_1 + 0 + 0$$

or

$$K_1 = \frac{3}{8}$$

For the second pole we multiply (9.3.15) by $s + 3$ to get

$$[(s+3)F(s)] = \frac{(s+2)(s+4)}{(s+1)(s+5)} = K_1 \frac{s+3}{s+1} + K_2 + K_3 \frac{s+3}{s+5}$$

and substituting $s = -3$ in the whole equation will result in

$$[(s+3)F(s)]_{s=-3} = \frac{1}{4} = K_2$$

Similarly, for the last pole multiply (9.3.15) by $s+5$

$$[(s+5)F(s)] = \frac{(s+2)(s+4)}{(s+5)(s+3)} = K_1 \frac{s+1}{s+1} + K_2 \frac{s+5}{s+5} + K_3$$

Substituting $s = -5$ results in

$$[(s+5)F(s)]_{s=-5} = \frac{3}{8} = K_3$$

and the decomposition is finished

$$F(s) = \frac{3}{8}\frac{1}{s+1} + \frac{1}{4}\frac{1}{s+3} + \frac{3}{8}\frac{1}{s+5}$$

Using the table we get the inversion

$$f(t) = [\frac{3}{8}e^{-t} + \frac{1}{4}e^{-3t} + \frac{3}{8}e^{-5t}]u(t) \tag{9.3.16}$$

The above steps can be written as a formula

$$K_i = [(s-p_i)F(s)]_{s=p_i} \tag{9.3.17}$$

where the symbols on the right side mean that we

1. first cancel the term $s - p_i$ against the same term in the denominator of F(s) and
2. only then substitute $s = p_i$.

This method works equally well for complex conjugate poles, with the additional advantage that we need not calculate the residue of the complex conjugate pole because we know that its residue is complex conjugate. Considering, for instance, our example in equation (9.3.3), we have

$$F_2(s) = \frac{3}{2}\frac{s^2+9}{(s+1)(s+1+j)(s+1-j)}$$

$$= \frac{K_1}{s+1} + \frac{K_2}{s+1+j} + \frac{\overline{K_2}}{s+1-j} \tag{9.3.18}$$

Then

$$K_1 = [(s+1)F(s)]_{s=-1} = \frac{3}{2}\frac{(-1)^2+9}{(-1+1+j)(-1-1+j)} = 15$$

and

$$K_2 = [(s+1+j)F(s)]_{s=-1-j} = \frac{3}{2}\frac{(-1+j)^2+9}{(-1-j+1)(-1-j+1-j)} = -\frac{27}{4} - j\frac{3}{2}$$

If we are confident that we did not make a mistake in the evaluations, the decomposition is finished because the complex conjugate pole will have a complex conjugate residue:

$$F_2(s) = \frac{3}{2} \frac{s^2 + 9}{(s+1)(s+1+j)(s+1-j)}$$

$$= \frac{15}{s+1} + \frac{-\frac{27}{4} - j\frac{3}{2}}{s+1+j} + \frac{-\frac{27}{4} + j\frac{3}{2}}{s+1-j} \qquad (9.3.19)$$

We do not even need to write the complete decomposition. We can use the formula (9.3.10) for the pair of complex conjugate poles, identify $A = -\frac{27}{4}$, $B = -\frac{3}{2}$, $c = 1$, $d = 1$ and write the inversion

$$f(t) = [15e^{-t} + 2e^{-t}(-\frac{27}{4}\cos t - \frac{3}{2}\sin t)]u(t) \qquad (9.3.20)$$

The result corresponds to (9.3.8).

Example 2.

Find the time domain response of the function

$$F(s) = \frac{s^2 + 4}{(s^2 + 9)(s^2 + 4s + 5)}$$

by using the residue calculus. The function has complex conjugate poles $p_{1,2} = \pm j3$ and $p_{3,4} = -2 \pm j1$. The function can be rewritten in the form

$$F(s) = \frac{s^2 + 4}{(s + j3)(s - j3)(s + 2 + j1)(s + 2 - j1)}$$

The residue at $s = -j3$ is

$$K_1 = \frac{(-j3)^2 + 4}{(-j3 - j3)(-j3 + 2 + j1)(-j3 + 2 - j1)} = \frac{3 + j1}{48}$$

The residue at $s = -2 - j1$ is

$$K_2 = \frac{(-2 - j1)^2 + 4}{(-2 - j1 + j3)(-2 - j1 - j3)(-2 - j1 + 2 - j1)} = \frac{-1 + j5}{16}$$

The above calculations may be fairly lengthy if we do them by hand but many calculators have complex arithmetic and make the evaluation quite easy. We can now identify the terms for formula (9.3.10). For the first pole, $-j3$, we have $A_1 = \frac{1}{16}$, $B_1 = \frac{1}{48}$, $c_1 = 0$, $d_1 = 3$.

For the complex pole, $-2 - j1$, we have $A_2 = -\frac{1}{16}$, $B_2 = \frac{5}{16}$, $c_2 = 2$, $d_2 = 1$. Inserting into (9.3.10)

$$f(t) = \left\{ 2\left[\frac{1}{16}\cos 3t + \frac{1}{48}\sin|3t|\right] + 2e^{-2t}\left[-\frac{1}{16}\cos t + \frac{5}{16}\sin t\right] \right\} u(t)$$

9.4. DECOMPOSITION WITH MULTIPLE POLES

Before we go into the details of decompositions with multiple poles, we will discuss some practical points. Multiple poles are important theoretically, simply because they can exist, but finding them in higher order polynomials is not a trivial problem. If the multiplicity is high, we may not even be able to find the multiple root correctly. Fortunately, when we look at the situation from a practical point of view, we discover that multiple poles are extremely rare. In fact, in network applications, if we have the choice, we will *always* avoid situations when they could occur. Responses of networks with multiple poles are inferior to responses where we are free to assign the poles as simple ones. The only situation of some practical importance is the case of the ramp function, see Table 9.2.1, but even this function is somewhat artificial. As a result of these *practical* considerations we feel that the student can skip this section in his first reading and return to it when needed.

If the function has multiple poles of the type $(s - p_i)^m$, $m > 1$, then the decomposition is done differently and an example will show best what has to be done. Let the function have one pole of multiplicity $m = 3$ and some other simple poles, let us say two. Then we must assume the decomposition in the form

$$F(s) = \frac{N(s)}{(s - p_1)^3 (s - p_2)(s - p_3)} \tag{9.4.1}$$

$$= \frac{K_{1,3}}{(s - p_1)^3} + \frac{K_{1,2}}{(s - p_1)^2} + \frac{K_{1,1}}{(s - p_1)^1} + \frac{K_{2,1}}{s - p_2} + \frac{K_{3,1}}{s - p_3}$$

The first subscript corresponds to the subscript of the pole and the second indicates the power of the term $s - p_1$. For each pole of multiplicity m we must assume m terms with all the powers from m to 1.

The first step to find the coefficients is to multiply the (9.4.1) by the highest power of the pole we are considering. In the above case, multiply by $(s - p_1)^3$ to get

$$[(s - p_1)^3 F(s)] = \frac{N(s)}{(s - p_2)(s - p_3)} = K_{1,3} + K_{1,2}(s - p_1) + K_{1,1}(s - p_1)^2$$

$$+ K_{2,1}\frac{(s - p_1)^3}{s - p_2} + K_{3,1}\frac{(s - p_1)^3}{s - p_3} \tag{9.4.2}$$

If we now substitute $s = p_1$, we get

$$[(s - p_1)^3 F(s)]_{s=p_1} = \frac{N(p_1)}{(p_1 - p_2)(p_1 - p_3)} = K_{1,3} \tag{9.4.3}$$

and the remaining terms disappear because their numerators are zero.

In order to get $K_{1,2}$, we differentiate (9.4.2) with respect to s:

$$\frac{d}{ds}[(s-p_1)^3 F(s)] = \frac{d}{ds}[\frac{N(s)}{(s-p_2)(s-p_3)} = K_{1,2} + 2K_{1,1}(s-p_1)$$

$$+ [\,terms\,having\,(s-p_1)\,in\,the\,numerator\,]$$

(9.4.4)

Check that after the differentiation we have, indeed, all the remaining terms multiplied by $s - p_1$. If we now substitute $s = p_1$, the result will be

$$\frac{d}{ds}[(s-p_1)^3 F(s)]_{s=p_1} = K_{1,2}$$

(9.4.5)

Finally, to obtain $K_{1,1}$ we differentiate (9.4.4) to get

$$\frac{d^2}{ds^2}[(s-p_1)^3 F(s)] = 2K_{1,1} + [\,terms\,having\,(s-p_1)\,in\,the\,numerator\,]$$

Substituting $s = p_1$ provides

$$\frac{d^2}{ds^2}[(s-p_1)^3 F(s)]_{s=p_1} = 2K_{1,1}$$

or

$$K_{1,1} = \frac{1}{2}\frac{d^2}{ds^2}[(s-p_1)^3 F(s)]_{s=p_1}$$

The residues of the simple poles are obtained by the method in section 9.3.3.

In general, if the ith pole has multiplicity m, then the decomposition coefficient $K_{i,j}$, $m \geq j \geq 1$, will be

$$K_{i,j} = \frac{1}{(m-j)!}\frac{d^{m-j}}{ds^{m-j}}[(s-p_i)^m F(s)]_{s=p_i}$$

(9.4.6)

with the understanding that we

1. first cancel the term with which we multiply against the same term in the denominator,
2. then obtain the necessary derivative and
3. only then substitute the pole p_i for s.

As an example, consider the function

$$F(S) = \frac{1}{(s-1)^2 s^3}$$

with the poles $p_1 = 1$ and $p_2 = 0$. The decomposition will have the form

$$F(s) = \frac{K_{1,2}}{(s-1)^2} + \frac{K_{1,1}}{s-1} + \frac{K_{2,3}}{s^3} + \frac{K_{2,2}}{s^2} + \frac{K_{2,1}}{s}$$

(9.4.7)

For the first pole we have

$$(s-1)^2 F(s) = s^{-3} \qquad (9.4.8)$$

Thus

$$K_{1,2} = \frac{1}{(2-2)!}[(s-1)^2 F(s)]_{s=1} = s^{-3}\big|_{s=1} = 1$$

$$K_{1,1} = \frac{1}{(2-1)!}\frac{d}{ds}[(s-1)^2 F(s)]_{s=1} = -3s^{-4}\big|_{s=1} = -3$$

For the second pole we have

$$s^3 F(s) = (s-1)^{-2} \qquad (9.4.9)$$

$$K_{2,3} = \frac{1}{(3-3)!}[s^3 F(s)]_{s=0} = [(s-1)^{-2}]_{s=0} = 1$$

$$K_{2,2} = \frac{1}{(3-2)!}\frac{d}{ds}[(s^{-3} F(s)]_{s=0} = [-2(s-1)^{-3}]_{s=0} = 2$$

$$K_{2,1} = \frac{1}{(3-1)!}\frac{d^2}{ds^2}[(s^{-3} F(s)]_{s=0} = \frac{1}{2}[6(s-1)^{-4}]_{s=0} = 3$$

The complete decomposition is

$$F(s) = \frac{1}{(s-1)^2 s^3} = \frac{1}{(s-1)^2} + \frac{-3}{s-1} + \frac{1}{s^3} + \frac{2}{s^2} + \frac{3}{s}$$

The above steps are equally valid for complex conjugate poles. The decomposition coefficient of the conjugate pole will be the conjugate of the coefficient of the other pole. Once we have the term $K_{i,j}$, then the time domain function corresponding to this term is

$$L^{-1}[\frac{K_{i,j}}{(s-p_i)^j}] = K_{ij}\frac{t^{j-1}e^{p_i t}}{(j-1)!} \qquad (9.4.10)$$

9.5. LAPLACE TRANSFORM OPERATIONS

Operations with time functions are usually complicated while operations in the Laplace domain are almost always much simpler. In order to be able to take advantage of these simplifications, we now transform several mathematical operations into the Laplace domain. The transformations are summarized in Table 9.5.1.

9.5.1 Differentiation

Differentiation of time functions is expressed in the Laplace domain by multiplication by the s variable. For our network theory this is the most important operation.

Consider $f(t)$ for which we know the Laplace transform $F(s)$. We wish to express differentiation of the time function in terms of $F(s)$; this means that we are looking for

$$L[f'(t)] = \int_{0^-}^{\infty} f'(t)e^{-st}dt \qquad (9.5.1)$$

The integral is solved by parts by selecting $u' = f'$, $u = f$, $v = e^{-st}$, $v' = -se^{-st}$:

$$L[f'(t)] = f(t)e^{-st}\Big|_{0^-}^{\infty} + s\int_{0^-}^{\infty} f(t)e^{-st}dt$$

or

$$L[f'(t)] = sF(s) - f(0) \qquad (9.5.2)$$

where $f(0)$ is the value of the function at $t = 0$.

Table 9.5.1.

Laplace transform operations

No	f(t)	F(s)
1	$a \cdot f(t)$	$a \cdot F(s)$
2	$\dfrac{d}{dt}f(t)$	$s \cdot F(s) - f(0)$
3	$\dfrac{d^n}{dt^n}f(t)$	$s^n F(s) - \displaystyle\sum_{i=1}^{n} s^{n-i} f^{(i-1)}(0)$
		where $f^{(k)}(0) = \lim\limits_{t \to 0} \dfrac{d^k}{dt^k} f(t)$
4	$\displaystyle\int_0^t f(z)dz$	$\dfrac{F(s)}{s} + \dfrac{f^{(-1)}(0)}{s}$
		where $f^{(-1)}(0) = \lim\limits_{t \to 0} \displaystyle\int_0^t f(z)dz$
5	$f(kt)$	$\dfrac{1}{k}F\left(\dfrac{s}{k}\right)$
6	$f(t-T)$	$e^{-sT}F(s)$
7	$\displaystyle\int_0^t f_1(t-z)f_2(z)dz$	$F_1(s) \cdot F_2(s)$
8	$\lim\limits_{t \to 0} f(t)$	$\lim\limits_{s \to \infty} s \cdot F(s)$
9	$\lim\limits_{t \to \infty} f(t)$	$\lim\limits_{s \to 0} s \cdot F(s)$

The Laplace transform of the second derivative can be obtained similarly but it is easier to apply (9.5.2):

$$L[f''(t)] = s[sF(s) - f(0)] - f'(0)$$

or

$$L[f''(t)] = s^2F(s) - sf(0) - f'(0) \tag{9.5.3}$$

Continuing to higher order derivatives, we obtain a general formula

$$L[f^{(n)}(t)] = s^nF(s) - s^{n-1}f(0) - s^{n-2}f'(0) - \cdots$$
$$\cdots - sf^{(n-2)}(0) - f^{(n-1)}(0) \tag{9.5.4}$$

Let us now see how this is used in applications. The most important case is that of an initial voltage on the capacitor or an initial current through the inductor. The current flowing through a capacitor is given by

$$i = C\frac{dv}{dt}$$

Denote by I the Laplace transform of the current i and by V the Laplace transform of the voltage v. Let the initial voltage across the capacitor be E_0. The situation is sketched in Fig. 9.5.1a and b. Using (9.5.2)

$$I = C[sV - E_0]$$

or

$$I = sCV - CE_0 \tag{9.5.5}$$

Should the initial voltage change its polarity with respect to the + and − sign indicated at the terminals, the sign in (9.5.5) will change to +. In the Laplace transform, the product CE_0 represents the Dirac impulse with an area CE_0 instead of 1.

The voltage across the inductor is given by

$$v = L\frac{di}{dt}$$

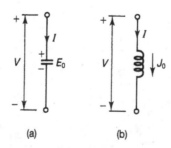

(a)   (b)

FIGURE 9.5.1   Initial conditions for: (a) Capacitor. (b) Inductor.

Let the initial current be J_0 with the direction shown in Fig. 9.5.1. Then in the Laplace domain

$$V = L[sI - J_0] = sLI - LJ_0 \qquad (9.5.6)$$

The product LJ_0 represents a Dirac impulse with the area LJ_0 instead of 1. Should the initial current flow in the opposite direction with respect to the + and − sign at the terminals, the sign in the formula is changed to +.

If there is no initial voltage on the capacitor or no initial current through the inductor, E_0 or J_0 are zero and

$$I = sCV$$
$$V = sLI \qquad (9.5.7)$$

We have already used these expressions in Chapter 7, but now we can see the theoretical connection. We will return to the above derivations in more detail in the next chapter.

9.5.2 Integration

Formulas for integration are derived similarly as in the case of differentiation

$$L[\int_{0^-}^{t} f(\tau)d\tau] = \int_{0^-}^{\infty} \left(\int_{0}^{t} f(\tau)d\tau \right) e^{-st} dt$$

Integrating by parts

$$L[\int_{0^-}^{t} f(\tau)d\tau] = \frac{-e^{-st}}{s} \int_{0^-}^{t} f(\tau)d\tau \Big|_{0^-}^{\infty} + \frac{1}{s} \int_{0^-}^{\infty} f(t)e^{-st} dt$$

or

$$L[\int_{0^-}^{t} f(\tau)d\tau] = \frac{F(s)}{s} + \frac{f^{(-1)}(0)}{s} \qquad (9.5.8)$$

The symbol $f^{(-1)}(0)$ denotes the integral of the function from 0^- to 0^+. This integral will normally be zero, except in the case of a Dirac impulse for which we have derived in section 9.2 that it has zero duration but, nevertheless, has an area equal to one. This is the case of the initial voltage on a capacitor or the initial current through the inductor. For the capacitor, we have in the time domain

$$v(t) = \frac{1}{C} \int_{0^-}^{t} i(t)dt$$

In the Laplace domain this will become

$$V = \frac{1}{C}[\frac{I}{s} + \frac{CE_0}{s}]$$

or

$$V = \frac{1}{sC} + \frac{E_0}{s} \tag{9.5.9}$$

The reader should note that this could be obtained simply by separating algebraically V in formula (9.5.5). There is, however, one important point to remember: the influence of the initial voltage across the capacitor is now expressed by $\dfrac{E_0}{s}$, which is equivalent to a *step* of voltage having a height E_0, see Table 9.2.1, formula 2.

The expression for inductor current can be found similarly but we will proceed differently here. We start with (9.5.6) and separate I algebraically:

$$I = \frac{V}{sL} + \frac{J_0}{s} \tag{9.5.10}$$

The initial condition is now taken into account by a step function having the amplitude J_0. If the initial voltage across the capacitor and the initial current through the inductor are zero, then the expressions simplify to

$$V = \frac{I}{sC}$$
$$I = \frac{V}{sL} \tag{9.5.11}$$

These expressions were also introduced in Chapter 7.

9.5.3   Initial and Final Value Theorems

It is sometimes advantageous to know how the function $f(t)$ behaves at times $t \to 0$ and $t \to \infty$. Formulas for both cases are derived by starting with (9.5.1):

$$L[f'(t)] = \int_{0^-}^{\infty} f'(t)e^{-st}dt = sF(s) - f(0)$$

Take the limit $s \to 0$:

$$\lim_{s\to\infty} \int_{0^-}^{\infty} f'(t)e^{-st}dt = \lim_{s\to\infty}[sF(s) - f(0)]$$

Since $e^0 = 1$, on the left side remains only $f'(t)$ under the integral sign and the solution is $f(t)$; inserting the limits

$$\lim_{t\to\infty}[f(t) - f(0)] = \lim_{s\to 0}[sF(s) - f(0)]$$

Canceling $f(0)$ on both sides leads to the *final value theorem:*

$$\lim_{t\to\infty} f(t) = \lim_{s\to 0} sF(s) \tag{9.5.12}$$

Another formula is obtained by starting again with (9.5.1) but assuming that the limit is $s \to \infty$:

$$\lim_{s \to \infty} \int_{0^-}^{\infty} f'(t)e^{-st}dt = \lim_{s \to \infty}[sF(s) - f(0)]$$

Since $e^{-\infty} = 0$, the integral is zero and the expression reduces to

$$0 = \lim_{s \to \infty}[sF(s) - f(0)]$$

Transferring $f(0)$ on the left side and realizing that this may be written as a limit of $f(t)$ for $t \to 0$ we obtain the *initial value theorem:*

$$\lim_{t \to 0} f(t) = \lim_{s \to \infty} sF(s) \qquad (9.5.13)$$

Note that formulas (9.5.12) and (9.5.13) have the same form, except for the limits. If the limit of t is 0, the limit of s is ∞ and vice versa.

Both theorems can give considerable insight into the behavior of functions. As an example, consider a Laplace domain function

$$F(s) = \frac{N(s)}{D(s)} = \frac{s^2 + ...}{s^5 + ...}$$

where the dots indicate that there are other terms with powers of s less than the one indicated. The initial value theorem indicates that

$$\lim_{t \to 0} f(t) = \lim_{s \to \infty} \frac{s(s^2 + ...)}{s^5 + ...} = 0$$

since the highest power in the numerator is less than the highest power in the denominator. How will the derivative of $f(t)$ behave at $t \to 0$? We already know that the derivative is obtained by multiplying the Laplace domain function by s. Therefore

$$f'(0) = \lim_{s \to \infty} \frac{s \cdot s(s^2 + ...)}{s^5 + ...}$$

Here one multiplication by s comes from the formula, the second from the differentiation. The numerator has the highest power 4, the denominator 5 and for extremely large s the limit will be zero. What about the second derivative? Multiply again by s. Now we have

$$f''(0) = \lim_{s \to \infty} \frac{s^2 \cdot s(s^2 + ...)}{s^5 + ...}$$

The highest powers in the numerator and denominator are equal, resulting in a nonzero limit of the second order derivative at $t \to 0$. We know that the time domain response starts at the origin and, in addition, has the first derivative equal to zero; this means that it must start horizontally from the same point. We cannot say anything else about the form of the curve, but even this gives some insight into the time domain behavior of the function. Note that all we need to know about the function are the highest powers of s in the numerator and denominator.

9.5.4 Scaling in the Time Domain

Assume that we have the function $f(t)$ and know its Laplace transform $F(s)$. We wish to know how scaling of time will influence the transform. We are looking for a formula for the scaled situation $f(kt)$. Using the definition

$$L[f(kt)] = \int_{0^-}^{\infty} f(kt)e^{-st}dt$$

and changing the variable by selecting

$$kt = \tau$$

$$dt = \frac{d\tau}{k}$$

we arrive at

$$L[f(kt)] = \frac{1}{k}\int_{0^-}^{\infty} f(\tau)e^{-(s/k)\tau}d\tau = \frac{1}{k}\int_{0^-}^{\infty} f(t)e^{-(s/k)t}dt$$

In the last step we returned to the original time variable because the use of a different letter cannot change the integral. The expression is similar to (9.1.3) with the only exception that we now have s/k instead of s. This means that we can write the formula

$$L[f(kt)] = \frac{1}{k}F(\frac{s}{k}) \qquad (9.5.14)$$

9.5.5 Delay

One of the important questions is: What will happen with the transform $F(s)$ if instead of $f(t)$ we consider $f(t-T)$, a function delayed (shifted to the right) by the amount of T. Using formula (9.1.3) we are looking for

$$L[f(t-T)] = \int_{0^-}^{\infty} f(t-T)e^{-st}dt$$

Substitute a new variable

$$\tau = t - T$$

Then

$$L[f(t-T)] = \int_{-T}^{\infty} f(\tau)e^{-st(\tau+T)}d\tau$$

Recalling that all functions start, by definition, at $t = 0$, we can modify the lower limit to 0 and write

$$L[f(t-T)] = e^{-sT}\int_{0^-}^{\infty} f(\tau)e^{-s\tau}d\tau$$

To write the final formula, we replace τ by t, because the change of the letter cannot change the value of the integral. But then the integral is nothing but $F(s)$, (9.1.3), and

$$L[f(t-T)] = e^{-sT} F(s) \qquad (9.5.15)$$

A few explanations will clarify the importance of this formula. The symbol $f(t-T)$ means that in the time domain every occurrence of t is replaced by $(t-T)$. For instance, let the parabola

$$f(t) = t^2 \quad \text{for } t \geq 0$$
$$= 0 \quad \text{for } t < 0 \qquad (9.5.16)$$

be shifted to the right by $T = 5$. Then the delayed parabola is defined by

$$f(t-T) = (t-T)^2 \quad \text{for } t-5 \geq 0$$
$$= 0 \qquad \text{for } t-5 < 0 \qquad (9.5.17)$$

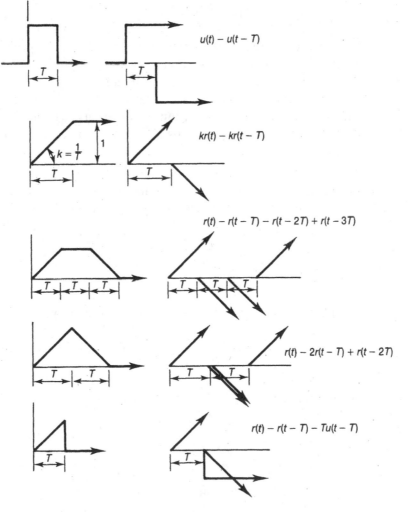

FIGURE 9.5.2   Composed functions.

As we know from section 9.2, the Laplace transform of (9.5.16) is $\frac{2}{s^3}$. According to the formula we just derived, the Laplace transform of (9.5.17) is $e^{-5s}\frac{2}{s^3}$.

In most application-oriented studies, only a few functions are needed: the Dirac impulse, the unit step, the ramp function, the sine and cosine functions, possibly multiplied by e^{-ct}. Many other functions can be composed of several of the above functions, properly shifted in time. Fig. 9.5.2 gives some of them.

Linear networks have the superposition property: if several signals are applied to the network, we can consider each of them separately, solve for the response of the network to each of them and add the responses to get the final answer. For this reason, it is not even necessary to consider shifts of input signals. Take, for instance, the trapezoidal impulse in Fig. 9.5.2, composed of four equal ramp functions. All we have to do is solve for one of them and plot the result. The responses to the other three ramp functions will be the same curves, only shifted to the right by appropriate amounts and either added or subtracted. This will give the response to the trapezoidal pulse.

9.6. SIGNALS AND NETWORKS

The most important property of the Laplace transform is the ease with which we can calculate the response of any linear network to any given input signal. Recall what we learned in Chapter 7. We introduced the impedances and admittances, we stated that the independent signal will be a symbol, to be specified later, and we learned how to get the solution and the output. Except for the case of initial conditions, which will be covered in the next chapter, we already know all the steps how to find the output in terms of the network elements and the yet unspecified input signal.

For time domain evaluation, all we have to do is replace the symbol of the independent source by *the Laplace transform of the signal* actually applied. Afterwards we must go ahead with the decompositions and inversions as we have learned in this chapter. Since the whole Chapter 10 is dedicated to time domain solutions, we will use only one small example here.

Example 1.

Consider the RL voltage divider in Fig. 9.6.1. Using E as a symbol, we get the output

$$V_{out}(s) = \frac{sL}{sL+R}E = \frac{s}{s+1}E$$

If the signal is a unit step, then we consult Table 9.2.1 and read that $E = \frac{1}{s}$. Inserting for the symbol of the independent source we get the expression

$$V_{out,\,step}(s) = \frac{s}{s+1}\frac{1}{s} = \frac{1}{s+1}$$

The inverse is taken from Table 9.2.1, formula 5, and is

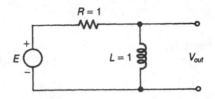

FIGURE 9.6.1 *RL* voltage divider.

$$v_{out,step}(t) = e^{-t}u(t).$$

Suppose that instead of the unit step we wish to apply a cosine function, $e(t) = 2\cos 2t$. Consulting Table 9.2.1 we see that $E = \dfrac{2s}{s^2+4}$. Substituting for the source symbol

$$V_{out,cos}(s) = \frac{s}{s+1}\frac{2s}{s^2+4}$$

To get the time domain response we must proceed with the decomposition

$$\frac{2s^2}{(s+1)(s^2+4)} = \frac{K}{s+1} + \frac{As+D}{s^2+4}$$

Cross-multiplication provides the equation for the numerators:

$$2s^2 = Ks^2 + 4K + As^2 + As + Bs + B$$

which leads to the system of equations

$$K + A = 2$$

$$A + B = 0$$

$$4K + B = 0$$

The solution is $A = \dfrac{8}{5}$, $B = -\dfrac{8}{5}$, $K = \dfrac{2}{5}$ and thus the decomposition is

$$V_{out,cos}(s) = \frac{\dfrac{2}{5}}{s+1} + \frac{\dfrac{8}{5}s - \dfrac{8}{5}}{s^2+4}$$

To be able to use formula 12 from Table 9.2.1 we identify, in addition, $c = 0$ and $d = 2$, and get the time domain response

$$V_{out,cos}(t) = \{\frac{2}{5}e^{-t} + \frac{8}{5}\cos 2t - \frac{8}{5}\sin 2t\}u(t).$$

PROBLEMS CHAPTER 9

P.9.1 Find the partial fraction decompositions and get the time domain responses for the following functions:

(a) $F(s) = \dfrac{s+12}{s(s+4)}$

(b) $F(s) = \dfrac{3s+2}{s^2+s}$

(c) $F(s) = \dfrac{6s^2+10s+1}{s(s^2+3s+2)}$

(d) $F(s) = \dfrac{2s}{s+2}$

(e) $F(s) = \dfrac{2(s^2+2s+3)}{(s+1)(s^2+1)}$

(f) $F(s) = \dfrac{2s+1}{s+3}$

(g) $F(s) = \dfrac{s+1}{(s+1+j)(s+1-j)}$

(h) $F(s) = \dfrac{(s+3)(s+5)}{(s+1)(s^2+2s+2)}$

(i) $F(s) = \dfrac{(s+3)(s+5)}{(s^2+2s+2)(s^2+1)}$

(j) $F(s) = \dfrac{s^2+3s+1}{s(s^2+2s+5)}$

(k) $F(s) = \dfrac{(s+1)(s+3)(s+5)}{s(s+2)(s+4)(s+6)}$

(l) $F(s) = \dfrac{s^2+3s+5}{(s+1)(s+2)(s+3)}$

(m) $F(s) = \dfrac{s^3+3s+7}{(s+1)(s^2+4s+8)}$

(n) $F(s) = \dfrac{5s^2+1}{s(s^2+1)}$

P.9.2 Functions of time, obtained by the Laplace transform, start at $t = 0$. This is formally expressed by multiplication of the function by $u(t)$, the unit step. For instance, $F(s) = 4/(s + 2)$ leads to $f(t) = 4e^{-2t}u(t)$. If we differentiate the time function, we must treat the expression as a product of two functions and the derivative is

$$f'(t) = -8e^{-2t}u(t) + 4e^{-2t}\delta(t)$$

Any function multiplied by $\delta(t)$ results in the Dirac impulse being multiplied by the value of the function at $t = 0$. In the above case

$$f'(s) = -8e^{-2t}u(t) + 4\delta(t)$$

Using the formula for differentiation we also obtain

$$sF(s) = \dfrac{4s}{s+2} = 4 - \dfrac{8}{s+2}$$

and the time function is

$$f'(t) = 4\delta(t) - 8e^{-2t}u(t),$$

the same result. Using the above explanation find the inverse of each of the functions below and differentiate the time functions with respect to t. Afterwards form the product $sF(s)$ and find the inverse.

(a) $F(s) = \dfrac{s+12}{s(s+4)}$

(b) $F(s) = \dfrac{2}{(s+1)(s+2)}$

(c) $F(s) = \dfrac{1}{(s+2)(s+3)(s+4)}$

(d) $F(s) = \dfrac{s+1}{s(s^2+2s+2)}$

P.9.3 Using the limit theorems, find where will the time domain functions of problems in P.9.1 start at $t = 0$ and end at $t \to \infty$.

P.9.4 In the problems P.9.1 multiply the functions by $1/s$, calculate the residues and find time domain responses. Skip problems (a), (b), (c), (j), (k) and (n), because they lead to a double pole at the origin.

P.9.5 Divide the functions (a), (b), (c), (j) and (n) in P.9.1. by s, find the partial fraction decompositions and the time domain responses. In problem P.9.4 you skipped problems which lead to a double pole at the origin. Find the partial fraction decompositions and the inverses for these functions.

P.9.6 The following problems have multiple poles. Using the theory of Section 4, find the partial fraction decompositions and the time domain responses.

(a) $F(s) = \dfrac{1}{(s+1)^3(s+2)^2}$

(b) $F(s) = \dfrac{s^2+1}{(s+1)^3(s+2)}$

(c) $F(s) = \dfrac{7s^2+14s+9}{(s+1)^2(s+2)^2}$

10 TIME DOMAIN

INTRODUCTION

This chapter will cover the theory to obtain mathematical expressions for time domain responses of linear networks. Let us stress here the words *linear* and *theory*, because the results will be mathematical expressions. In practice, most time domain responses are obtained numerically on computers and a brief introduction to it is given in Chapter 15, where we use some numerical methods.

Laplace traansformation is the tool we will use. It has the advantage that there is very little difference from what we have learned in the section on frequency domain in Chapter 8. All we have to add are special independent sources corresponding to initial voltages on capacitors and initial currents through inductors. Everything we have learned so far can be used for time domain analysis as well.

We start separately with networks composed of resistors and capacitors, first without and then with initial conditions. Next we cover networks with resistors and inductors, also first without and then with initial conditions. Finally, we go to networks which have all the linear components we have discussed so far in this book.

10.1. NETWORKS WITH CAPACITORS

We have discussed capacitors and their properties in several earlier chapters but let us repeat some of the basic conclusions.

A capacitor is an element which can store electrons and in such case we say that it has an initial voltage, E_0. The current, $i_C(t)$, flowing through the linear capacitor, and the voltage, $v_C(t)$, appearing across it, are related by the equation

$$i_C(t) = C \frac{dv_C(t)}{dt} \tag{10.1.1}$$

We also derived in Chapter 9, Eq. (9.5.2), that the derivative of a time function, $f(t)$, is expressed in the Laplace domain by

$$L\left[\frac{df(t)}{dt}\right] = sF(s) - f(0)$$

Since Eq. (10.1.1) asks for the derivative of the voltage and since our initial voltage is E_0, the derivative of the voltage across the capacitor will be, in the Laplace domain,

$$L[\frac{dv(t)}{dt}] = sV - E_0$$

Using Eq. (10.1.1), we get in the Laplace domain

$$I_C = C[sV_C - E_0]$$

and finally

$$I_C = sCV_C - CE_0 \tag{10.1.2}$$

This fundamental expression was already derived in Eq. (9.5.5). The next two subsections will consider separately the cases when the capacitor is without and with the initial voltage (initial condition).

10.1.1. Capacitors Without Initial Conditions

If the capacitor has no initial voltage, then $E_0 = 0$ and Eq. (10.1.2) simplifies to

$$I_C = sCV_C \tag{10.1.3}$$

This was already introduced in Chapter 7 without giving any reasons. We also used a special name, *admittance*, for the product

$$Y_C = sC \tag{10.1.4}$$

We can invert Eq. (10.1.3) and obtain

$$V_C = \frac{1}{sC} I_C \tag{10.1.5}$$

The name *impedance* was introduced for the expression

$$Z_C = \frac{1}{sC} \tag{10.1.6}$$

It is worth noting again that at zero frequency, $s = 0$, the impedance becomes infinitely large and the capacitor behaves as an open circuit for dc. At infinitely high frequency it behaves as a short circuit. We have used these expressions many times before. We also stated that E (for an independent voltage source), or J (for an independent current source) will be symbols, to be replaced later by an appropriate expression. In the time domain, the expression will be replaced by the Laplace transform of the signal, and all the necessary transformations have been collected in Table 9.2.1. We will now use the previous derivations to obtain time domain responses of various networks.

We will start with the RC and RL dividers, already analyzed in Section 8.3. The first one is in Fig. 10.1.1. The output voltage was derived in equation (8.3.1) and selecting $R = 1\ \Omega$, and $C = 1\ F$ we simplify it to

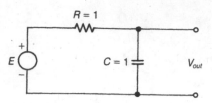

FIGURE 10.1.1 RC voltage divider.

$$V_{out}(s) = \frac{1}{s+1}E$$

We can now study the time domain responses by substituting the Laplace transform of the signal for the symbol E. Suppose that we wish to know the response to the Dirac impulse. Consulting Table 9.2.1 we find in formula 1 that $L[\delta(t)] = 1$ and the output becomes

$$V_{out,Dirac}(s) = \frac{1}{s+1}$$

No partial fraction decomposition is needed and formula 5 of Table 9.2.1 gives the answer

$$v_{out,Dirac}(t) = e^{-1}u(t),$$

where multiplication by $u(t)$ formally makes the signal zero for $t < 0$. The response is plotted in Fig. 10.1.2. Suppose that instead of the Dirac impulse we wish to use the unit step. Table 9.2.1 tells us (formula 2) that $L[u(t)] = \frac{1}{s}$ and thus

$$V_{out,step}(s) = \frac{1}{s+1}\frac{1}{s}$$

Using the residue method of Chapter 9 we obtain the decomposition

$$V_{out,step}(s) = \frac{1}{s} - \frac{1}{s+1}$$

and the inverse is again found by using Table 9.2.1:

$$v_{out,step}(t) = [1 - e^{-t}]u(t)$$

This response is plotted in Fig. 10.1.3.

The RC voltage divider, considered in Chapter 8, is redrawn in Fig. 10.1.4. We derived for it the Eq. (8.3.5) and for the indicated elements it simplifies to

$$V_{out}(s) = \frac{s}{s+1}E$$

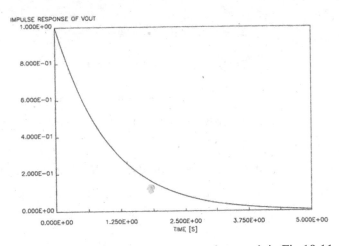

FIGURE 10.1.2 Impulse response of network in Fig 10.11.

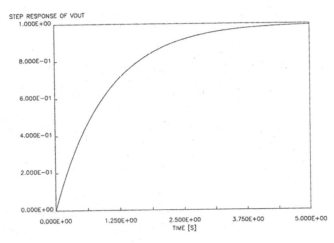

FIGURE 10.1.3 Step resonse of network in Fig 10.1.1.

Suppose that we want to find the time domain response for the Dirac impulse. Substituting $E = 1$ we have

$$V_{out,Dirac}(s) = \frac{s}{s+1}$$

Since the numerator and denominator have the same degrees, we must first divide the denominator into the numerator to obtain

$$V_{out,Dirac}(s) = 1 - \frac{s}{s+1}$$

For the inversion we consult Table 9.2.1 where we find that

$$V_{out,Dirac}(s) = \delta(t) - e^{-t}u(t)$$

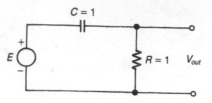

FIGURE 10.1.4   CR voltage divider.

The plot is in Fig. 10.1.5, but the Dirac impulse at the time origin cannot be plotted. Suppose that we would like to know the step response of this network. Substituting $E = \dfrac{1}{s}$ we get

$$V_{out,step}(s) = \frac{s}{s+1}\frac{1}{s} = \frac{1}{s+1}$$

and the time domain response is

$$V_{out,step}(s) = e^{-t}u(t)$$

which happens to be the same function as plotted in Fig. 10.1.2.

As we already know, the admittance form is suitable for nodal analysis and the impedance form for mesh analysis. We now consider three somewhat more complicated networks and find their time responses.

Example 1.

Find the output voltage V_2 for the network in Fig. 10.1.6 by keeping the elements and the source as variables. Afterwards substitute unit values for all elements and find the step response. Using admittances and the symbol J for the independent current source, we get the nodal equations

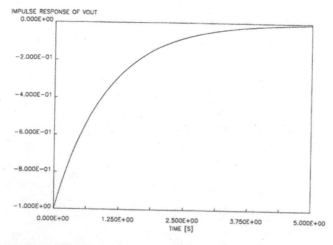

FIGURE 10.1.5   Impulse response of network in Fig 10.1.4.

$$(G_1 + sC)V_1 - sCV_2 = J$$
$$- sCV_1 + (G_2 + sC)V_2 = 0$$

In matrix form

$$\begin{bmatrix} G_1 + sC & -sC \\ -sC & G_2 + sC \end{bmatrix}\begin{bmatrix} V_1 \\ V_2 \end{bmatrix} = \begin{bmatrix} J \\ 0 \end{bmatrix}$$

The determinant is equal to

$$D = (G_1 + sC)(G_2 + sC) - s^2 C^2 = sC(G_1 + G_2) + G_1 G_2$$

and applying the Cramer's rule

$$V_2 = \frac{sC}{sC(G_1 + G_2) + G_1 G_2} J(s)$$

Substituting unit values for the elements

$$V_2(s) = \frac{s}{2s + 1} J$$

For the unit step this changes to

$$V_{2,step}(s) = \frac{1}{2(s + 0.5)}$$

and the time response is

$$v_{2,step}(t) = \frac{1}{2} e^{-0.5t} u(t)$$

This is one exponential and the plot will have a similar shape as in Fig. 10.1.2.

Example 2.

Use the mesh method for the network in Fig. 10.1.7 and find the voltage across the capacitor C_2 in terms of the input voltage E. Afterwards find the Dirac impulse response. The mesh equations are

$$(R_1 + \frac{1}{sC_1})I_1 - R_1 I_2 = E$$

$$-R_1 I_1 + (R_1 + R_2 + \frac{1}{sC_2})I_2 = 0$$

In matrix form

$$\begin{bmatrix} R_1 + \dfrac{1}{sC_1} & -R_1 \\ -R_1 & R_1 + R_2 + \dfrac{1}{sC_2} \end{bmatrix}\begin{bmatrix} I_1 \\ I_2 \end{bmatrix} = \begin{bmatrix} E \\ 0 \end{bmatrix}$$

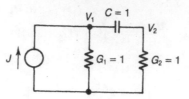

FIGURE 10.1.6   Network with one capacitor.

Inserting numerical values the matrix becomes

$$\begin{bmatrix} 1+\dfrac{1}{2s} & -1 \\ -1 & 2+\dfrac{1}{s} \end{bmatrix} \begin{bmatrix} I_1 \\ I_2 \end{bmatrix} = \begin{bmatrix} E \\ 0 \end{bmatrix}$$

The determinant is

$$D = (1+\frac{1}{2s})(2+\frac{1}{s}) - 1 = 1 + \frac{2}{s} + \frac{1}{2s^2}$$

and the current I_2 is

$$I_2 = \frac{1}{1+\dfrac{2}{s}+\dfrac{1}{2s^2}} E$$

To simplify the result we multiply both the numerator and denominator by $2s^2$ to get

$$I_2 = \frac{2s^2}{2s^2+4s+1} E$$

The output voltage will be.

$$V_{out} = \frac{1}{sC_2} I_2 = \frac{2s}{2s^2+4s+1} E$$

We will return to this result in Example 3.

Example 3.

Consider again the network in Fig. 10.1.7, but apply first the source transformation and then use nodal analysis. Find the output voltage and response to the Dirac impulse.

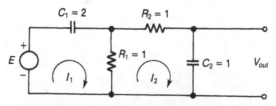

FIGURE 10.1.7   Network with two meshes.

The transformation is shown in Fig. 10.1.8. The nodal equations are

$$(sC_1 + G_1 + G_2)V_1 - G_2V_2 = sC_1E$$

$$-G_2V_1 + (G_2 + sC_2)V_2 = 0$$

Substituting numerical values and writing in matrix form

$$\begin{bmatrix} 2s+2 & -1 \\ -1 & 1+s \end{bmatrix} \begin{bmatrix} V_1 \\ V_2 \end{bmatrix} = \begin{bmatrix} 2sE \\ 0 \end{bmatrix}$$

The determinant is

$$D = (2s+2)(1+s) - 1 = 2s^2 + 4s + 1$$

and the output voltage is

$$V_2 = \frac{2s}{2s^2 + 4s + 1} E$$

The result is the same as in Example 2. Continuing for Dirac impulse we set $E = 1$ and

$$V_{2,Dirac}(s) = \frac{2s}{2s^2 + 4s + 1} = \frac{s}{s^2 + 2s + 0.5}$$

The decomposition is

$$V_{2,Dirac}(s) = \frac{1.20711}{s+1.7011} - \frac{0.20711}{s+0.29289}$$

and the time domain response is

$$v_{2,Dirac}(t) = \left\{ 1.20711e^{-1.7011t} - 0.20711e^{-0.29289t} \right\} u(t)$$

The response is plotted in Fig. 10.1.9.

Let us now summarize the results of this section. In the derivations keep the independent source as a symbol E or J. For the capacitor with no initial charge ($E_0 = 0$) use its admittance, $Y_c = sC$, or its impedance, $Z_c = \frac{1}{sC}$ depending on the formulation method. Derive the desired output and only then substitute the Laplace transform of the input signal for the symbol E or J. Find the partial fraction expansion and use Table 9.2.1 to obtain the time domain response.

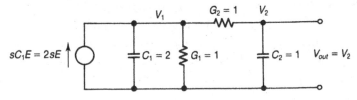

FIGURE 10.1.8 Network Fig. 10.1.8 modified for nodal formulation.

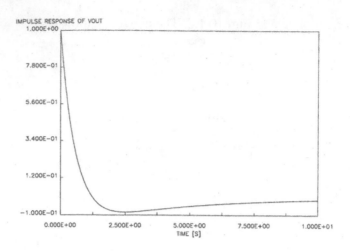

IMPULSE RESPONSE OF VOUT

FIGURE 10.1.9   Impulse response of network in Fig. 10.1.7.

10.1.2. Capacitors with Initial Conditions

If the capacitor has an initial voltage, the situation is as shown in Fig. 10.1.10a. The initial voltage, E_0, can have signs either as shown, or reversed, depending on how we charged the capacitor before we connected it into the network. Due to the influences coming from the network, there will be, in addition, a voltage, $v(t)$, across it and a current, $i(t)$, flowing through it. In agreement with our general definition, this current is taken as positive when it flows from + to – , as in the figure.

We derived in (10.1.2) how a capacitor with an initial voltage is described in the Laplace domain. We also defined its admittance, $Y_c = sC$. Substitute the admittance into (10.1.2) and rewrite in the form

$$I_C + CE_0 = Y_C V_C \tag{10.1.7}$$

This is an equation in currents, so every term in it must be a current. Without the initial voltage $I_C = Y_C V_C$. With the initial condition we have another component, CE_0, flowing into the capacitor as well. This can be interpreted as shown in Fig. 10.1.10b. The additional current acts as a special independent current source. Recall what we said in Chapter 9 about the Dirac impulse, $\delta(t)$. It is a strange impulse, simultaneously infinitely large and infinitely narrow, but, curiously, having an area equal to 1. We derived for it that $L[\delta(t)] = 1$. There is no s in the transform, similarly as in our current source due to the initial condition. The only difference is that this source is CE_0 times larger, or has an area CE_0. We thus conclude that the initial voltage across the capacitor, E_0, is expressed in the Laplace domain by a current source in parallel with the capacitor and its signal is a Dirac impulse with an area CE_0. Should the + and – sign of the initial voltage be reversed, the direction of the equivalent source is reversed as well.

Let us now return to equation (10.1.2). Simple algebraic operations lead to

$$V_C = \frac{1}{sC} I_C + \frac{E_0}{s} \tag{10.1.8}$$

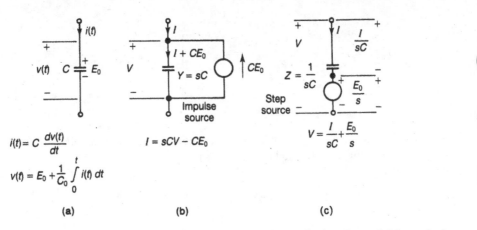

FIGURE 10.1.10   Capacitor with initial voltage: (a) Symbol. (b) Equivalent for nodal formulation. (c) Equivalent for mesh formulation.

We can modify this expression by using the impedance, $Z_C = \dfrac{1}{sC}$,

$$V_C = Z_C I_C = \frac{E_0}{s}$$

This is an equation in voltages and every term must describe a voltage. The overall voltage across the capacitor is equal to the sum of two components, one due to the current through it and one due to the initial condition. The orientations are as shown in Fig. 10.1.10c. Going back to the Laplace transform theory, we see that the additional voltage is a *step source* having an amplitude E_0.

The figures greatly simplify the way we take initial conditions into consideration. If the capacitor has an initial voltage, all we have to do is draw an additional independent source into the network. If we are using nodal equations, we take advantage of the Fig. 10.1.10b, because the initial condition is in the form of an independent current source. If we use mesh formulation, we use the form with the independent voltage source, Fig. 10.1.10c. The Kirchhoff voltage and current laws apply the same way as before. It is also becoming clear why we have used the letter E_0 to indicate the initial condition: In the Laplace transform, this initial condition forms an independent source, and as such must be transferred to the right hand side of the equation before we can proceed with the solution.

We will now take examples with initial conditions and set up the equations. Consider first the RC voltage divider, now with an initial condition as shown in Fig. 10.1.11a. Using the results summarized in Fig. 10.1.10 we can redraw this network either as shown in Fig. 10.1.11b, suitable for mesh formulation, or as shown in Fig. 10.1.11c, suitable for nodal formulation. From now on the analysis proceeds exactly the same way as before. For mesh analysis we write

$$I\left(R + \frac{1}{sC}\right) = E - \frac{E_0}{s}$$

or

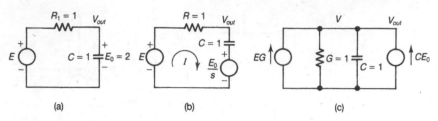

(a)   (b)   (c)

FIGURE 10.1.11 RC network with initial condition: (a) Orginal. (b) Equivalent for mesh formulation. (c) Equivalent for nodal formulation.

$$I = \frac{sC}{sCR+1}\left(E - \frac{E_0}{s}\right) = \frac{sCE - CE_0}{sCR+1}$$

The output voltage is

$$V_{out}(s) = E - RI = E - R\frac{sCE - CE_0}{sCR+1}$$

If we substitute numerical values we get

$$V_{out}(s) = \frac{E + E_0}{s+1}$$

At this point we must decide what we will use for the external voltage source. If we select a unit step, the result is

$$V_{out,step}(s) = \frac{1}{s(s+1)} + \frac{E_0}{s+1}.$$

Partial fraction decomposition of the first term provides

$$V_{out,step}(s) = \frac{1}{s} - \frac{1}{(s+1)} + \frac{E_0}{s+1},$$

and for $E_0 = 2$ the final result is

$$V_{out,step}(s) = \frac{1}{s} + \frac{1}{(s+1)}.$$

Consulting table 9.2.1 we get the inverse

$$v_{out,step}(s) = (1 + e^{-t})u(t).$$

This result is plotted in Fig. 10.1.12.

If we consider nodal formulation, then we first transform the external source into a current source, as shown in Fig. 10.1.11, and write one nodal equation

$$V(G + sC) = EG + CE_0,$$

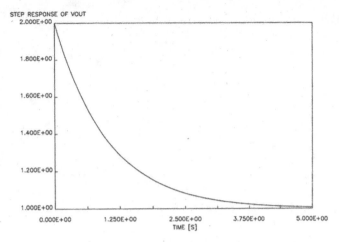

FIGURE 10.1.12 step response of network in Fig. 10.1.11.

which, after the substitution of element values, results in

$$V = \frac{E + E_0}{s+1},$$

the same as above.

Our next problem will be the other possible RC divider, sketched in Fig. 10.1.13a. If we consider first the mesh formulation, the network is redrawn as in Fig. 10.1.13b, for which the single mesh equation is

$$(R + \frac{1}{sC})I = E - \frac{E_0}{s},$$

the current is

$$I = \frac{sC}{sCR+1}(E - \frac{E_0}{s}).$$

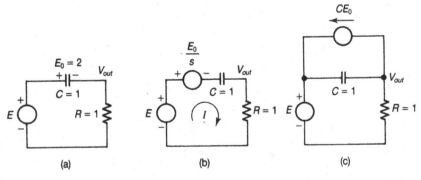

FIGURE 10.1.13 CR network with initial condition: (a) Orginal. (b) Equivalent for mesh formulation. (c) Equivalent with impules source.

and the output voltage is

$$V_{out} = RI = \frac{sCR}{sCR+1}(E - \frac{E_0}{s})$$

Inserting element values

$$V_{out} = \frac{sE - E_0}{s+1}$$

For the unit step $E = \frac{1}{s}$ and since $E_0 = 2$, we get

$$V_{out,step} = \frac{-1}{s+1}$$

The time domain response is

$$v_{out,step}(t) = -e^{-t}u(t)$$

If we wish to use the current source equivalent for the initial condition, then we have to analyze the network in Fig. 10.1.13c, but here we have a dilemma: we cannot transform the voltage source on the left into a current source, because the capacitor is now with its own current source. However, with a bit of thinking there is always a way. What we can do is take the network in Fig. 10.1.13b, and transform both sources simultaneously into one current source. This is shown in Fig. 10.1.14 for which we now write

$$(G + sC)V = sC(E - \frac{E_0}{s})$$

with the result

$$V = \frac{sE - E_0}{s + \frac{G}{C}},$$

the same as above.

The above networks have shown that there is nothing new in terms of network analysis. There are some additional independent sources, due to the initial conditions, but all the methods we learned in the previous chapters remain valid. What is even more important, the analysis remains exactly the same if we move to more complicated networks. One of them will be analyzed now.

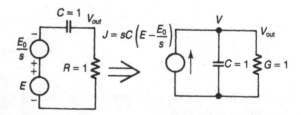

FIGURE 10.1.14   Transformation of network in Fig. 10.1.13(c).

Example 4.

Consider the network in Fig. 10.1.15a. It has no independent external source but all three capacitors have initial conditions. The network is redrawn in b with equivalent current sources, suitable for nodal formulation, and in c with voltage sources, suitable for mesh formulation. We will use both. For Fig. 10.1.15b the nodal equations are

$$(sC_1 + G_1)V_1 - G_1 V_2 + 0V_3 = C_1 E_{01}$$

$$-G_1 V_1 + (G_1 + sC_2 + sC_3)V_2 - sC_3 V_3 = -C_2 E_{02} + C_3 E_{03}$$

$$0V_1 - sC_3 V_2 + (sC_3 + G_2)V_3 = -C_3 E_{03}$$

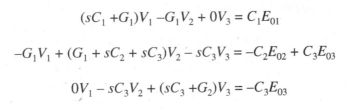

(a)

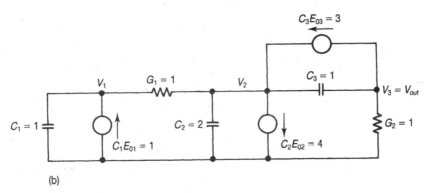

(b)

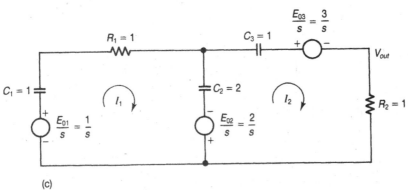

(c)

FIGURE 10.1.15 Network with initial conditions (a) Orginal. (b) Transformed for nodal formulation. (c) Transformed for mesh formulation.

In matter form

$$\begin{bmatrix} sC_1+G_1 & -G_1 & 0 \\ -G_1 & G_1+sC_2+sC_3 & -sC_3 \\ 0 & -sC_3 & sC_3+G_2 \end{bmatrix}\begin{bmatrix} V_1 \\ V_2 \\ V_3 \end{bmatrix} = \begin{bmatrix} C_1E_{01} \\ -C_2E_{02}+C_3E_{03} \\ -C_3E_{03} \end{bmatrix}$$

To continue we insert numerical values:

$$\begin{bmatrix} s+1 & -1 & 0 \\ -1 & 3s+1 & -s \\ 0 & -s & s+1 \end{bmatrix}\begin{bmatrix} V_1 \\ V_2 \\ V_3 \end{bmatrix} = \begin{bmatrix} 1 \\ -1 \\ -3 \end{bmatrix}$$

After a few intermediate steps, we find the determinant of the matrix to be

$$D = s(2s^2 + 6s + 4)$$

Using the Cramer's rule for V_3, we find, after a few steps, that the numerator is

$$N_3 = s(-10s - 12)$$

We can cancel $2s$ in the numerator against the same term in the denominator and get

$$V_3 = \frac{-5s-6}{s^2+3s+2}$$

The same network can also be solved by the mesh formulation and the other representation of the initial voltages, as shown in Fig. 10.1.15c. The equations are

$$(\frac{1}{sC_1}+R_1+\frac{1}{sC_2})I_1 - \frac{1}{sC_2}I_2 = \frac{E_{01}+E_{02}}{s}$$

$$-\frac{1}{sC_2}I_1 + (\frac{1}{sC_2}+\frac{1}{sC_3}+R_2)I_2 = \frac{-E_{02}-E_{03}}{s}$$

Using numerical values

$$\begin{bmatrix} \dfrac{2s+3}{2s} & -\dfrac{1}{2s} \\ -\dfrac{1}{2s} & \dfrac{2s+3}{2s} \end{bmatrix}\begin{bmatrix} I_1 \\ I_2 \end{bmatrix} = \begin{bmatrix} \dfrac{3}{s} \\ -\dfrac{5}{s} \end{bmatrix}$$

The determinant of the matrix is

$$D = \frac{s^2+3s+2}{s^2}$$

and, using the Cramer's rule, the numerator for I_2 is

$$N_2 = \frac{-5s-6}{s^2}$$

As a result, the current of the second mesh is

$$I_2 = \frac{-5s-6}{s^2+3s+2}$$

The output voltage, V_{out}, is taken across the resistor R_2 and is numerically equal to the current I_2. This is in agreement with our previous solution, obtained by the nodal method. The function can be decomposed into

$$V_3 = -\frac{1}{s+1} - \frac{4}{s+2}$$

and consulting Table 9.2.1 the inverse is

$$v_3(t) = \{-1e^{-t} - 4e^{-2t}\}u(t)$$

This is plotted in Fig. 10.1.16.

Example 5.

Consider the network in Fig. 10.1.17a and its Laplace domain equivalent in b. Find the response of the network for $j(t) = \sin t$. The nodal equations are

$$(G_1 + sC)V_1 - sCV_2 = J(s) - CE_0$$

$$-sCV_1 + (sC + G_2)V_2 = CE_0$$

In matrix form

$$\begin{bmatrix} 1+s & -s \\ -s & 2+s \end{bmatrix} \begin{bmatrix} V_1 \\ V_2 \end{bmatrix} = \begin{bmatrix} J-2 \\ 2 \end{bmatrix}$$

The denominator is $D = 3s + 2$, $N_2 = Js + 2$ and

$$V_2 = \frac{Js+2}{3s+2}$$

The Laplace transform of the input signal is $J = \dfrac{1}{s^2+1}$ and thus

$$V_2 = \frac{\frac{2}{3}s^2 + \frac{1}{3}s + \frac{2}{3}}{(s^2+1)(s+\frac{2}{3})}$$

We need the decomposition

$$\frac{\frac{2}{3}s^2 + \frac{1}{3}s + \frac{2}{3}}{(s^2+1)(s+\frac{2}{3})} = \frac{K}{s+\frac{2}{3}} + \frac{As+B}{s^2+1}$$

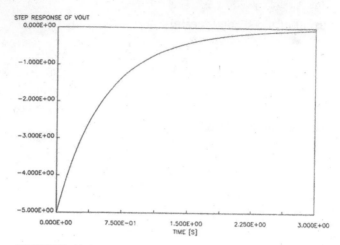

FIGURE 10.1.16   Response of network in Fig. 10.1.15.

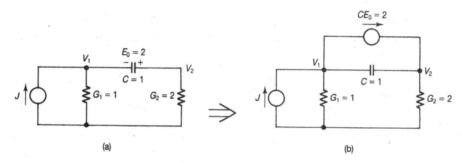

FIGURE 10.1.17   Network with initial voltage: (a) Original. (b) Nodal formulation equivalent.

Cross-multiplying and comparing coefficients at equal powers we get the system

$$K + A = \frac{2}{3}$$

$$\frac{2}{3}A + B = \frac{1}{3}$$

$$K + \frac{2}{3}B = \frac{2}{3}$$

Its solution is, $K = \dfrac{20}{39}$, $A = \dfrac{6}{39}$ and $B = \dfrac{9}{39}$. The decomposition is

$$V_2 = \frac{20}{39}\frac{1}{s+\dfrac{2}{3}} + \frac{\dfrac{6}{39}s + \dfrac{9}{39}}{s^2 + 1}$$

Using formulas from Table 9.2.1 we get the time domain response

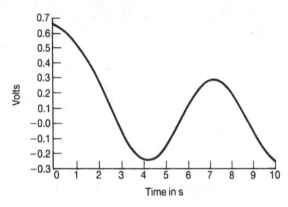

FIGURE 10.1.18 Response of network Fig 10.1.17 to j(t) = sin t.

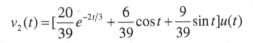

$$v_2(t) = [\frac{20}{39}e^{-2t/3} + \frac{6}{39}\cos t + \frac{9}{39}\sin t]u(t)$$

It is plotted in Fig. 10.1.18.

10.2. NETWORKS WITH INDUCTORS

The voltage across an inductor and the current flowing through it are coupled by the fundamental equation

$$v_L(t) = L\frac{di_L(t)}{dt} \qquad (10.2.1)$$

The current creates a magnetic field which stores energy and tries to continue the flow of current even if the source of current has changed. We speak about initial current through the inductor and denote it by J_0. If we proceed similarly as in the case of the capacitor and use the expression for the Laplace transform of the derivative, we get

$$V_L = sLI_L - LJ_0 \qquad (10.2.2)$$

This fundamental expression relates, in the Laplace transform, the voltage across the inductor, V_L, the current through the inductor, I_L, and the initial current, J_0. It was already derived in (9.5.6). We will consider separately the cases when the inductor is without and with the initial current.

10.2.1. Inductors Without Initial Conditions

If no initial current flows through the inductor, $J_0 = 0$ and (10.2.2) simplifies to

$$V_L = sLI_L . \qquad (10.2.3)$$

The product sL is an impedance

$$Z_L = sL \tag{10.2.4}$$

If we invert (10.2.3), we obtain

$$I_L = \frac{1}{sL} V_L \tag{10.2.5}$$

and the word *admittance* is used for the ratio

$$Y_L = \frac{1}{sL} \tag{10.2.6}$$

We use the impedance when writing mesh equations and the admittance in nodal equations. If $s = 0$ then the impedance is zero and the inductor acts as a short circuit at dc. If s is very large, approaching infinity, then the impedance becomes very large and the inductor acts as an open circuit.

Since RL dividers lead to the same equations as the RC dividers (see our derivations in Section 8.3), we will not consider them here. Instead we will take somewhat more complicated networks.

Example 1.

Analyze the network in Fig. 10.2.1 using the mesh and nodal equations. Find the time domain response for a Dirac impulse and unit step. The mesh equations are

$$(R_1 + sL_1)I_1 - R_1 I_2 = E$$

$$-R_1 I_1 + (R_1 + R_2 + sL_2)I_2 = 0$$

Using numerical values, the matrix equation is

$$\begin{bmatrix} 1+s & -1 \\ -1 & 2+2s \end{bmatrix} \begin{bmatrix} I_1 \\ I_2 \end{bmatrix} = \begin{bmatrix} E \\ 0 \end{bmatrix}$$

The determinant is

$$D = 2s^2 + 4s + 1$$

and

$$I_2 = \frac{1}{2s^2 + 4s + 1} E$$

For the output voltage across the inductor we must multiply the current I_2 by the impedance $Z_L = sL_2 = 2s$. The result is

$$V_{out}(s) = sLI_2 = \frac{2s}{2s^2 + 4s + 1} E$$

Finding the roots of the denominator, substituting for the Dirac impulse $E = 1$ and finding the residues:

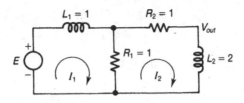

FIGURE 10.2.1 Network with two inductors.

$$V_{out,Dirac}(s) = \frac{1.20711}{s+1.70711} - \frac{0.20711}{s+0.29289}$$

The time domain response is

$$v_{out,Dirac} = [1.20711e^{-1.70711t} - 0.20711e^{-0.29289t}]u(t)$$

For unit step we substitute $E = \frac{1}{s}$ and the function becomes

$$V_{out,step} = \frac{2s}{2s^2+4s+1}\frac{1}{s} = \frac{2}{2s^2+4s+1} = \frac{0.70711}{s+0.29289} - \frac{0.70711}{s+1.70711}$$

The time domain response is

$$v_{out,step} = [0.70711e^{0.29289t} - 0.70711e^{-1.70711t}]u(t)$$

For nodal formulation we first transform the voltage source into a current source as shown in Fig. 10.2.2. The equations are

$$(G_1 + G_2 + \frac{1}{sL_1})V_1 - G_2V_2 = \frac{E}{sL_1}$$

$$-G_2V_1 + (G_2 + \frac{1}{sL_2})V_2 = 0$$

Substituting numerical values and writing in matrix form

$$\begin{bmatrix} 2+\frac{1}{s} & -1 \\ -1 & 1+\frac{1}{2s} \end{bmatrix}\begin{bmatrix} V_1 \\ V_2 \end{bmatrix} = \begin{bmatrix} \frac{E}{s} \\ 0 \end{bmatrix}$$

The determinant is $D = \frac{2s^2+4s+1}{2s^2}$ and the numerator is $N_2 = \frac{E}{s}$. The output voltage, V_2, is the same as obtained above.

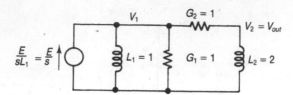

FIGURE 10.2.2   Network Fig. 10.2.1. prepared for nodal formulation.

Example 2.

Dependent sources are handled the same way as before, as seen from the analysis of the network in Fig. 10.2.3.

The equations are

$$(1+\frac{1}{2s})V_1 -\frac{1}{2s}V_2 = J$$

$$-\frac{1}{2s}V_1 +(\frac{1}{2s}+2)V_2 +1.5V_1 = 0$$

and in matrix form

$$\begin{bmatrix} \dfrac{2s+1}{2s} & -\dfrac{1}{2s} \\ \dfrac{3s-1}{2s} & \dfrac{4s+1}{2s} \end{bmatrix}\begin{bmatrix} V_1 \\ V_2 \end{bmatrix}=\begin{bmatrix} J \\ 0 \end{bmatrix}$$

The determinant is $D = \dfrac{8s+9}{4s}$, $N_2 = \dfrac{1-3s}{2s}J$ and the output voltage is

$$V_{out} = \frac{-6s+2}{8s+9}J$$

If the source is a Dirac impulse, then we must first divide the two polynomials to obtain

$$V_{out,Dirac} = -0.75+\frac{1.09375}{s+1.125}$$

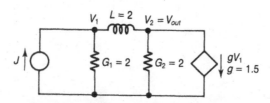

FIGURE 10.2.3   Network with a VC.

The time domain response is

$$v_{out,Dirac} = -0.75\delta(t) + 1.09375e^{-1.125t}u(t)$$

10.2.2. Inductors With Initial Conditions

An inductor may have an initial current, J_0, in addition to a given voltage $v(t)$ across it and a current $i(t)$ flowing through it. As always, positive direction of the current $i(t)$ is from + to $-$, but initial current can flow in either direction, depending on the previous state of the network. Our reference direction is as shown in Fig. 10.2.4a.

In the Laplace domain, an inductor with an initial current is described by the equation (10.2.2). If we substitute $Z_L = sL$, then

$$V_L = Z_L I_L - LJ_0 \tag{10.2.7}$$

It indicates that the voltage across the inductor is formed from two components, one due to the current flowing through the impedance and another due to the initial current J_0. This is interpreted as shown in Fig. 10.2.4b. The voltage source is a Dirac impulse having an area LJ_0. Should the direction of the initial current J_0 be reversed, the + and $-$ signs of the equivalent voltage source will be reversed as well.

Let us now return to equation (10.2.7) and separate the current on the left side of the equation

$$I_L = \frac{1}{sL}V_L + \frac{J_0}{s} \tag{10.2.8}$$

If we now insert $Y_L = \dfrac{1}{sL}$ we can write

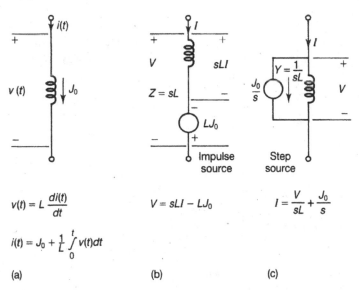

$$v(t) = L\frac{di(t)}{dt} \qquad\qquad V = sLI - LJ_0 \qquad\qquad I = \frac{V}{sL} + \frac{J_0}{s}$$

$$i(t) = J_0 + \frac{1}{L}\int_0^t v(t)dt$$

(a)   (b)   (c)

FIGURE 10.2.4   Inductor with initial current: (a) Symbol. (b) Equivalent for mesh formulation. (c) Equivalent for nodal formulation.

$$I_L - \frac{J_0}{s} = Y_L V_L$$

The equation indicates that two currents are flowing into the admittance, one due to the voltage V_L across the inductor and one due to the initial current, J_0. This is interpreted as shown in Fig. 10.2.4c. Going back to the Laplace transform theory, we see that the source is a step having an amplitude J_0.

Similarly as in the case of the capacitor, the figures considerably simplify our work. All we have to do is redraw the network using one of the two possible representations. For mesh equations we use the form with the voltage source, for nodal equations the form with the current source and apply KVL or KCL for the modified network. It is also becoming clear why we have used the letter J_0 to indicate the initial condition: In the Laplace transform, this initial condition forms an independent source and in the equations all such sources must go to the right hand side before we can go ahead with the solution. We will use examples to clarify application.

Example 3.

Find the unit step response of the network in Fig. 10.2.5a using the mesh and nodal formulation.

For mesh formulation we redraw the network as shown in Fig. 10.2.5b and obtain

$$\begin{bmatrix} 2+2s & -2s \\ -2s & 1+2s \end{bmatrix} \begin{bmatrix} I_1 \\ I_2 \end{bmatrix} = \begin{bmatrix} E+4 \\ -4 \end{bmatrix}$$

The determinant is $D = 6s + 2$, $N_2 = 2(Es - 4)$ and

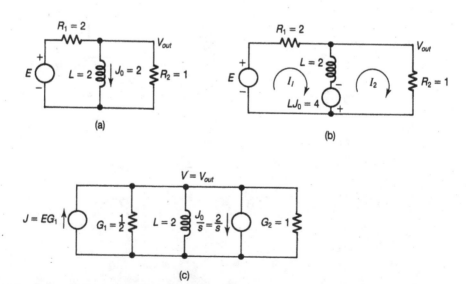

(a)   (b)

(c)

FIGURE 10.2.5   Network with initial current: (a) Original. (b) Transformed for mesh formulation. (c) Transformed for nodal formulation.

$$I_2 = \frac{E_s - 4}{3s + 1}$$

The output voltage is $V_{out} = R_2 I_2$ and since $R_2 = 1$ the above result applies for V_{out} as well. For nodal formulation we redraw the network as in Fig. 10.2.5c. Only one equation is needed

$$(\frac{1}{2} + \frac{1}{2s} + 1)V = \frac{E}{2} - \frac{2}{s}$$

and the output voltage is

$$V_{out} = \frac{Es - 4}{3s + 1},$$

thc same as above. For unit step we substitute $E = \frac{1}{s}$ and

$$V_{out,step} = -\frac{3}{3s + 1} = -\frac{1}{s + \frac{1}{3}}$$

The time domain response is

$$v_{out,step} = -e^{-0.33333t}u(t)$$

Example 4.

Consider Fig. 10.2.6a with three inductors, each having an initial current. The network has no independent external source. Solve using both mesh and nodal methods.

For mesh analysis we redraw the network as shown in Fig. 10.2.6b. The equations are

$$(sL_1 + sL_2 + R_1)I_1 - sL_2 I_2 = -L_1 J_{01} - L_2 J_{02}$$

$$-sL_2 I_2 + (sL_2 + sL_3 + R_2)I_2 = L_2 J_{02} - L_3 J_{03}$$

Inserting numerical values and writing in matrix form

$$\begin{bmatrix} 3s+1 & -2s \\ -2s & 3s+1 \end{bmatrix}\begin{bmatrix} I_1 \\ I_2 \end{bmatrix} = \begin{bmatrix} -5 \\ 1 \end{bmatrix}$$

The determinant of the matrix is

$$D = 5s^2 + 6s + 1$$

Using Cramer's rule we find

$$I_2 = \frac{-7s + 1}{5s^2 + 6s + 1}$$

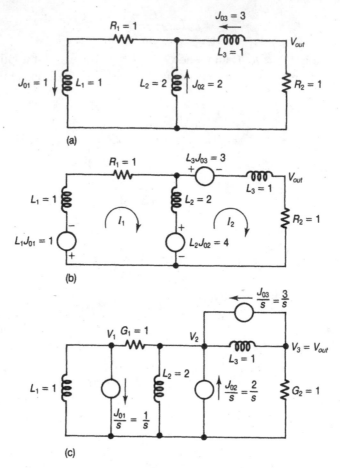

FIGURE 10.2.6 Network with initial currents: (a) Schematic. (b) Transformed for mesh formulation. (c) Transformed for nodal formulation.

The output voltage is

$$V_{out} = R_2 I_2 = \frac{-7s+1}{5s^2+6s+1}$$

Finding the roots and the partial fraction decomposition

$$V_{out} = \frac{0.6}{s+0.2} - \frac{2}{s+1}$$

and the time domain output is

$$v_{out} = [0.6e^{-0.2t} - 2e^{-t}]u(t)$$

If we use the representation with current sources, then the same network can be redrawn as shown in Fig. 10.2.6c. Applying the nodal equations

$$(\frac{1}{sL_1}+G_1)V_1 - G_1V_2 + 0V_3 = -\frac{J_{01}}{s}$$

$$-G_1V_1 + (G_1 + \frac{1}{sL_2} + \frac{1}{sL_3})V_2 - \frac{1}{sL_3}V_3 = \frac{J_{02}}{s} + \frac{J_{03}}{s}$$

$$0V_1 - \frac{1}{sL_3}V_2 + (\frac{1}{sL_3}+G_2)V_3 = -\frac{J_{03}}{s}$$

If we substitute numerical values, then the system will be

$$
\begin{bmatrix}
\dfrac{s+1}{s} & -1 & 0 \\[2mm]
-1 & \dfrac{2s+3}{2s} & -\dfrac{1}{s} \\[2mm]
0 & -\dfrac{1}{s} & \dfrac{s+1}{s}
\end{bmatrix}
\begin{bmatrix}
V_1 \\ V_2 \\ V_3
\end{bmatrix}
=
\begin{bmatrix}
-\dfrac{1}{s} \\[2mm]
\dfrac{5}{s} \\[2mm]
-\dfrac{3}{s}
\end{bmatrix}
$$

The determinant of the matrix is

$$D = \frac{5s^2+6s+1}{2s^3}$$

If we wish to get only the output voltage, $V_{out} = V_3$, then the numerator is

$$N_3 = \frac{-7s+1}{2s^3}$$

and the output voltage is

$$V_{out} = \frac{-7s+1}{5s^2+6s+1},$$

the same as above.

The examples have shown that a proper choice of the equivalent source can save us a lot of work; Fig. 10.2.6b leads to only two equations, Fig. 10.2.6c leads to three. If the network has only inductors and resistors, it will be generally advantageous to select mesh formulation with voltage sources because we will avoid fractions of the type $\dfrac{1}{s}$. It can pay to first decide which of the methods leads to a smaller set of equations.

Another fact which is good to remember (although it was not derived here) is that for networks combined of only resistors and capacitors, or resistors and inductors, without any dependent sources, all poles are real, irrespective of the size of the network.

We will now make several conclusions and recommendations:

1. When writing the system equations, keep the symbols for external sources, E and/or J, throughout all calculations, up to the point when partial fraction

decomposition can be applied. Only then substitute the actual Laplace domain expression for the input signal.

2. Decomposition requires factorization of the denominator polynomial. For polynomials of first and second order, this can be done symbolically, with elements kept as variables. If the polynomial is of order 3 or more, only numerical solutions are feasible and a computer should be used to find them.

3. Responses are formed by two types of components. Those which die out with time are called *transients*. If the input is a periodic signal (sine or cosine), then the output contains a component of the same periodic function, modified in amplitude and in phase. This is called the *steady state* component.

We also make a few observations.

(a) The unit step seems to be the most useful signal; it is simple and smaller networks can be handled by hand.

(b) The Dirac impulse is even simpler, but is an artificial signal. It is the derivative of the unit step and therefore the response of the network to the Dirac impulse will be the derivative of the response to a unit step.

(c) Finding the time domain response of a network to a periodic input signal (sine or cosine) is a laborious process even for small networks.

In the next section we will summarize the develpment

10.3. RLC CIRCUITS

In this section we will study time domain responses of series and parallel tuned circuits. In order to simplify the derivations, we will set fixed values for the capacitor and inductor; only the resistor will be changing.

For the series tuned circuit, Fig. 10.3.1, we have

$$(R + \frac{1}{sC} + sL)I = E$$

and

$$I = \frac{1}{L} \frac{s}{s^2 + s\frac{R}{L} + \frac{1}{LC}} E$$

If we insert the element values $L = 1\ H$, $C = 1\ F$, select unit step $E = \frac{1}{s}$ and calculate the voltage across the capacitor, we get

$$V_{C,step} = \frac{1}{sC} I = \frac{1}{s(s^2 + sR + 1)}$$

In dealing with time responses it is always useful to find out beforehand where the response will be at $t = 0$ and where it will get for $t \to \infty$. For this we use the limiting theorems derived in (9.5.12) and (9.5.13). At the time $t = 0$

$$v_{C,step}(0) = \lim_{s \to \infty} sV = \lim_{s \to \infty} \frac{1}{s^2 + sR + 1} = 0$$

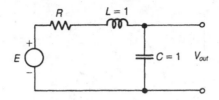

FIGURE 10.3.1 Series tuned circuit with L = 1 H and C = 1 F.

At $t \rightarrow \infty$ we use the other formula.

$$v_{C,step}(\infty) = \lim_{s=0} sV = \lim_{s=0} \frac{1}{s^2 + sR + 1} = 1$$

and the response will start at 0 and end at 1.

The roots of the quadratic polynomial are

$$p_{1,2} = -\frac{R}{2} \pm \left[\frac{R^2}{4} - 1\right]^{1/2}$$

Depending on the value of the discriminant (expression in the square brackets), these roots can be either distinct real, or coincide as one double real root, or they can be complex conjugate. We will consider all cases.

CASE A.

If $\frac{R^2}{4} - 1 > 0$, the polynomial has two real roots. Selecting, for instance, $R = 4$, the roots become

$$p_1 = -0.268$$
$$p_2 = -3.732$$

and

$$V_{C,step} = \frac{1}{s(s - p_1)(s - p_2)} = \frac{K_1}{s} + \frac{K_2}{s - p_1} + \frac{K_3}{s - p_2}$$

where

$$K_1 = 1$$
$$K_2 = -1.07735$$
$$K_3 = 0.07735$$

The time response is

$$v_{C,step}(t) = [1 - 1.07735e^{-0.268\,t} + 0.07735e^{-3.732\,t}]u(t)$$

and is plotted in Fig. 10.3.2, full curve.

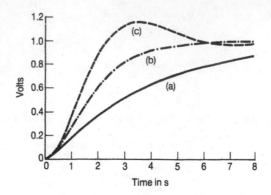

FIGURE 10.3.2   Unit step responses of the circuit in Fig. 10.3.1 for: (a) R = 4 Ω (full curve). (b) R = 2 Ω (dotted curve). (c) R = 1 Ω (dashed curve).

CASE B.

If $R = 2$, then

$$V_{C,step} = \frac{1}{s(s+1)^2} = \frac{1}{s} - \frac{1}{(s+1)^2} - \frac{1}{s+1}$$

The decomposition for the double root is obtained by the theory explained in Section 9.4. The inverse for the double pole is in formula 6, Table 9.2.1 and the time domain response is

$$v_{C,step}(t) = [1 - te^{-t} - e^{-t}]u(t).$$

It is plotted in Fig. 10.3.2, dotted curve.

CASE C.

If $R < 2$, then

$$V_{C,step} = \frac{1}{s(s^2 + sR + 1)} = \frac{1}{s} - \frac{s+R}{(s+\frac{R}{2})^2 + d^2}$$

where

$$d = \left[1 - \frac{R^2}{4}\right]^{1/2}$$

We identify for formula 12 in Table 9.2.1 $A = 1$, $B = R$, $c = \dfrac{R}{2}$, d as given above and $\dfrac{B - Ac}{d} = \dfrac{R}{2d}$. The inverse is

$$v_{C,step}(t) = [1 - e^{-\frac{R}{2}t}\cos dt - \frac{R}{2d}e^{-\frac{R}{2}t}\sin dt]u(t)$$

If we select $R = 1$, then $c = 0.5$, $d = 0.866$ and

$$v_{C,step}(t) = [1 - e^{-0.5t} (\cos 0.866\, t + 0.577 \sin 0.866\, t)]u(t),$$

see the dashed curve in Fig. 10.3.2.

The shape of the curves depends on the value of the resistor; the smaller the resistor, the more pronounced is the oscillatory behavior of the network. As a matter of interest, if we select $R = 0.1$, then the time domain response will be as shown in Fig. 10.3.3. If we make the resistor zero, the expression for the voltage across the capacitor reduces to

$$V_{C,step} = \frac{1}{s(s^2 + 1)} = \frac{1}{s} - \frac{s}{s^2 + 1},$$

the inverse will be

$$v_{C,step}(t) = 1 - \cos t$$

and will oscillate permanently.

Another practical case is a parallel tuned circuit shown in Fig. 10.3.4. The output voltage is given by

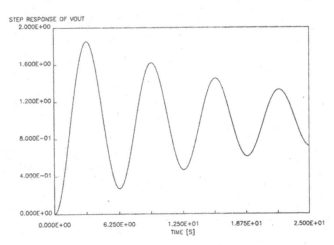

FIGURE 10.3.3   Step response of network Fig. 10.3.1 for $R = 0.1\ \Omega$.

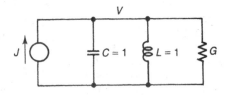

FIGURE 10.3.4   Parallel tuned circuit.

$$V = \frac{1}{C} \frac{s}{s^2 + s\dfrac{1}{RC} + \dfrac{1}{LC}} J$$

The equation is very similar to the equations discussed above. If we use $C = 1\ F$ and $L = 1\ H$, it simplifies to

$$V = \frac{s}{s^2 + s\dfrac{1}{R} + 1} J$$

The difference from the previous case is that the middle term of the denominator is now divided by R, instead of multiplied.

PROBLEMS CHAPTER 10

P.10.1   Analyze the network by first keeping all elements and the source as variables and obtain the voltage V_2. Afterwards substitute numerical values for elements and find the time responses for (a) the Dirac impulse and (b) for unit step input.

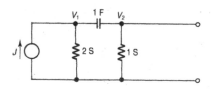

P.10.2   Do the same as in P.10.1.

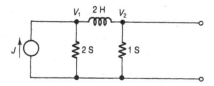

P.10.3   Obtain the output voltage by keeping the source as a variable. Obtain time domain responses for input being (a) Dirac impulse, (b) unit step, (c) $\sin 2t$, and (d) $\cos 2t$.

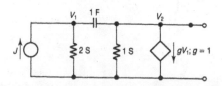

P.10.4   Do the same as in P.10.3.

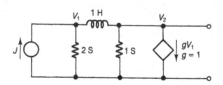

P.10.5   Do the same as in P.10.3.

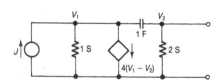

P.10.6   Use mesh formulation to calculate the the output voltage in terms of E, then let the input be the Dirac impulse, the unit step and $\sin 2t$.

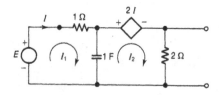

P.10.7   Use mesh formulation and the voltage-source equivalent for the initial condition. Calculate the output voltage for the source being (a) Dirac impulse, (b) unit step, (c) $\sin 2t$. Also transform the voltage source into a current source, use the current source representation for the initial condition and do the same in nodal analysis.

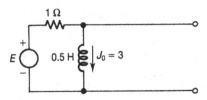

P.10.8 Do the same as in P.10.7.

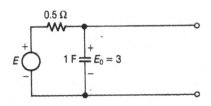

P.10.9 Find the voltage V_2 for the source being (a) Dirac impulse and (b) unit step.

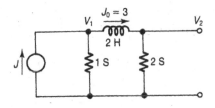

P.10.10 Do the same as in P.10.9.

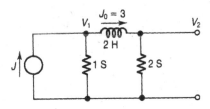

P.10.11 Do the same as in P.10.9.

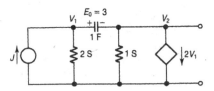

P.10.12 Do the same as in P.10.9.

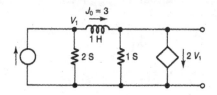

P.10.13 Use mesh analysis to obtain the voltage across the output resistor. Use sin $3t$ for the input.

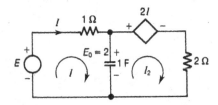

P.10.14 Do the same as in P.10.13.

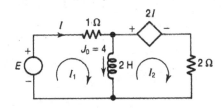

P.10.15 Use nodal formulation to obtain the output response for (a) no source, (b) $j(t) = \delta(t)$, (c) $j(t) = u(t)$.

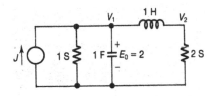

P.10.16 Find the output voltage using mesh formulation. Let the source be (a) no source, (b) $e(t) = \delta(t)$.

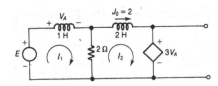

P.10.17 Use nodal formulation and calculate the output voltage for (a) no input, (b) $j(t) = \delta(t)$.

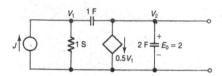

P.10.18 Derive the expression for the output voltage. This is a third-order network with a real root at $s = -1$. Find (a) the Dirac impulse and (b) unit step response.

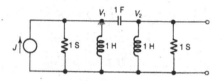

P.10.19 Find the output response for unit step. Mesh analysis is advantageous in this network.

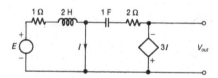

P.10.20 Find the output voltage for (a) no source, (b) $e(t) = \delta(t)$. It is advantageous to use nodal formulation.

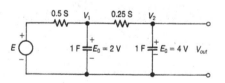

P.10.21 Find the output voltage using mesh and nodal formulation. Use two cases: (a) no source, (b) $e(t) = u(t)$.

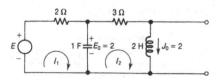

P.10.22 Find the time domain response of this network at the output node.

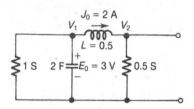

P.10.23 Find the output voltage for (a) no source, (b) $e(t) = u(t)$. Use both mesh and nodal formulations.

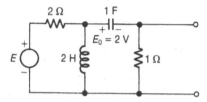

P.10.24 Find the output voltage for (a) no source, (b) $e(t) = \delta(t)$.

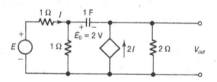

P.10.25 Do the same as in P.10.24.

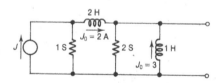

P.10.26 Keep both E and E_0 as variables and find the output voltage. Afterwards solve for (a) $E = 0$ and $E_0 = 2$, (b) $E = \frac{1}{s}$ and $E_0 = 0$, (c) $E = \frac{1}{s}$ and $E_0 = 2$. Note that nodal formulation is advantageous in this case.

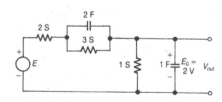

11

NETWORK FUNCTIONS

INTRODUCTION

The concept of network functions is one of the most practical ones in the theory of linear networks. Network functions are expressed in terms of the Laplace domain variable and contain all information about the investigated output of the network. It is possible to go from there to frequency or time domain analysis. In addition, we can establish stability properties and also derive some other fundamental rules, if we decompose the network function into its poles and zeros.

11.1. DEFINITION OF NETWORK FUNCTIONS

Network functions are usually considered in cases where the network has one independent source, E or J, and one output. The functions are obtained in the Laplace domain and initial conditions are ignored.

In order to introduce the concept, we take the simplest possible network, an R-L voltage divider shown in Fig. 11.1.1. The output voltage is equal to

$$V_{out} = \frac{2}{s+2} E$$

where E can be any input signal. If we do not take a specific signal but rather divide the equation by E, we get a transfer function

$$T = \frac{V_{out}}{E} = \frac{2}{s+2}$$

This ratio is one of the possible network functions; specifically, for this example, it is a voltage transfer function

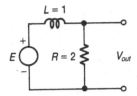

FIGURE 11.1.1 LR voltage divider.

Had we chosen a larger network, the output would be in the form

$$V_{out} = \frac{N}{D} E \qquad (11.1.1)$$

where N and D are polynomials in the variable s. Dividing by E we get the network function.

$$T = \frac{V_{out}}{E} = \frac{N}{D} \qquad (11.1.2)$$

One network can have many network functions, depending on where we apply the independent source and where we take the output. In order to be able to define the function, we must satisfy two conditions:

1. The network must be linear
2. There are zero initial conditions on capacitors and inductors.

If these conditions are satisfied, then we define the network function as the ratio

$$\text{Network function} = \frac{\text{output}}{\text{input}} \qquad (11.1.3)$$

The output variable, V or I, is *always* in the numerator while the input variable, E or J, is in the denominator, similarly as in (11.1.2).

Depending on the type of the input source (voltage or current) and the type of the output (voltage or current), we can define the following network functions:

1. Voltage transfer function:

$$T_V = \frac{V_{out}}{E} \qquad (11.1.4)$$

The source is an independent voltage source, the output is either the voltage of one node or the difference of two nodal voltages. The voltage transfer function is a dimensionless quantity.

2. Current transfer function:

$$T_I = \frac{I_{out}}{J} \qquad (11.1.5)$$

The source is an independent current source, the output is the current through any element. The current transfer function is a dimensionless quantity.

3. Transfer impedance:

$$Z_{TR} = \frac{V_{out}}{J} \qquad (11.1.6)$$

The source is an independent current source, the output is the voltage of any node or the difference of two nodal voltages. The transfer impedance has the dimension of an impedance.

4. Transfer admittance:

$$Y_{TR} = \frac{I_{out}}{E}$$ (11.1.7)

The source is an independent voltage source, the output is the current through any element. Transfer admittance has the dimension of an admittance.

5. Input impedance:

$$Z_{in} = \frac{V_{in}}{J}$$ (11.1.8)

The output variable is the voltage across the source J. The input impedance is an impedance.

6. Input admittance:

$$Y_{in} = \frac{I_{in}}{E}$$ (11.1.9)

The output variable is the current flowing through the voltage source. The input admittance is an admittance.

The input impedance and input admittance are *always* related by the expression

$$Z_{in} = \frac{1}{Y_{in}}$$ (11.1.10)

However, such relation *does not* apply for the transfer impedance and admittance. The reader should remember that

$$\dot{Z}_{TR} \ IS \ NOT \ \frac{1}{Y_{TR}}$$

because they use different variables, see (11.1.6) and (11.1.7).

Let us now look again at the definition of the network function and give it still another explanation. Suppose that instead of dividing (11.1.1) by E we leave the source on the right but substitute instead $E = 1$. The result will be the same. However, we learned that the Laplace transform of the Dirac impulse is equal to 1 and we can thus say that the network function describes simultaneously the response of the network to the Dirac impulse. Once we have the network function, we can always get the output by multiplying the network function by E or J and use it for frequency domain or time domain analysis. We have covered a similar subject already in Chapter 7, only now we give it the above additional meaning. Several examples will provide some additional practice.

Example 1.

Find the Z_{TR} and T_V for the network in Fig. 11.1.2.

The nodal equations are:

$$(G_1 + sC_1)V_1 - sC_1V_2 = J$$
$$-(sC_1 + g)V_1 + (sC_1 + sC_2 + G_2)V_2 = 0$$

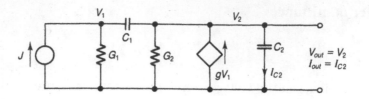

FIGURE 11.1.2   Network with a *CC*.

Although it should already be clear, the dependent source transconductance must stay on the left and the independent source on the right. In matrix form

$$\begin{bmatrix} (G_1 + sC_1) & -sC_1 \\ -(g + sC_1) & (sC_1 + sC_2 + G_2) \end{bmatrix}\begin{bmatrix} V_1 \\ V_2 \end{bmatrix} = \begin{bmatrix} J \\ 0 \end{bmatrix}$$

The determinant of the system is

$$D = s^2 C_1 C_2 + s(C_1 G_1 + C_1 G_2 + C_2 G_1 - C_1 g) + G_1 G_2$$

Using Cramer's rule we evaluate V_2 and also define $I_{out} = sC_2 V_2$. The transfer functions are:

$$Z_{TR} = \frac{V_2}{J} = \frac{sC_1 + g}{D}$$

and

$$T_I = \frac{I_{out}}{J} = \frac{s^2 C_1 C_2 + sC_2 g}{D}$$

At this point we make a very important observation: All network functions of any given network have the same denominator. This clearly follows from the solution by the Cramer's rule where the determinant of the matrix turns out to be the denominator of every network function.

Example 2.

Find V_{out} and I_{out} for the network in Fig. 11.1.3.

This example will show that the numerator and denominator may be functions of both s and $1/s$. Although in this case it would be easier to transform the voltage source into a current source and apply nodal formulation, we will use the mesh method. The system matrix is

$$\begin{bmatrix} (R_1 + \dfrac{1}{sC_1}) & -\dfrac{1}{sC_1} \\ -\dfrac{1}{sC_1} & (R_2 + \dfrac{1}{sC_1} + \dfrac{1}{sC_2}) \end{bmatrix}\begin{bmatrix} I_1 \\ I_2 \end{bmatrix} = \begin{bmatrix} E \\ 0 \end{bmatrix}$$

The determinant of the matrix is

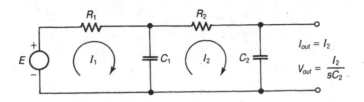

FIGURE 11.1. 3 Network with two capacitors.

$$D = \frac{1}{s^2 C_1 C_2} + \frac{1}{s}(\frac{R_1}{C_1} + \frac{R_1}{C_2} + \frac{R_2}{C_1}) + R_1 R_1$$

$$= \frac{s^2 C_1 C_2 R_1 R_2 + s(R_1 C_1 + R_1 C_2 + R_2 C_2) + 1}{s^2 C_1 C_2}$$

The output current is $I_{out} = I_2$ and the output voltage is $V_{out} = \dfrac{I_2}{sC_2}$. Applying Cramer's rule for I_2 we obtain

$$I_2 = I_{out} = \frac{\dfrac{E}{sC_1}}{D}$$

The output current is

$$I_{out} = I_2 = \frac{sC_2 E}{s^2 C_1 C_2 R_1 R_2 + s(R_1 C_1 + R_1 C_2 + R_2 C_2) + 1}$$

and the network functions are

$$Y_{TR} = \frac{I_{out}}{E} = \frac{I_2}{E} = \frac{sC_2}{s^2 C_1 C_2 R_1 R_2 + s(R_1 C_1 + R_1 C_2 + R_2 C_2) + 1}$$

and

$$T_V = \frac{V_{out}}{E} = \frac{I_2}{sC_2} = \frac{1}{s^2 C_1 C_2 R_1 R_2 + s(R_1 C_1 + R_1 C_2 + R_2 C_2) + 1}$$

Example 3.

For the network in Fig. 11.1.4 find the indicated output voltage and current.

The nodal matrix equation can be written by inspection:

$$\begin{bmatrix} (G_1 + G_2 + sC) & -G_2 \\ -G_2 & (G_2 + \dfrac{1}{sL}) \end{bmatrix} \begin{bmatrix} V_1 \\ V_2 \end{bmatrix} = \begin{bmatrix} J \\ 0 \end{bmatrix}$$

The denominator of the transfer functions is

$$D = sCG_2 + (G_1 G_2 + \frac{C}{L}) + \frac{G_1 + G_2}{sL}$$

Using Cramer's rule, we obtain

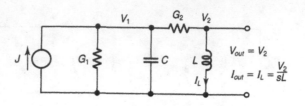

FIGURE 11.1. 4   GLC network.

$$V_{out} = \frac{G_2 J}{sCG_2 + (G_1 G_2 + \frac{C}{L}) + \frac{G_1 + G_2}{sL}}$$

and since $I_{out} = \frac{V_{out}}{sL}$, we also get

$$I_{out} = \frac{\frac{G_2}{sL}}{sCG_2 + (G_1 G_2 + \frac{C}{L}) + \frac{G_1 + G_2}{sL}} J$$

Similarly as we have done before we remove the fractions by multiplying the numerator and denominator by sL. The results are

$$Z_{TR} = \frac{sLG_2}{s^2 LCG_2 + s(LG_1 G_2 + C) + G_1 + G_2}$$

$$T_I = \frac{G_2}{s^2 LCG_2 + s(LG_1 G_2 + C) + G_1 + G_2}$$

Example 4.

Find the transfer function T_V for the network in Fig. 11.1.5. Use mesh analysis and numerical values of the elements.

Since the direction of the driving current of the CV is the same as the direction of I_1, it must be true that

$$I_{R,2} = I_1 - I_2$$

The mesh equations are:

$$(3 + s)I_1 - (2 + s)I_2 = E$$
$$-(2 + s)I_1 + (2 + 3s)I_2 + r(I_1 - I_2) = 0$$

Here we used the letter r of the CV to clarify the equation. Substituting $r = 1$ and writing in matrix form

$$\begin{bmatrix} (3+s) & -(2+s) \\ -(1+s) & (1+3s) \end{bmatrix} \begin{bmatrix} I_1 \\ I_2 \end{bmatrix} = \begin{bmatrix} E \\ 0 \end{bmatrix}$$

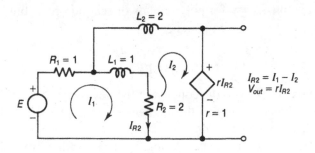

FIGURE 11.1. 5 Network with a *CV*.

The determinant of the system is

$$D = 2s^2 + 7s + 1$$

Since in this example the output voltage is given by $V_{out} = r(I_1 - I_2) = I_1 - I_2$, we must find both I_1 and I_2:

$$I_1 = \frac{1+3s}{D} E$$

$$I_2 = \frac{1+s}{D} E$$

Then

$$V_{out} = \frac{2s}{D} E$$

and the transfer function is

$$T_V = \frac{V_{out}}{E} = \frac{2s}{2s^2 + 7s + 1}$$

Let us summarize what we have learned in this section.

1. Network functions are defined only for linear networks with zero initial conditions.
2. To obtain the network function, connect the required source at the input, but specify it only as *E* or *J*.
3. Solve the network to find the desired output and divide by *E* or *J* to obtain the network function. The input variable is always in the denominator.
4. Should the numerator and/or denominator have fractions of the type $\frac{1}{s}$, remove them by simultaneously multiplying *D* and *N* by an appropriate factor.
5. The network function can be used to obtain frequency domain response: replace the source symbol by the source phasor and set $s = j\omega$ where ω is the angular frequency of the signal.
6. The network function can be used to obtain time domain response: first multiply the equation by the source symbol, replace the source symbol by the Laplace transform of the actual input signal, find the poles and get the partial fraction expansion. For the inverse use Table 9.2.1.

Finding poles and zeros is easy for polynomials of second order, but higher orders can be handled only numerically by a computer.

11.2. POLES, ZEROS AND STABILITY

In the first section we introduced the concept of the network function. In all cases, the final results were rational functions in s. If the elements were kept as variables, then each power of s was accompanied by some combination of the elements. When the network elements were given numerically, the final network function was a ratio of two polynomials in s with constant coefficients. Roots of the denominator are the *poles*, roots of the numerator are the *zeros*. We also discovered that although a network may have many network functions, *all* will have the same denominator. The numerator depends on the type of the input (E or J), on the place where the source is applied and where we take the output.

From this discussion we see that the denominator expresses some fundamental properties of the network whereas the numerator is related to a particular input and output. A network function will have the general form

$$F(s) = \frac{N(s)}{D(s)} = \frac{A_M s^M + \cdots + A_1 s^1 + A_0}{B_N s^N + \cdots + B_1 s^1 + B_0} = \frac{\sum_{i=0}^{M} A_i s^i}{\sum_{i=0}^{N} B_i s^i} \tag{11.2.1}$$

Since it is known from mathematics that a polynomial of order n has exactly n roots, we can also rewrite the network function in the form

$$F(s) = K_0 \frac{(s - z_1)(s - z_2)....(s - z_M)}{(s - p_1)(s - p_2)....(s - p_N)} = K_0 \frac{\prod_{i=1}^{M}(s - z_i)}{\prod_{i=1}^{N}(s - p_i)} \tag{11.2.2}$$

Where

$$K_0 = \frac{A_M}{B_N} \tag{11.2.3}$$

is equal to the ratio of the coefficients at the highest powers of s. Most of the poles and zeros of practical networks are complex,

$$z_i = \alpha_i + j\beta_i$$
$$p_i = \gamma_i + j\delta_i \tag{11.2.4}$$

with $j = \sqrt{-1}$

The step from (11.2.1) to (11.2.2) may be difficult, but the two forms are equivalent, at least theoretically. If the coefficients of the polynomials are numbers, we can find the roots by numerical methods. If the coefficients are functions of the element variables, then the situation is much more difficult and we will remember the following *practical* advice:

1. The root of the first order polynomial $P(s) = as + b = 0$ is $s = -\dfrac{b}{a}$.

2. The roots of a second order polynomial $P(s) = as^2 + bs + c = 0$ are found by means of the formula

$$s_{1,2} = \frac{-b \pm \sqrt{b^2 - 4ac}}{2a}$$

The coefficients may be either numbers or element symbols.

3. For polynomials with constant coefficients of order 3 or more use a root-finding computer program.

4. Do not try to find the roots of polynomials of order 3 or more if the coefficients are not numerically given.

In these few sentences are hidden many problems of network design: only very few practical networks can be fully solved by hand. The rest must use special computer methods which are the subject of advanced courses.

The positions of the zeros and poles can be plotted in the complex plane. Zeros are marked as small circles, poles as crosses. Thus the transfer function

$$F = 5 \frac{s^2 + 4}{s^3 + 4s^2 + 6s + 4} = 5 \frac{(s + j2)(s - j2)}{(s + 2)(s + 1 + j1)(s + 1 - j1)} \tag{11.2.5}$$

has the plot shown in Fig. 11.2.1. The multiplicative constant 5 cannot be determined from the pole-zero plot. If the polynomial of the numerator or denominator has a complex root, then there exists *always* another, complex conjugate root, having the same real part but opposite sign of the imaginary part.

Suppose that instead of the function (11.2.5) we have another function with multiple poles, for instance

$$F = \frac{5(s^2 + 4)}{s^4 + 6s^3 + 14s^2 + 16s + 8} = \frac{5(s + j2)(s - j2)}{(s + 2)^2(s + 1 + j1)(s + 1 - j1)} \tag{11.2.6}$$

The pole $s = -2$ has multiplicity 2 but the positions of the poles are the same as in the previous example. To indicate that the pole has multiplicity n, we write a small n next to this pole. The same will apply to multiple zeros.

The concept of poles and zeros can also be related to matrix algebra. Since the denominator is found as the determinant of the system matrix, and the pole is its root, then any value of s which makes the determinant equal to zero is the *pole*. The numerator polynomial is obtained by Cramer's rule by replacing one column of the matrix by the right hand side. Since the numerator polynomial is found as the determinant of this modified matrix, then any value of s which makes the determinant of the modified matrix zero is also the *zero* of the transfer function.

Poles and zeros are mathematical abstractions, valid only for linear networks. They cannot be easily related to elements or how the network is constructed. The poles are clearly more fundamental than the zeros because all network functions of any given network will have the same poles. For this reason, they are sometimes called *natural frequencies* of the network. The zeros depend on the place where the source is connected to the network and on the place where we take the output. If the zeros are purely imaginary (lying on the vertical axis), then they do have a clear practical meaning: they completely suppress a signal occurring at their frequency.

Importance of the poles can be best understood by considering the Dirac impulse as the input. This signal has finite area and thus delivers a finite amount of energy to the network.

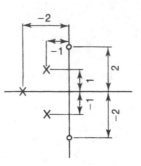

FIGURE 11.2.1   Pole-zero plot.

We will say that the network is stable if its response to the Dirac impulse, a signal with finite energy, does not grow for large t. Should it grow, then we will say that the network is unstable. Networks composed of passive components like R, L, C cannot be unstable but stability problems can occur in networks with sources of energy, like dependent sources or amplifiers.

Assume that we know the poles. If all poles are simple, then the network function can be decomposed into partial fractions

$$F(s) = \sum_{i=1}^{N} \frac{K_i}{s - p_i} \qquad (11.2.7)$$

and the Laplace transform inverse is

$$f(t) = \sum_{k=1}^{N} K_i e^{p_i t} \qquad (11.2.8)$$

Consider now just one term in (11.2.7) and $s = p$. If the pole is negative real, then it lies on the negative real axis, to the left of the origin. The time response of such term is an exponential, $f(t) = e^{-pt}$, and decreases with time. It will become practically zero for large t. This is a *stable* situation. On the other hand, if the pole lies on the positive real axis, the response will be $f(t) = e^{pt}$, an exponential which grows very rapidly; we say that such term is unstable.

For a complex conjugate pair the exact response depends on the residue, but the type of the function will be $f(t) = e^{\gamma t} \cos(\delta t + \phi)$. If the real part of the pole, γ, is negative, we will have a decaying oscillatory behavior which will become practically zero for large t. If the real part is positive, the oscillations will grow without bound. We can also have a special case when the real part of the pole is zero. In such case we will get oscillations which neither grow nor decay. Such situation can occur either if there are no resistors in the network or if all the losses in the resistors are somehow restored from other sources of energy (battery).

If the poles are multiple, the functions become more complicated but the real part must be negative in order to get a response which decreases with time. A multiple pole on the imaginary axis will always lead to an unstable behavior.

We can now summarize:

1. A stable network must have all poles (simple or multiple) in the left half plane.
2. Only simple poles may lie on the imaginary axis; in such case the network will oscillate with constant amplitude.

The positions of the zeros did not come into our stability considerations. They do not influence stability but they do influence amplitudes of the exponentials and phase shifts of the sinusoids.

Example 1.

Find the poles and zeros of the function

$$T = \frac{2s}{2s^2 + 7s + 1}$$

and determine whether the function is stable or not.

The poles are the roots of the denominator,

$$2s^2 + 7s + 1 = 0$$

Using the formula a quadratic polynomial we find that the roots are $p_1 = -0.14922$ and $p_2 = -3.3508$. They are real, in the left half plane and thus the network function is stable. The function can be rewritten as

$$T = \frac{2s}{2(s + 0.14922)(s + 3.3508)}$$

Note that we must not forget the coefficient 2 which multiplied the highest power of s in the denominator.

Example 2.

The network in Fig. 11.2.2a has all elements given by their values, except for the voltage controlled voltage source which has the gain A. Find the value of A for which the network will become unstable.

The network is redrawn in Fig. 11.2.2b with transformed sources so that we can use nodal analysis. The equations are

$$(2 + s)V_1 - V_2 = E + AsV_2$$

$$-V_1 + (1 + s)V_2 = 0$$

Transferring the term with V_2 to the left and leaving E on the right leads to the following matrix equation

$$\begin{bmatrix} 2 + s & -(1 + sA) \\ -1 & 1 + s \end{bmatrix} \begin{bmatrix} V_1 \\ V_2 \end{bmatrix} = \begin{bmatrix} E \\ 0 \end{bmatrix}$$

The determinant of the system matrix is $D = s^2 + s(3 - A) + 1$. Normally, to find poles, we set the polynomial equal to zero and find the roots. In this problem we only have to determine the value of A for which the poles will move on the imaginary axis. This will happen when the middle term of the polynomial will become zero, $A = 3$. Then the equation reduces to

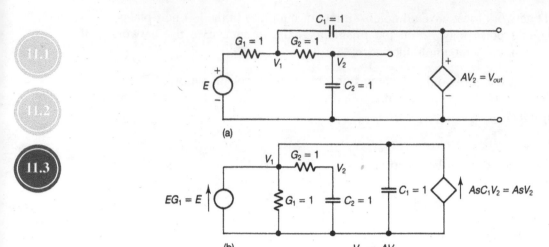

FIGURE 11.2.2   Network with a VV: (a) Original, (b) Transformed for nodal formulation.

$$s^2 + 1 = 0$$

and the poles are $p_{1,2} = \pm j1$. The network will be stable for $A < 3$ and unstable for $A \geq 3$.

11.3.   FREQUENCY RESPONSES FROM POLES AND ZEROS

If the network function poles and zeros are known, it is very easy to evaluate various frequency domain responses. We will derive the necessary expressions.

The network function in terms of poles and zeros has the form

$$F == K_0 \frac{(s - z_1)(s - z_2) \cdots (s - z_M)}{(s - p_1)(p - p_2) \cdots (s - p_N)} \qquad (11.3.1)$$

Where

$$\begin{aligned} z_i &= \alpha_i + j\beta_i \\ p_i &= \gamma_i + j\delta_i \end{aligned} \qquad (11.3.2)$$

For evaluation in the frequency domain we have always substituted $s = j\omega$, which is a point on the imaginary axis. A zero, $\alpha + j\beta$, is also a point in the complex plane and both points are in Fig. 11.3.1. Considering one term in (11.3.1) we must form the difference

$$(s - z) = (0 + j\omega) - (\alpha + j\beta) = -a + j(\omega - \beta) \qquad (11.3.3)$$

The absolute value is

$$A = \sqrt{\alpha^2 - (\omega - \beta)^2} \qquad (11.3.4)$$

We can also calculate the angle ϕ

$$\phi = arctan \frac{\omega - \beta}{-\alpha} \qquad (11.3.5)$$

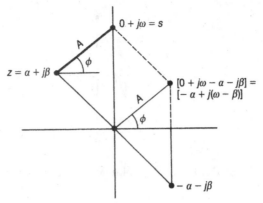

FIGURE 11.3.1 Subtraction of two vectors.

Thus for one term in (11.3.1) we can write the equivalence

$$s - z = A e^{j\varphi} \tag{11.3.6}$$

The same steps apply for the denominator terms $s - p$. For the amplitude response, the absolute values (distances) from zeros will be in the numerator and absolute values from the poles in the denominator. The formula becomes

$$|F(j\omega)| = K_0 \frac{A_{z,1} A_{z,2} \cdots A_{z,M}}{A_{p,1} A_{p,2} \cdots A_{p,N}} \tag{11.3.7}$$

The angles appear as exponents of e. The exponents are added separately in the numerator and in the denominator, but those in the denominator can be brought to the numerator by changing their signs. Thus the overall phase will be

$$\Phi = (\phi_{z,1} + \phi_{z,2} + \cdots \phi_{z,M}) - (\phi_{p,1} + \phi_{p,2} + \cdots \phi_{z,N}) \tag{11.3.8}$$

Inserting from (11.3.4) we get

$$|F| = K_0 \left[\frac{\prod_{i=1}^{M} [\alpha_i^2 + (\omega - \beta_i)^2]}{\prod_{i=1}^{N} [\gamma_i^2 + (\omega - \delta_i)^2]} \right]^{1/2} \tag{11.3.9}$$

This formula gives the *amplitude response*. Similarly for the phases we use (11.3.5) and get

$$\Phi = \sum_{i=1}^{M} \arctan \frac{\beta_i - \omega}{\alpha_i} - \sum_{i=1}^{N} \arctan \frac{\delta_i - \omega}{\gamma_i} \tag{11.3.10}$$

This expression is the *phase response*.

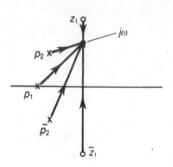

FIGURE 11.3.2   Amplitude response from pole-zero plot.

It is advanatageous to get some insight. Consider Fig. 11.3.2 where we have three poles in the left half plane and two zeros on the imaginary axis. In addition, consider any $s = j\omega$. This will be a point on the imaginary axis, as shown. Each expression in (11.3.9) is nothing but the distance of this point from the poles and zeros. Products of the distances from zeros are in the numerator of (11.3.9) products of the distances from the poles in the denominator.

The phase characteristic could be found similarly by adding phase angles from the zeros and subtracting those from the poles. Because of the difficulties of measuring the angles, this would be rarely done.

It is interesting to add the following facts without deriving them. If we want to transfer, without distortion, a signal composed of several sinusoidal components, then we need a network whose amplitude response is constant for all frequencies of the signal. It is relatively easy to see how the amplitude response deviates from the desired constant response. Simultaneously, the phase response must be a straight line passing through the origin. It is, unfortunately, much more difficult to interpret significance of the deviations in the phase response. Because of these difficulties with the phase response, another, much better measure is often used. It is called the *group delay*

$$\tau(\omega) = -\frac{d\Phi}{d\omega} \qquad (11.3.11)$$

The minus sign in front of the expression makes the group delay positive for most pole-zero configurations. Differentiating (11.3.10) with respect to ω we get

$$\tau(\omega) = \sum_{i=1}^{M} \frac{\alpha_i}{\alpha_i^2 + (\omega - \beta_i)^2} - \sum_{i=1}^{N} \frac{\gamma_i}{\gamma_i^2 + (\omega - \delta_i)^2} \qquad (11.3.12)$$

What should the group delay response be in order not to distort signals? We stated that the ideal phase response should be a straight line passing through the origin. Since the group delay is the derivative of the phase and derivative of a straight line is a constant, we conclude that for transfer without distortion the group delay should be constant for all frequencies of interest.

Formulas (11.3.9), (11.3.10) and (11.3.12) are very easy to program but finding the poles and zeros for a larger network is next to impossible without a computer.

Example 1.

Fig. 11.3.3 shows a third order filter which has the name Butterworth filter. Derive the transfer impedance for this filter. You will get in the denominator a polynomial of third order which happens to have a real pole at $s = -1$. Divide the denominator by the term $s + 1$ to lower the degree and find the remaining two poles. Using nodal formulation we get the system equation

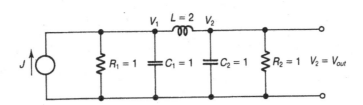

$$\begin{bmatrix} 1+s+\dfrac{1}{2s} & -\dfrac{1}{2s} \\ -\dfrac{1}{2s} & 1+s+\dfrac{1}{2s} \end{bmatrix} \begin{bmatrix} V_1 \\ V_2 \end{bmatrix} = \begin{bmatrix} J \\ 0 \end{bmatrix}$$

The determinant (after a few steps) is

$$D = \frac{s^3 + 2s^2 + 2s + 1}{s}$$

and the numerator is

$$N_2 = \frac{J}{2s}$$

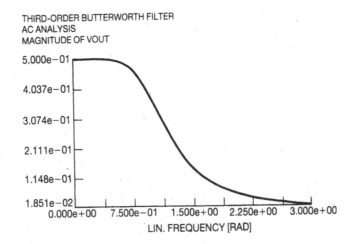

FIGURE 11.3.3 Third-order Butterworth filter.

THIRD-ORDER BUTTERWORTH FILTER
AC ANALYSIS
MAGNITUDE OF VOUT

FIGURE 11.3.4 Amplitude response of filter in Fig.11.3.3.

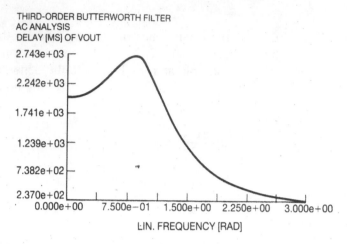

THIRD-ORDER BUTTERWORTH FILTER
AC ANALYSIS
DELAY [MS] OF VOUT

FIGURE 11.3.5   Group delay of the network in Fig. 11.3.3.

The network function is

$$Z_{TR} = \frac{V_{out}}{J} = \frac{1}{2(s^3 + 2s^2 + 2s + 1)}$$

Since we were given the real pole, we divide the denominator by the term $(s + 1)$. The result is the polynomial $s^2 + s + 1$ which has the roots

$$s_{1,2} = \frac{-1 \pm \sqrt{3}}{2}$$

We have plotted the amplitude response in Fig. 11.3.4, and the group delay in Fig. 11.3.5.

PROBLEMS CHAPTER 11

P.11.1   Display the poles and zeros of the following functions in the complex plane. Determine whether the functions are stable or unstable. Also sketch the amplitude, phase and group delay responses.

(a) $F(s) = \dfrac{1}{s+1}$

(b) $F(s) = \dfrac{1}{(s+1)(s+2)}$

(c) $F(s) = \dfrac{s+1}{s+2}$

(d) $F(s) = \dfrac{s+3}{(s+2)(s+4)}$

(e) $F(s) = \dfrac{s+3}{s^2 - 3s + 2}$

(f) $F(s) = \dfrac{s-1}{(s+2)(s^2 - 5s + 3)}$

P.11.2   If a complex conjugate pair of poles is close to the imaginary axis, then there is a peak in the amplitude response. The group delay also has a peak in the vicinity of such a pole. Make a sketch for the following pole pairs. Note that we are changing real parts of the poles without changing the imaginary parts.

(a) $F_1(s) = \dfrac{1}{(s+1+j1)(s+1-j1)}$

(b) $F_2(s) = \dfrac{1}{(s+0.75+j1)(s+0.75-j1)}$

(c) $F_3(s) = \dfrac{1}{(s+0.5+j1)(s+0.5-j1)}$

(d) $F_3(s) = \dfrac{1}{(s+0.25+j1)(s+0.25-j1)}$

P.11.3 It is not possible to discover from frequency domain responses whether the function is stable or unstable. Make an estimate on the amplitude response of the following pairs of functions; the second is always an unstable function.

(a) $F_1(s) = \dfrac{1}{s+1}$ and $F_2(s) = \dfrac{1}{s-1}$

(b) $F_1(s) = \dfrac{s-1}{s+2}$ and $F_2(s) = \dfrac{s+1}{s-2}$

(c) $F_1(s) = \dfrac{s-1}{s+1}$ and $F_2(s) = \dfrac{s+1}{s-1}$

P.11.4 Plot the group delay responses of the functions in P.11.3.

P.11.5 The so called Butterworth filters have poles spread equally on a unit circle. The numerator is a constant equal to 1, the coordinates of the poles are

(a) one pole: $p_1 = 1.0$
(b) two poles:
$p_{1,2} = -0.707 \pm j0.707$
(c) three poles:
$p_1 = -1$
$p_{2,3} = -0.5 \pm j0.866$
(d) four poles:
$p_{1,2} = -0.383 \pm j0.924$
$p_{3,4} = -0.92388 \pm 0.38268$

Plot the poles and see how they are placed on the unit circle. Make an estimate of the amplitude and group delay responses.

P.11.6 Another type of simple filters are the Chebychev filters. They have equal ripples in the magnitude or dB function, in the case given here equal to 0.5 dB. Let the numerator constant be equal to 1 and use the poles

(a) two poles:
$p_{1,2} = -0.713 \pm j1.004$
(b) three poles:
$p_1 = -0.626$
$p_{2,3} = -0.313 \pm j1.022$
(c) four poles:
$p_{1,2} = -1.75 \pm j1.016$
$p_{3,4} = -0.423 \pm j0.421$

Plot the poles and make an estimate on the amplitude responses.

P.11.7(a)–(d) Networks which have only resistors and capacitors, or networks which have only resistors and inductors, have network functions with simple poles, all lying on the negative real axis. Show that this is true for the four networks. Find the input impedance, input admittance, transfer admittance and current transfer functions.

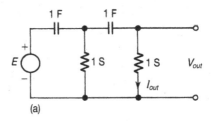

(a)

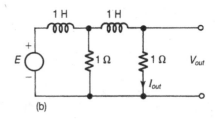

(b)

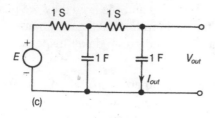

(c)

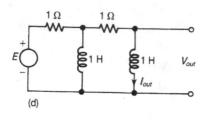

(d)

P.11.8 Network having resistors, inductors and capacitors can have complex conjugate poles or only real poles, depending on element values. They can never be unstable and all poles will have neg ative real parts. Select any RLC network without dependent sources and confirm, at least on your examples, that this is true.

12

ACTIVE NETWORKS

INTRODUCTION

In the low frequency regions we can avoid the use of inductors by building RC-active networks. Although this book will not go into the details of the design of such networks, we will present a practical method for their analysis. It is based on nodal formulation and on the fact that practically all active networks use voltage sources whose one terminal is grounded.

The amplifiers are almost always operational amplifiers and their idealized equivalent, with infinite gain, is a convenient abstraction. Unfortunately, infinite gain cannot be handled by computers. In order to avoid difficulties, we introduce a modification which makes it possible to deal with ideal or nonideal operational amplifiers in a unified way in hand or computer applications.

Analysis is first explained on networks with finite gain amplifiers, then on networks with ideal operational amplifiers and finally on networks with nonideal operational amplifiers.

12.1. AMPLIFIERS AND OPERATIONAL AMPLIFIERS

Amplifiers used in active networks are equivalent to voltage controlled voltage sources with one output terminal grounded. Fig. 12.1.1a shows the equivalent VV, Fig. 12.1.1b show the usual symbol. If the input voltage is V_{in}, then the output is

$$V_0 = AV_{in} \qquad (12.1.1)$$

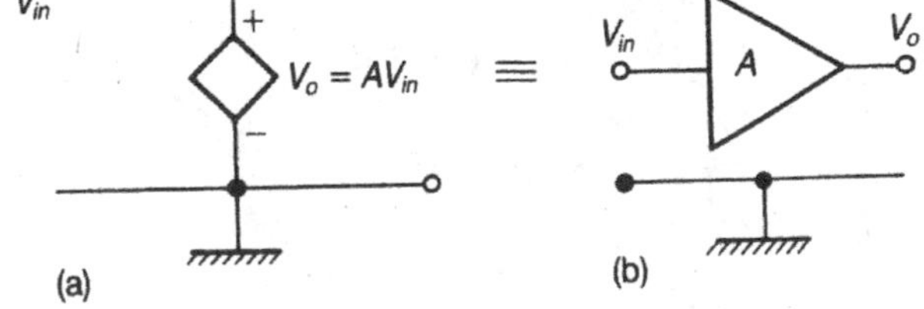

FIGURE 12.1.1 Amplifier with gain A: (a) Equivalent VV. (b) Usual symbol.

where A is the gain of the amplifier. If $A > 0$ the amplifier is said to be noninverting, if $A < 0$, then it is inverting. On the right side of Fig. 12.1.1 is the usual symbol for this type of amplifier. This symbol is used when the input as well as the output voltage are measured with respect to ground. We also note that no current is flowing into the amplifier at the input terminal.

Differential amplifiers are used more often. Fig. 12.1.2a shows the equivalent VV, Fig. 12.1.2b the usual symbol. It is understood that V_+ denotes the non-inverting terminal and V_- the inverting terminal. Both voltages are measured with respect to ground. The output of the device is

$$V_0 = A(V_+ - V_-) \tag{12.1.2}$$

measured with respect to ground. The other terminal of the internal voltage controlled voltage source is grounded. The symbol for the differential amplifier is on the right of Fig. 12.1.2. The gain of the differential amplifier, A, is always assumed to be positive.

It should be clear that grounding the negative terminal reduces the differential amplifier to the amplifier shown in Fig. 12.1.1. Should we ground the positive terminal, the differential amplifier becomes an inverting amplifier.

For the purposes of analysis, it is convenient to define an idealized element, the *ideal operational amplifier*. Its definition is the same as in (12.1.2) but the gain of the amplifier is assumed to be infinite. In some cases this is a simplification, in others a complication. If we keep the gain, A, during the analysis and later want to see what the result would be for $A \to \infty$, we must apply mathematical limiting, a process which is more complicated than simple substitution. In addition, infinity cannot be treated by a computer. For these reasons, we modify the operational amplifier and define the inverted gain B.

$$B = -\frac{1}{A} \tag{12.1.3}$$

This modifies (12.1.2) to

$$V_+ - V_- + BV_0 = 0 \tag{12.1.4}$$

Equations (12.1.3) and (12.1.4) will be used extensively in the following sections. Their advantage lies in the fact that at any step of the analysis, even at the very beginning, we can substitute $B = 0$ and convert a nonideal operational amplifier into an ideal one.

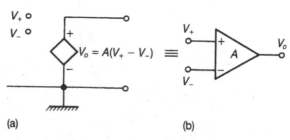

(a)   (b)

FIGURE 12.1.2   Differential amplifier with gain A: (a) Equivalent differential VV. (b) Usual symbol.

12.2. PRINCIPLES OF ACTIVE NETWORK ANALYSIS

As we have seen in the previous section, amplifiers are dependent voltage sources with one output terminal grounded. No current flows into the amplifier at its input terminals. This fact can be used to simplify analysis. In order to demonstrate the principle of the method to be described, we take two examples.

Consider the simple network in Fig. 12.2.1a; it has one independent voltage source and all passive elements are conductances. If we wish to use nodal analysis, then we must first apply voltage splitting, done in Fig. 12.2.1b. It is then convenient to redraw the network as in Fig. 12.1.1c, with the sources on both sides. Finally we can apply transformation of the voltage sources into current sources and get the final form suitable for nodal formulation. The equations will be

$$(G_1 + G_2 + G_3)V_1 - G_3V_2 = EG_1$$

$$-G_3V_1 - (G_3 + G_4 + G_5)V_2 = EG_5$$

Now suppose that we proceed somewhat differently by arguing that we really do not need to write the KCL equation at the node where the voltage is known. In Fig. 12.2.2 is the same network as in Fig. 12.2.1a but the node where the voltage source is connected to the network is marked by a large cross; this means that we will not write the KCL equation at this node. Writing them for the remaining two nodes results in

$$(G_1 + G_2 + G_3)V_1 - G_3V_2 - EG_1 = 0$$

$$-G_3V_1 + (G_3 + G_4 + G_5)V_2 - EG_5 = 0$$

As always we transfer the independent sources (terms with the symbol E) on the right side and we get exactly the same equations as above.

The rule which skips writing the KCL at the nodes with voltage sources can be used in cases when *the voltage sources have one node grounded*. Fortunately, this happens to be true for practically all active networks, and the rule is valid for dependent sources as well.

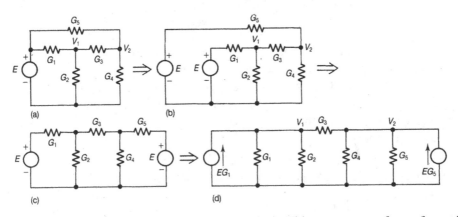

FIGURE 12.2.1 Developing method for nodal analysis with a sequence of transformations (a) to (d).

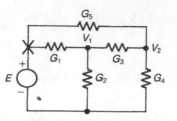

FIGURE 12.2.2 Elimination of a node for KCL equations.

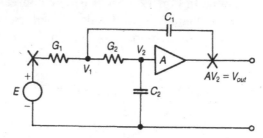

FIGURE 12.2.3 Sallen–Key active network.

Consider the network in Fig. 12.2.3. We must mark voltages at all nodes of the network. Going from the left of the figure we mark the first node with E, a known voltage. The other voltages are unknown and we mark them V_1 and V_2. We do not know the voltage of the last node, but we know something about it nevertheless: it will be A times the voltage which appears at the input of the amplifier. We thus mark this last node with the voltage AV_2. The KCL are written at the two nodes which have not been marked. The steps to write KCL are the same as we have used already many times. The equations are

$$(G_1 + G_2 + sC_1)V_1 - (G_2 + sAC_1)V_2 = EG_1$$

$$-V_1G_2 + (G_2 + sC_2)V_2 = 0$$

and in matrix form

$$\begin{bmatrix} (G_1 + G_2 + sC_1) & -(G_2 + sC_1A) \\ -G_2 & (G_2 + sC_2) \end{bmatrix} \begin{bmatrix} V_1 \\ V_2 \end{bmatrix} = \begin{bmatrix} EG_1 \\ 0 \end{bmatrix} \qquad (12.2.1)$$

We postpone solution and discussion to Section 6. At this moment we are interested in the methods how to set up the correct set of equations for various types of amplifiers. We will start with finite-gain amplifiers, because their analysis is somewhat simpler. They will be followed by networks with ideal operational amplifiers and finally with operational amplifiers which are not ideal. At this moment we note that it is advantageous to work with conductances and not resistances.

12.3.   NETWORKS WITH FINITE-GAIN AMPLIFIERS

The first network we consider is in Fig. 12.3.1 and we will keep all admittances as symbols. This has certain advantages: any G in the result can be replaced by sC and many configurations can thus be obtained from one analysis. The first and last nodes are marked by crosses because of the voltage sources. The voltages are written at the nodes: E at the first node, then from left to right V_1, V_2 and AV_2 at the rightmost node. Only two equations need be written:

$$(G_1 + G_2 + G_3 + G_5)V_1 - G_3V_2 - G_2AV_2 = G_1E$$

$$-G_3V_1 + (G_3 + G_4 + G_6)V_2 - G_6AV_2 = 0$$

In matrix form

$$\begin{bmatrix} (G_1 + G_2 + G_3 + G_5) & -(AG_2 + G_3) \\ -G_3 & (G_3 + G_4 + G_6(1-A)) \end{bmatrix} \begin{bmatrix} V_1 \\ V_2 \end{bmatrix} = \begin{bmatrix} G_1E \\ 0 \end{bmatrix}$$

The transfer function if found by first solving for V_2 and then multiplying the result by A to obtain the V_{out}. The voltage transfer function is

$$T_V = \frac{AG_1G_3}{(G_1 + G_2 + G_3 + G_5)[G_3 + G_4 + G_6(1-A)] - G_3(AG_2 + G_3)}$$

The second network is shown in Fig 12.3.2. It has one voltage source and two amplifiers. The voltages and crosses are in the figure; we must write the KCL equations for the two uncrossed nodes only:

$$(G_1 + sC_2)V_1 - sC_2A_2V_2 = EG_1$$

$$-G_2A_1V_1 + (G_2 + sC_1)V_2 = EsC_1$$

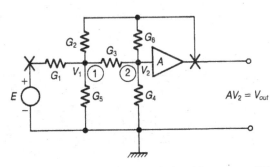

FIGURE 12.3.1 Active network with one amplifier.

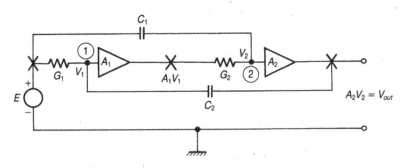

FIGURE 12.3.2 Active network with two amplifiers.

In matrix form:

$$\begin{bmatrix} (G_1 + sC_2) & -sC_2 A_2 \\ -G_2 A_1 & (G_2 + sC_1) \end{bmatrix} \begin{bmatrix} V_1 \\ V_2 \end{bmatrix} = \begin{bmatrix} EG_1 \\ EsC_1 \end{bmatrix}$$

Using Cramer's rule we find the determinant, the numerator N_2 and the output voltage $V_{out} = A_2 V_2$; the voltage transfer function is

$$T_V = \frac{A_2(s^2 C_1 C_2 + sC_1 G_1 + G_1 G_2 A_1)}{s^2 C_1 C_2 + s[C_1 G_1 + C_2 G_2(1 - A_1 A_2)] + G_1 G_2}$$

The third network is shown in Fig. 12.3.3, again with two amplifiers. The process of selecting the nodes for which we write the KCL equations should now be sufficiently clear. The voltages are written into the figure and nodal equations are written only for the two nodes which were not crossed out:

$$(G_1 + G_2 + sC_1 + sC_3)V_1 - G_1 V_2 + sC_3 A_1 V_2 - G_2 A_1 A_2 V_2 = EsC_1$$

$$-G_1 V_1 + (G_1 + sC_2)V_2 = 0$$

In matrix form:

$$\begin{bmatrix} (G_1 + G_2 + sC_1 + sC_3) & -(G_1 + sC_3 A_1 + G_2 A_1 A_2) \\ -G_1 & (G_1 + sC_2) \end{bmatrix} \begin{bmatrix} V_1 \\ V_2 \end{bmatrix} = \begin{bmatrix} EsC_1 \\ 0 \end{bmatrix}$$

Using Cramer's rule we find D, N_2 for the voltage V_2 and get the output from $V_{out} = A_1 A_2 V_2$. The transfer function is

$$T = \frac{sC_1 G_1 A_1 A_2}{s^2(C_1 + C_3)C_2 + s[C_1 G_1 + C_2 G_1 + C_2 G_2 + C_3 G_1(1 - A_1)] + G_1 G_2(1 - A_1 A_2)}$$

The next network is in Fig. 12.3.4. All the necessary information regarding the voltages and places where the KCL are written has already been marked. The equations are:

$$(G_1 + G_4 + sC_1)V_1 - sC_1 A_1 V_1 - G_4 A_2 V_2 = EG_1$$

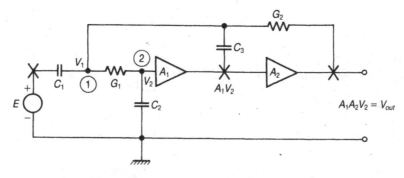

FIGURE 12.3.3   Active network with two amplifiers.

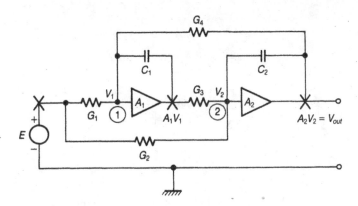

FIGURE 12.3.4   Active network with two amplifiers.

$$-G_3 A_1 V_1 \ (G_2 + G_3 + sC_2)V_2 - sC_2 A_2 V_2 = EG_2$$

The system matrix equation is

$$\begin{bmatrix} (G_1 + G_4 + sC_1(1 - A_1)) & -A_2 G_4 \\ A_1 G_3 & (G_2 + G_3 + sC_2(1 - A_2)) \end{bmatrix} \begin{bmatrix} V_1 \\ V_2 \end{bmatrix} = \begin{bmatrix} EG_1 \\ EG_2 \end{bmatrix}$$

Using standard steps we find

$$D = s^2 C_1 C_2 (1 - A_1)(1 - A_2) + s[C_1(1 - A_1)(G_2 + G_3) + C_2(1 - A_2)(G_1 + G_4)]$$

$$+ (G_1 + G_4)(G_2 + G_3) - A_1 A_2 G_3 G_4$$

$$N = A_2[sC_1 G_2(1 - A_1) + G_2(G_1 + G_4) + A_1 G_1 G_3]$$

The output is $V_{out} = A_2 V_2$ and the transfer function is

$$T = \frac{V_{out}}{E} = \frac{N}{D}$$

It should be noted here that the above networks were selected for the purpose of practicing the analysis method and not as examples of active networks which could be used for design. Many other considerations must be taken into account before we can be sure that a given network is a good one. In particular, it is necessary to study sensitivity of the network function to changes in element values. This important subject is covered in Chapter 15.

Example 1.

The network in Fig. 12.3.5 has two amplifiers, the first having the gain $A_1 = -1$. Find the output voltage by keeping the gain of the second amplifier, A_2, as a variable. In the result substitute $A_2 = 2$ and determine whether the network is stable. Also find the value of A_2 for which the network will become unstable. The voltages are written at each node; note especially that the voltage at the output of the first amplifier is $-V_1$. The nodal equations for the two nodes without crosses are

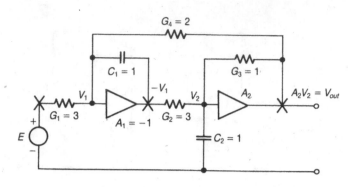

FIGURE 12.3.5   Active network with two amplifiers.

$$(5 + s)V_1 + sV_1 - 2A_2V_2 = 3E$$

$$3V_1 + (4 + s)V_2 - A_2V_2 = 0$$

In matrix form

$$\begin{bmatrix} 2s+5 & -2A_2 \\ 3 & 4-A_2+s \end{bmatrix}\begin{bmatrix} V_1 \\ V_2 \end{bmatrix} = \begin{bmatrix} 3E \\ 0 \end{bmatrix}$$

The determinant is, after a few simple steps,

$$D = 2s^2 + s(13 - 2A_2) + 20 + A_2$$

The numerator is $N_2 = -9E$ but we must not forget that the output is $V_{out} = A_2V_2$. The transfer function is

$$T_V = \frac{-9A_2}{2s^2 + s(13 - 2A_2) + 20 + A_2}$$

Substituting $A_2 = 2$ results in the transfer function

$$T_V = \frac{-18}{2s^2 + 9s + 22}$$

The poles are $p_{1,2} = -2.25 + j2.437$. Since the real parts of the poles are negative, the network is stable.

The network will become unstable when the poles move on the imaginary axis and will have zero real part. This will happen when the middle term of the denominator becomes zero: $13 - 2A_2 = 0$. The network will be unstable for $A_2 = \frac{13}{2} = 6.5$.

Example 2.

Find the voltage transfer function for the network in Fig. 12.3.6. Also determine its poles and zeros and sketch the amplitude and group delay responses. This network leads to somewhat more complicated steps but the style of analysis is the same as for the previous networks. Write the KCL equations for the nodes without crosses yourself. In matrix form this will lead to

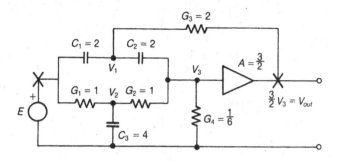

FIGURE 12.3.6 Active network wiyh a Twin-T.

$$
\begin{bmatrix}
4s+2 & 0 & -(2s+3) \\
0 & 4s+2 & -1 \\
-2s & -1 & 2s+\dfrac{7}{6}
\end{bmatrix}
\begin{bmatrix}
V_1 \\
V_2 \\
V_3
\end{bmatrix}
=
\begin{bmatrix}
2sE \\
E \\
0
\end{bmatrix}
$$

The determinant is

$$
D = (4s+2)(2s+\frac{7}{6}) - 2s(2s+3)(4s+2) - (4s+2)
$$

$$
= (4s+2)[(4s+2)(2s+\frac{7}{6}) - (4s^2+6s) - 1]
$$

The step which took the term $(4s + 2)$ in front of the square bracket is quite crucial for further analysis. After a few additional steps the denominator becomes

$$
D = 4(4s+2)(s^2 + \frac{2}{3}s + \frac{1}{3})
$$

For output voltage we must first find V_3 by evaluating the determinant

$$
N_3 =
\begin{vmatrix}
4s+2 & 0 & 2sE \\
0 & 4s+2 & E \\
-2s & -1 & 0
\end{vmatrix}
= (4s+2)(4s^2+1)E
$$

The output is $V_{out} = \dfrac{3}{2}V_3$ and the overall voltage transfer function is

$$
T = \frac{3}{2}\frac{(4s+2)(4s^2+1)}{4(4s+2)(s^2+\frac{2}{3}s+\frac{1}{3})}
$$

The network is a third order network but we can cancel the term $(4s + 2)$ in the numerator against the same term in the denominator and get the final transfer function

$$
T = \frac{3}{8}\frac{4s^2+1}{s^2+\frac{2}{3}s+\frac{1}{3}}
$$

12.1
12.2
12.3
12.4
12.5
12.6

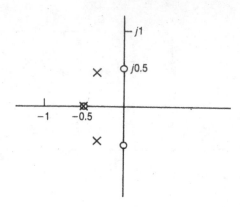

FIGURE 12.3.7   Pole-zero plot for network Fig 12.3.6.

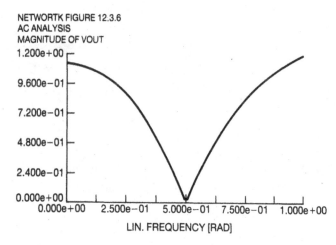

NETWORTK FIGURE 12.3.6
AC ANALYSIS
MAGNITUDE OF VOUT

FIGURE 12.3.8   Amplitude response for network Fig 12.3.6.

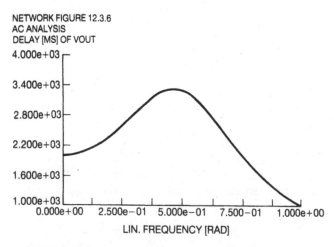

NETWORK FIGURE 12.3.6
AC ANALYSIS
DELAY [MS] OF VOUT

FIGURE 12.3.9   Group delay for network Fig. 12.3.6.

The network has two purely imaginary zeros at $Z_{1,2} = \pm j\frac{1}{2}$ and two complex conjugate poles $P_{1,2} = \frac{-1 \pm j\sqrt{2}}{3}$. In addition, it has a real zero at the same location as a pole at $s = -\frac{1}{2}$. The pole-zero plot is in Fig. 12.3.7 and cancellation of the pole and zero is indicated by plotting the circle and the cross at the same location. The amplitude response is in Fig. 12.3.8, the group delay is in Fig. 12.3.9.

12.4. NETWORKS WITH IDEAL OPAMPS

In this section we will learn how to write equations for networks with ideal operational amplifiers. We have already discussed that it is not convenient to use infinite gain and that it is much more practical to define the inverted gain,

$$B = -\frac{1}{A} \tag{12.4.1}$$

which simplifies the operational amplifier equation to

$$V_+ - V_- + BV_0 = 0 \tag{12.4.2}$$

Since all the voltages in (12.4.2) are unknown, all are placed on the left side of the equal sign.

Assume now that the gain is increased and becomes infinitely large. This makes $B = 0$ and (12.4.2) simplifies to

$$V_+ - V_- = 0 \tag{12.4.3}$$

which is equivalent to

$$V_+ = V_- \tag{12.4.4}$$

This says that the plus terminal and the minus terminal are *at the same potential*. If both terminals are floating, then whatever voltage, V_i, we write at one of these terminals, the same voltage is written at the other terminal. Sometimes the terminal V_+ is grounded; this means that the other terminal is at ground potential and $V_- = 0$.

We will apply the above conclusions to several networks, first to Fig. 12.4.1. The voltages are written into the figure and the nodes with voltage sources are marked by crosses. Since the + terminal is grounded, the − terminal of an ideal operational amplifier must be at ground potential and this is indicated by writing the zero voltage at node 1. Only one KCL equation is needed for this network, written for the node which has zero voltage:

$$(G_1 + G_2)\, 0 - G_2 V_{out} = EG_1$$

We retained the term which is multiplied by zero but it clearly need not be written at all. The transfer function is

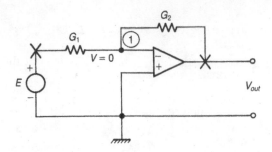

FIGURE 12.4.1   Network with ideal OPAMP.

$$T = \frac{V_{out}}{E} = -\frac{G_1}{G_2} = -\frac{R_2}{R_1} \qquad (12.4.5)$$

The network is an inverting device with the gain defined by the ratio of the two resistors. However, the network is *not* equivalent to an amplifier with the same gain because a current flows from the voltage source through the conductances G_1 and G_2 to the output. If, for instance, a capacitor is used instead of G_2, the transfer function becomes

$$T = \frac{V_{out}}{E} = -\frac{G_1}{sC_2}$$

and the network becomes an inverting integrator (inverting because of the minus sign and integrator because of the s in the denominator).

The second fundamental network is shown in Fig. 12.4.2. Only one node need be considered for the KCL, the others are marked by crosses:

$$(G_1 + G_2)E - G_2 V_{out} = 0$$

The transfer function is

$$T = \frac{V_{out}}{E} = 1 + \frac{G_1}{G_2} \qquad (12.4.6)$$

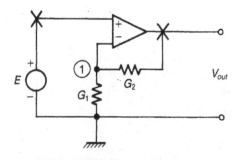

FIGURE 12.4.2   Realizing positive gain with an OPAMP.

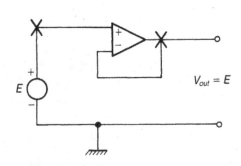

FIGURE 12.4.3   Unity gain buffer.

$$T = \frac{V_{out}}{E} = 1 + \frac{G_1}{G_2} \qquad (12.4.6)$$

It is a noninverting amplifier (no current flows from the voltage source E) and the gain is defined by the ratio of the two conductances. Let $G_1 = 0$ as a special case. This removes G_1 from the figure and in such case the other conductance is meaningless; we can replace it by a short circuit, as shown in Fig. 12.4.3. The network becomes a unity gain buffer, an amplifier with a gain equal to one.

A more complicated network is shown in Fig. 12.4.4. We have already considered this network in Section 2 because the OPAMP with the two conductances G_4 and G_5 is nothing but an amplifier with finite gain, as we established in the previous case. We proceed as before: cross out the nodes where there are voltage sources and write voltages at all nodes. This has been done in the figure. Since the OPAMP is ideal, the terminals V_+ and V_- must be at the same potential and we take this fact into account by writing *the same voltage* at these two nodes. The KCL are written at the nodes denoted by the numbers in circles:

$$(G_1 + G_2 + G_3 + sC_1)V_1 - G_3V_2 - sC_1V_{out} = EG_1$$

$$- G_3V_1 + (G_3 + sC_2)V_2 = 0$$

$$(G_4 + G_5)V_2 - G_5V_{out} = 0$$

In matrix form:

$$\begin{bmatrix} (G_1 + G_2 + G_3 + sC_1) & -G_3 & -sC_1 \\ -G_3 & (G_3 + sC_2) & 0 \\ 0 & (G_4 + G_5) & -G_5 \end{bmatrix} \begin{bmatrix} V_1 \\ V_2 \\ V_{out} \end{bmatrix} = \begin{bmatrix} EG_1 \\ 0 \\ 0 \end{bmatrix}$$

We will leave the problem of finding the transfer function to the reader.

Example 1.

Find the input impedance of the network in Fig. 12.4.5.

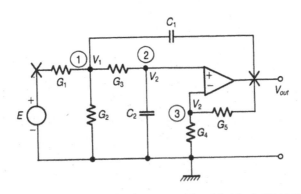

FIGURE 12.4.4   Active network with ideal OPAMP.

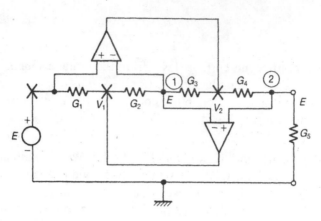

FIGURE 12.4.5   Active network with two ideal OPAMPs.

$$E(G_2 + G_3) - V_1 G_2 - V_2 G_3 = 0$$

$$E(G_4 + G_5) - G_4 V_2 = 0$$

In matrix form:

$$\begin{bmatrix} G_2 & G_3 \\ 0 & G_4 \end{bmatrix} \begin{bmatrix} V_1 \\ V_2 \end{bmatrix} = \begin{bmatrix} E(G_2 + G_3) \\ E(G_4 + G_5) \end{bmatrix}$$

The determinant of the system is $D = G_2 G_4$. In order to calculate the input impedance, we must know the current flowing into the network. It is

$$I_{in} = (E - V_1)G_1$$

and we must find V_1 by replacing the first column in the matrix by the right hand side. The result is

$$V_1 = E - \frac{E G_3 G_5}{G_2 G_4}$$

Inserting into the expression for the current

$$I_{in} = E \frac{G_1 G_3 G_5}{G_2 G_4}$$

and the input impedance is

$$Z_{in} = \frac{E}{I_{in}} = \frac{G_2 G_4}{G_1 G_3 G_5}$$

If we now replace one of the admittances in the numerator, for instance G_2, by the admittance of a capacitor, sC_2, then

$$Z_{in} = \frac{sC_2 G_4}{G_1 G_3 G_5}$$

This expression represents the impedance of an inductor with an equivalent inductance

$$L = \frac{C_2 G_4}{G_1 G_3 G_5}.$$

The network acts as an inductor with one terminal grounded.

Example 2.

Find the transfer function for the network in Fig. 12.4.6. All the necessary voltages have been written into the figure. The nodes where we do not write the KCL are marked by the crosses, the nodes where we must write the equations are indicated by numbers in circles.

The matrix equation is

$$
\begin{bmatrix}
(G_1 + G_4 + sC_1) & -G_4 & -(sC_1 + G_1) \\
(G_2 + G_3) & 0 & -G_2 \\
(G_5 + sC_2) & -sC_2 & 0
\end{bmatrix}
\begin{bmatrix}
V_1 \\
V_2 \\
V_{out}
\end{bmatrix}
=
\begin{bmatrix}
0 \\
0 \\
EG_5
\end{bmatrix}
$$

and the transfer function is

$$T = \frac{V_{out}}{E} = \frac{G_4 G_5 (G_2 + G_3)}{s^2 C_1 C_2 G_3 + s C_2 G_1 G_3 + G_2 G_4 G_5}$$

Example 3.

Fig. 12.4.7 shows a practical active network with three ideal operational amplifiers. Write the nodal equations and put them into matrix form. Try to solve using Cramer's rule and determinant expansion, as explained in Chapter 3. Also find the numerators N_2, N_3, N_4.

All the preliminary steps needed for analysis are marked in the figure: nodes where we do not write KCL are marked by crosses, nodes where we must write the KCL are marked by numbers in circles. We directly write the matrix form

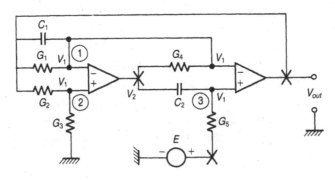

FIGURE 12.4.6   Active network with two ideal OPAMPs.

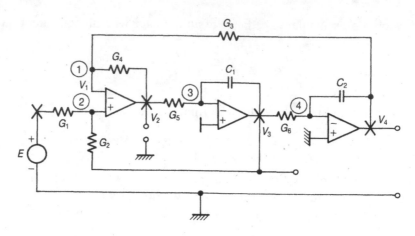

FIGURE 12.4.7 Active network with three ideal OPAMPs.

$$\begin{bmatrix} (G_3 + G_4) & -G_4 & 0 & -G_3 \\ (G_1 + G_2) & 0 & -G_2 & 0 \\ 0 & -G_5 & -sC_1 & 0 \\ 0 & 0 & -G_6 & -sC_2 \end{bmatrix} \begin{bmatrix} V_1 \\ V_2 \\ V_3 \\ V_4 \end{bmatrix} = \begin{bmatrix} 0 \\ EG_1 \\ 0 \\ 0 \end{bmatrix}$$

The solutions are:

$$D = s^2 C_1 C_2 G_4 (G_1 + G_2) + sC_2 G_2 G_5 (G_3 + G_4) + G_3 G_5 G_6 (G_1 + G_2)$$

$$N_2 = s^2 C_1 C_2 G_1 (G_3 + G_4) E$$

$$N_3 = -sC_2 G_1 G_5 (G_3 + G_4) E$$

$$N_4 = G_1 G_5 G_6 (G_3 + G_4) E$$

12.5. NETWORKS WITH NONIDEAL OPAMPS

In this section we are coming to situations where we can no longer assume that the OPAMPs are ideal. In order to make an easy comparison with the method of the previous section, we will consider the same networks.

Once the amplifiers are nonideal we cannot make any assumptions about the various voltages in the network. All are unknown except where we apply the independent voltage sources. We assign consecutive subscripts to the voltages of all nodes, cross out the nodes which are connected to voltage sources and write the KCL for the remaining nodes. This does not differ from all previous steps. Finally, we use, for each operational amplifier, equation (12.1.4) in which we substitute actual nodal voltages for the symbols V_+, V_- and V_0.

Let us take the network in Fig. 12.5.1. The input terminals of the OPAMP are no longer at the same potential and we assign voltage V_1 to the internal node. The KCL for this node is

$$(G_1 + G_2)V_1 - G_2V_{out} = EG_1$$

For the OPAMP we substitute V_1 for V_-, 0 for V_+, and V_{out} for V_0. The second equation is

$$-V_1 + BV_{out} = 0$$

Calculate V_{out} and obtain the transfer function

$$T = \frac{V_{out}}{E} = \frac{G_1}{-G_2 + B(G_1 + G_2)}$$

If we set $B = 0$, then the expression simplifies to the result we obtained in the previous section.

The second network is in Fig. 12.5.2; only one KCL need be written, followed by the equation expressing the properties of the OPAMP.

$$(G_1 + G_2)V_1 - G_2V_{out} = 0$$

$$E - V_1 + BV_{out} = 0$$

The solution is

$$T = \frac{V_{out}}{E} = \frac{G_1 + G_2}{G_2 - (G_1 + G_2)B}$$

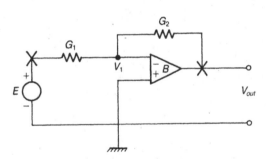

FIGURE 12.5.1   Network Fig. 12.4.1.with nonideal OPAMP.

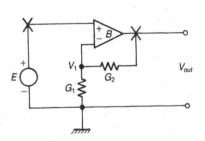

FIGURE 12.5.2   Network Fig. 12.4.2. with nonideal OPAMP.

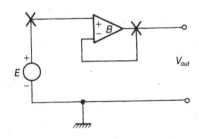

FIGURE 12.5.3   Unity gain buffer nonideal OPAMP.

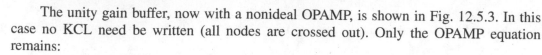

If we set $B = 0$, the expression reduces to (12.4.6).

The unity gain buffer, now with a nonideal OPAMP, is shown in Fig. 12.5.3. In this case no KCL need be written (all nodes are crossed out). Only the OPAMP equation remains:

$$E - V_{out} + BV_{out} = 0$$

The property of this network with a nonideal OPAMP is described by the transfer function

$$T = \frac{V_{out}}{E} = \frac{1}{1 - B}$$

The next case is the network in Fig. 12.5.4. We must write three KCL equations at the nodes marked by the voltages V_1, V_2, and V_3 and attach the equation for the OPAMP:

$$(G_1 + G_2 + G_3 + sC_1)V_1 - G_3V_2 - sC_1V_{out} = EG_1$$

$$-G_3V_1 + (G_3 + sC_2)V_2 = 0$$

$$(G_4 + G_5)V_3 - G_5V_{out} = 0$$

$$V_2 - V_3 + BV_{out} = 0$$

In matrix form:

$$
\begin{bmatrix}
(G_1 + G_2 + G_3 + sC_1) & -G_3 & 0 & -sC_1 \\
-G_3 & (G_3 + sC_2) & 0 & 0 \\
0 & 0 & (G_4 + G_5) & -G_5 \\
0 & 1 & -1 & B
\end{bmatrix}
\begin{bmatrix}
V_1 \\
V_2 \\
V_3 \\
V_{out}
\end{bmatrix}
=
\begin{bmatrix}
EG_1 \\
0 \\
0 \\
0
\end{bmatrix}
$$

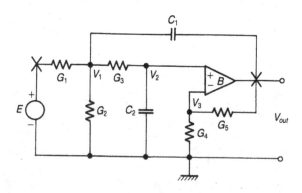

FIGURE 12.5.4   Network Fig. 12.4.4.with nonideal OPAMP.

Example 1.

Write the nodal equations and the necessary matrix equation for the network in Fig. 12.5.5, now with nonideal OPAMPs.

Only two nodal equations need be written, followed by the equations of the two OPAMPs:

$$-G_2V_1 + (G_2 + G_3)V_2 - G_3V_3 = 0$$

$$- G_4V_3 + (G_4 + G_5)V_4 = 0$$

$$E - V_2 + B_1V_3 = 0$$

$$- V_2 + V_4 + B_2V_1 = 0$$

In matrix form:

$$
\begin{bmatrix}
-G_2 & (G_2 + G_3) & -G_3 & 0 \\
0 & 0 & -G_4 & (G_4 + G_5) \\
0 & -1 & B_1 & 0 \\
B_2 & -1 & 0 & 1
\end{bmatrix}
\begin{bmatrix}
V_1 \\
V_2 \\
V_3 \\
V_4
\end{bmatrix}
=
\begin{bmatrix}
0 \\
0 \\
-E \\
0
\end{bmatrix}
$$

Example 2.

The network in Fig. 12.5.6 has two nonideal OPAMPs. Write the nodal equations and put them into matrix form.

All the preliminary operations have been written into the figure. We give only the matrix form:

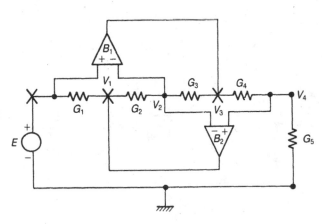

FIGURE 12.5.5   Network Fig. 12.4.5.with nonideal OPAMPs.

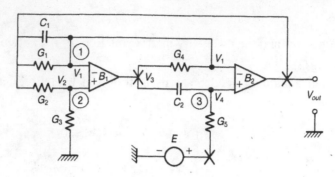

FIGURE 12.5.6   Network Fig. 12.4.6.with nonideal OPAMPs.

$$
\begin{bmatrix}
(G_1 + G_4 + sC_1) & 0 & -G_4 & 0 & -(G_1 + sC_1) \\
0 & (G_2 + G_3) & 0 & 0 & -G_2 \\
0 & 0 & -sC_2 & (G_5 + sC_2) & 0 \\
-1 & 1 & B_1 & 0 & 0 \\
-1 & 0 & 0 & 1 & B_2
\end{bmatrix}
\begin{bmatrix}
V_1 \\ V_2 \\ V_3 \\ V_4 \\ V_{out}
\end{bmatrix}
=
\begin{bmatrix}
0 \\ 0 \\ EG_5 \\ 0 \\ 0
\end{bmatrix}
$$

Example 3.

Write the nodal equations for the network in Fig. 12.5.7 and put them in matrix form. The OPAMPS are nonideal.

All the preliminary steps have been applied to the figure. The nodal equations are

$$(G_3 + G_4)V_1 - G_4V_3 - G_3V_7 = 0$$

$$(G_1 + G_2)V_2 - G_2V_5 = EG_1$$

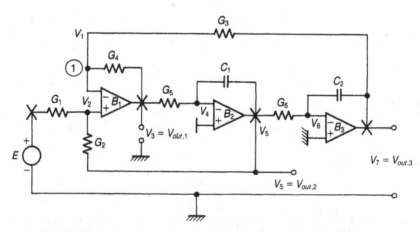

FIGURE 12.5.7   Network Fig. 12.4.7.with nonideal OPAMPs.

$$- G_5 V_3 + (G_5 + sC_1)V_4 - sC_1 V_5 = 0$$

$$- G_6 V_5 + (G_6 + sC_2)V_6 - sC_2 V_7 = 0$$

$$-V_1 + V_2 + B_1 V_3 = 0$$

$$- V_4 + B_2 V_5 = 0$$

$$- V_6 + B_3 V_7 = 0$$

Because of the size, we give only the system matrix:

$$\begin{bmatrix}
(G_3 + G_4) & 0 & -G_4 & 0 & 0 & 0 & -G_3 \\
0 & (G_1 + G_2) & 0 & 0 & -G_2 & 0 & 0 \\
0 & 0 & -G_5 & (G_5 + sC_1) & -sC_1 & 0 & 0 \\
0 & 0 & 0 & 0 & -G_6 & (G_6 + sC_2) & -sC_2 \\
-1 & +1 & B_1 & 0 & 0 & 0 & 0 \\
0 & 0 & 0 & -1 & B_2 & 0 & 0 \\
0 & 0 & 0 & 0 & 1 & -1 & B_3
\end{bmatrix}$$

The last example shows that hand calculations are next to impossible if we have to consider several nonideal OPAMPs. For this reason, analysis is usually done with ideal OPAMPs and conclusions are taken from such results.

12.6. PRACTICAL USE OF RC ACTIVE NETWORKS

All previous sections of this chapter were devoted to methods of writing equations of active networks. In several cases we also solved the matrix equations. Some of the networks we considered are practical, some were invented solely for the purpose of teaching the method of writing the minimum set of equations directly. In this section we take a few practical networks and discuss typical ways how one uses the results of the analysis.

As an introduction consider Fig. 12.6.1a where we have the symbol for an amplifier with $A \geq 1$. Its practical realization by means of an ideal OPAMP and two resistors is in Fig. 12.6.1b. In active networks this is always used whenever we need $A > 1$.

The first network with some practical importance was the network in Fig. 12.2.3, system matrix equation (12.2.1). We now solve it. The determinant is

$$D = s^2 C_1 C_2 + s[C_2(G_1 + G_2) + C_1 G_2(1 - A)] + G_1 G_2$$

Using Cramer's rule we obtain the numerator for V_2 and $V_{out} = AV_2$. The transfer function is

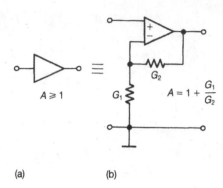

FIGURE 12.6.1   Amplifier with positive gain: (a) Symbol. (b) Realization with ideal OPAMP.

$$T = \frac{V_{out}}{E} = \frac{AG_1G_2}{s^2C_1C_2 + s[C_2(G_1+G_2)+C_1G_2(1-A)]+G_1G_2}$$

In applications we are interested in the Q and ω_0 and we now derive practical formulas how to get both these quantities directly from the second-order transfer function. The denominator polynomial is of the type

$$D = Us^2 + Vs + W$$

where U, V and W are functions of network elements. Rewrite as

$$D = U(s^2 + \frac{V}{U}s + \frac{W}{U})$$

and compare with the general form introduced earlier [equation (8.4.5)]:

$$D = s^2 + \frac{\omega_0}{Q}s + \omega_0^2$$

We can identify

$$\omega_0^2 = \frac{W}{U} \tag{12.6.1}$$

and

$$\frac{\omega_0}{Q} = \frac{V}{U} \tag{12.6.2}$$

Substituting ω_0 from (12.6.1) we obtain

$$Q = \frac{\sqrt{UW}}{V}$$

Equations (12.6.1)–(12.6.3) simplify our work. For the above example we immediately have

$$\omega_0^2 = \frac{G_1 G_2}{C_1 C_2}$$

and

$$Q = \frac{\sqrt{C_1 C_2 G_1 G_2}}{C_2 G_1 + C_2 G_2 + C_1 G_2 (1 - A)}$$

These expressions can be used for the design of the network. Suppose that we wish to realize $\omega_0 = 1$ and control Q by the gain of the amplifier only. We can select, more or less arbitrarily, $G_1 = G_2 = 1$ and $C_1 = C_2 = 1$. Then

$$Q = \frac{1}{3 - A}$$

or

$$A = \frac{3Q - 1}{Q}$$

Should we wish to realize the network for $Q = 2$, then we need $A = 2.5$.

The next practical network is in Fig. 12.6.2. Writing the nodal equations:

$$(G_1 + sC_1)V_1 - sC_1 V_2 - G_1 V_{out} = 0$$

$$(G_2 + G_3)V_1 - G_2 V_{out} = EG_3$$

$$(G_4 + G_5 + sC_2)V_1 - 4V_2 = EsC_2$$

In matrix form

$$\begin{bmatrix} (G_1 + sC_1) & -sC_1 & -G_1 \\ (G_2 + G_3) & 0 & -G_2 \\ (G_4 + G_5 + sC_2) & -G_4 & 0 \end{bmatrix} \begin{bmatrix} V_1 \\ V_2 \\ V_{out} \end{bmatrix} = \begin{bmatrix} 0 \\ G_3 E \\ sC_2 E \end{bmatrix}$$

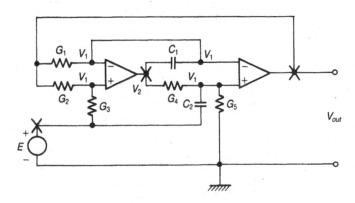

FIGURE 12.6.2 All-pass with OPAMPs.

The determinant of the matrix is the denominator of the transfer function:

$$D = s^2 C_1 C_2 G_2 + s C_1 G_2 G_5 + G_1 G_3 G_4$$

Using Cramer's rule for the output voltage we obtain the numerator:

$$N = E[s^2 C_1 C_2 G_2 - s C_1 G_3 G_5 + G_1 G_3 G_4]$$

Let us now select

$$G_2 = G_3 = G$$

The transfer function becomes

$$T = \frac{V_{out}}{E} = \frac{s^2 C_1 C_2 G - s C_1 G_5 G + G_1 G G_4}{s^2 C_1 C_2 G + s C_1 G_5 G + G_1 G G_4}$$

We note that the expressions in the numerator and denominator differ only by the sign of the middle terms. Networks which have this property are called *all-pass* networks because they do not influence the amplitude of the signal, only the phase (and group delay). Such networks are used in the design of filters. Applying our formulas we get

$$\omega_0^2 = \frac{G_1 G_4}{C_1 C_2}$$

and

$$Q = \frac{1}{G_5} \left[\frac{C_2 G_1 G_4}{C_1} \right]^{1/2}$$

Suppose that we select $C_1 = C_2 = 1$, $G_5 = 0.5$ and all other conductances $G_i = 1$; this results in $Q = 2$. Because the network is an all-pass, the amplitude response is equal to 1 for all frequencies, but the group delay is a function of frequency. Fig. 12.6.3 is the plot of this group delay.

We will consider one more practical network, this time with three operational amplifiers. It is shown in Fig. 12.6.4. The voltages at the input terminals of the OPAMPs are equal to zero; we write the nodal equations at these points. In matrix form:

$$\begin{bmatrix} -(G_3 + sC_1) & 0 & -G_2 \\ -G_4 & -sC_2 & 0 \\ 0 & -G_5 & -G_6 \end{bmatrix} \begin{bmatrix} V_1 \\ V_2 \\ V_3 \end{bmatrix} = \begin{bmatrix} EG_1 \\ 0 \\ 0 \end{bmatrix}$$

The denominator of all network functions is

$$D = -[s^2 C_1 C_2 G_6 + s C_2 G_3 G_6 + G_2 G_4 G_5]$$

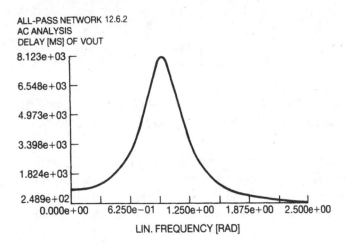

FIGURE 12.6.3 Group delay of network Fig. 12.6.2.

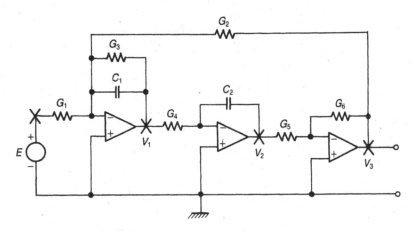

FIGURE 12.6.4 Network with three ideal operational amplifiers.

Solving for all three voltages at the outputs of the OPAMPs, we get the transfer functions

$$T_1 = \frac{V_1}{E} = \frac{sC_2 G_1 G_6}{D}$$

$$T_2 = \frac{V_2}{E} = \frac{-G_1 G_4 G_6}{D}$$

$$T_3 = \frac{V_3}{E} = \frac{G_1 G_4 G_5}{D}$$

Finally, using formulas (12.6.1) and (12.6.3), we obtain

$$\omega_0^2 = \frac{G_2 G_4 G_5}{C_1 C_2 G_6}$$

and

$$Q = \frac{1}{G_3}\left[\frac{C_1 G_2 G_4 G_5}{C_2 G_6}\right]^{1/2}$$

If we select, for instance, $C_1 = C_2 = 1$ and $G_2 = G_4 = G_5 = G_6 = 1$, then Q is controlled only by the conductance G_3. We thus have a very simple way to construct networks with various Q.

The considerations presented above are not sufficient for reliable design of active networks. It is also necessary to check how changes in element values influence the properties of the network. If small changes in the values result in large changes of the network responses, then such networks are not suitable for practical applications. These topics are generally lumped under the name of network sensitivity. This important subject is covered in Chapter 15.

PROBLEMS CHAPTER 12

P.12.1.(a)   The amplifier has unity gain, the other elements are variable. Derive the transfer function in terms of variable elements. Make $C_1 = C_2 = C$ and $G_1 = G_2 = G$ and find the expression for ω_0 and Q. What is the meaning of $Q = 0.5$ in terms of the poles of the network?

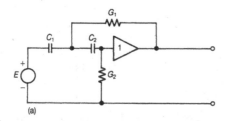

(a)

P.12.1.(b)   Do the same as in (a). What is the difference?

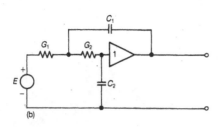

(b)

P.12.1.(c)   Do the same as in (a).

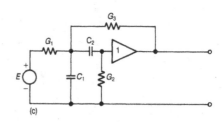

(c)

P.12.2(a)–(e)   The for the output in terms of variable elements.

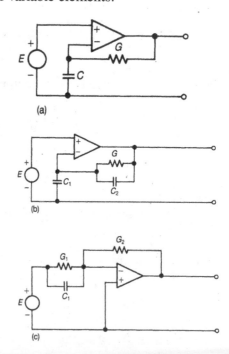

(a)

(b)

(c)

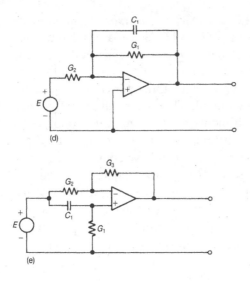

(d)

(e)

P.12.3 The operational amplifier in this configuration acts as a unity gain buffer. Derive the expression for the output in terms of variable elements. Afterwards make $G_1 = G_2 = G$ and $C_1 = C_2 = C$ and plot the poles. Next remove the operational amplifier by short-circuiting its input and output and derive the transfer function for equal G and C. Note that in this case you do not get a double pole.

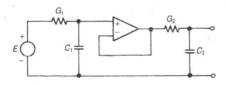

P.12.4 Analyze the network by keeping all elements as variables.

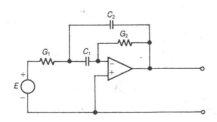

P.12.5 Derive the network by keeping all elements as variables.

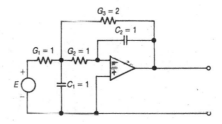

P.12.6 Use numerical values of elements and derive the transfer function What are the poles network.

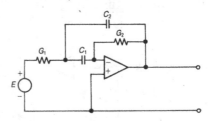

P.12.7 Use $C_1 = C_2 = 1$ but keep the conductances and the gain of the VV as variables. Derive the transfer function. Set $G_1 = G_2 = G$ and get the expression for ω_0 and Q? If we wish to make $Q = 2$, what gain of the amplifier must we select?

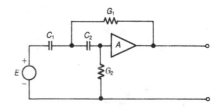

P.12.8 Derive the transfer function using given numerical values. Keep the gain, A, as a variable.

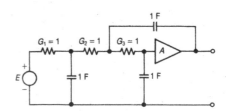

P.12.9 Use given element values but keep the amplifier gain, A, as a variable. Derive the transfer function.

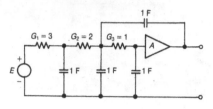

P.12.10 Find the input impedance in terms of variable elements.

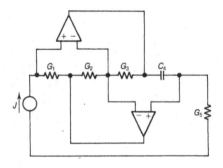

P.12.11 Find the transfer function by keeping the elements as variables. Also give expression for ω_0 and Q. Note that some conductances and the two capacitors have equal values.

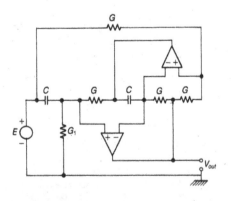

P.12.12 Derive the transfer function by keeping G_1 and G_2 as variables. Notice that the term $(1 + G_2)$ can be factored out of the denominator. Select $G_2 = 4G_1$ and notice the type of the numerator and denominator polynomial. They have numerically equal coefficients but the sign at the odd power of s differs. Such networks are called all-pass because they have constant frequency response (pass all frequencies equally) but do influence the group delay. Find the poles and zeros for $G_1 = 1$, show that the amplitude response is independent of frequency and sketch the group delay.

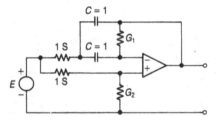

P.12.13 Use given numerical values but keep G_1 and G_2 as variables. Derive the transfer function.

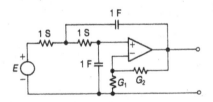

P.12.14 Keep the capacitors as variables and set $G_1 = G_2 = 1$. Find the transfer function.

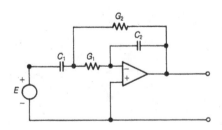

P.12.15 Set $G_1 = G_2 = G_3 = 1$ and keep the capacitors as variables. Find the transfer function.

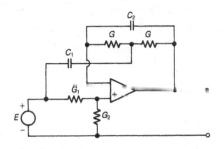

P.12.16 Let all conductances be equal to one and keep the capacitors as variables. Find the transfer function.

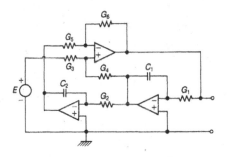

P.12.17 Find the transfer function by keeping all elements as variables. The operational amplifiers are ideal.

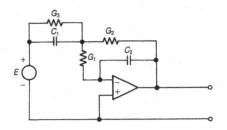

P.12.18 This network is an example how small active filters can be built. The dotted line separates two networks, in this case with identical structures. Because the second network is getting its input from the amplifier, the two networks do not influence each other. We can realize one of them for a given pair of complex conjugate poles and the other one for another pair. The overall response will be equal to the product of both transfer functions. In both networks set $G_2 = 1$ and $C_1 = 1$ but keep G_1 and C_2 and the gain A as variables. Derive the transfer function. Design the first network to realize the denominator $D_1 = s^2 + 2s + 2$, and the second network to realize $D_2 = s^2 + 2s + 5$. Plot the poles. Sketch the frequency domain response of each of the networks separately and then when they are connected together.

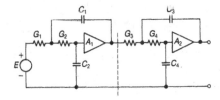

13 TWO-PORTS

INTRODUCTION

The theory of two-ports was invented many years ago as a method to analyze networks with practically zero computational resources: small networks are analyzed in a special way and their interconnections are handled by the two-port rules. Since the arrival of computers the two-port theory lost its importance. We added a brief explanation because it is mentioned in the next chapter where we study the transformers. Only impedance and admittance parameters are covered here, although several others also exist.

13.1. DEFINITION OF TWO-PORTS

In order to introduce two-port theory, consider an arbitrary network in which we select only four terminals and make them the only places where connections to other networks or elements can be made. The situation is sketched in Fig. 13.1.1. Instead of taking each terminal separately and measuring the voltages at these terminals with respect to ground, we group them into two pairs. Each pair is called a *port* and we thus have a *twoport* network. Anything can be inside the box but to make the theory practical we usually consider only a few components. It is a common practice to denote by V_1 the voltage at the left port and by V_2 the voltage at the right port, with + and − as indicated. There will also be currents and the directions are *always* assumed to be as shown in the figure. *Note the currents again*! At the + terminals they flow *away* from the terminal and *into* the network, at the − terminals they flow *into* the terminal and *out* of the network.

In our earlier chapters we always considered the current as positive when it was *leaving* the node, but here the currents go away from the top nodes and go to the nodes at the bottom of the figure. We were also considering the voltages with respect to ground, but here we are taking the voltage between two nodes. Why are there such two different ways of considering the voltages and currents? The reasons are historical. Two-port theory was developed many years ago, and at the beginning even the definition shown in Fig. 13.1.1

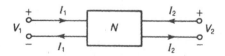

FIGURE 13.1.1   Two-port orientations of voltages and currents.

was not unified; in German literature the currents on the right were originally reversed. The directions shown in Fig. 13.1.1 are now accepted as standard in the two-port theory. In all other applications, voltages are measured with respect to ground and currents are considered as positive when leaving a node.

Since we have four variables, V_1, V_2, I_1 and I_2, we can consider any two of them as independent variables and the other two as dependent variables. Each dependent variable will, in general, depend on the two independent variables.

We can have several choices. For instance, we can select the currents I_1 and I_2 as independent variables; then the dependent variables will be V_1 and V_2. Other choices exist as well, but we will limit our explanations to only two of them, the impedance and admittance parameters.

13.1.1. Impedance Parameters

Let the currents be the independent variables. Then we can write two general equations

$$V_1 = z_{11} I_1 + z_{12} I_2$$
$$V_2 = z_{21} I_1 + z_{22} I_2$$

(13.1.1)

which couple the independent and dependent variables in a most general form. The subscripts are chosen to take advantage of standard matrix notation:

$$\begin{bmatrix} V_1 \\ V_2 \end{bmatrix} = \begin{bmatrix} z_{11} & z_{12} \\ z_{21} & z_{22} \end{bmatrix} \begin{bmatrix} I_1 \\ I_2 \end{bmatrix}$$

(13.1.2)

The network inside the rectangle in Fig. 13.1.1 is given and we cannot go inside it but we can attach sources and do measurements (or calculations). The steps will become clear from the following.

For the system (13.1.1) we can set $I_2 = 0$, which means that we *open circuit* the second port and attach a source at the first port. This simplifies (13.1.1) to

$$V_1 = z_{11} I_1$$
$$V_2 = z_{21} I_1$$

Dividing by I_1 we obtain

$$z_{11} = \left. \frac{V_1}{I_1} \right|_{I_2 = 0}$$

(13.1.3)

$$z_{21} = \left. \frac{V_2}{I_1} \right|_{I_2 = 0}$$

(13.1.4)

Similarly, setting $I_1 = 0$ (open-circuiting the first port) and placing the source to the second port we get

$$V_1 = z_{12} I_2$$
$$V_2 = z_{22} I_2$$

Dividing by I_2

$$z_{12} = \left.\frac{V_1}{I_2}\right|_{I_1 = 0} \tag{13.1.5}$$

$$z_{22} = \left.\frac{V_2}{I_2}\right|_{I_1 = 0} \tag{13.1.6}$$

The parameters z_{ij} have the dimensions of an impedance and are sometimes referred to as *two-port impedance parameters*. Since they were obtained by open-circuiting the ports, they are also known as *open circuit parameters*.

We will demonstrate the derivation of these parameters on the example of a *T*-network, shown in Fig. 13.1.2. First we attach a source at the first port. We can use either a voltage or a current source; in this case we take the voltage source as shown in Fig. 13.1.2a. Since $I_2 = 0$, we have

$$I_1 = \frac{V_1}{Z_1 + Z_3}$$

from which we get immediately

$$z_{11} = \frac{V_1}{I_1} = Z_1 + Z_3$$

The voltage at the middle node must be equal to V_2 since no current flows through Z_2 and there can be no voltage drop across it,

$$V_2 = I_1 Z_3$$

and

$$z_{21} = \frac{V_2}{I_1} = Z_3$$

In the next step we place the voltage source at the second port and opencircuit the first port, as shown in Fig. 13.1.2b. Proceeding the same way, we obtain

$$z_{12} = \frac{V_1}{I_2} = Z_3$$

$$z_{22} = \frac{V_2}{I_2} = Z_2 + Z_3$$

Collecting the terms in the form of a matrix

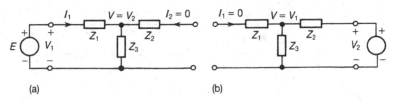

FIGURE 13.1.2 Z parameters for a T network: (a) For z_{11} and z_{21}. (b) For z_{12} and z_{22}.

$$\mathbf{Z} = \begin{bmatrix} Z_1 + Z_3 & Z_3 \\ Z_3 & Z_2 + Z_3 \end{bmatrix} \tag{13.1.7}$$

By setting Z_1 or Z_2 or both equal to zero, we get the open-circuit parameters for three additional networks, shown in Fig. 13.1.3 with their matrices. Note that in the simplest case, when only Z_3 is left, the open-circuit matrix becomes singular (the determinant of the matrix is zero). Another point to remember: if the network does not have dependent sources and is composed of only R, L, and C elements, then z_{12} is *always* equal z_{21}. Such networks are called *reciprocal*.

Example 1.

Find the open-circuit parameters for the network in Fig. 13.1.4.

It is advantageous to use current sources when deriving the z-parameters. We must proceed with two solutions, as given in Figs. 13.1.4a and b.

(a) Note how we numbered the nodes, to have the subscripts 1 and 2 for the required results. We must calculate V_1 and V_2. Using standard nodal formulation (write the KCL equations) we arrive at the system matrix

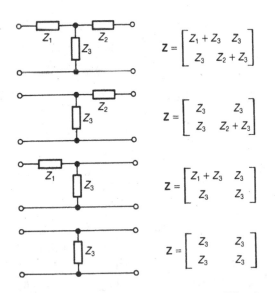

FIGURE 13.1.3 T network and its simplifications.

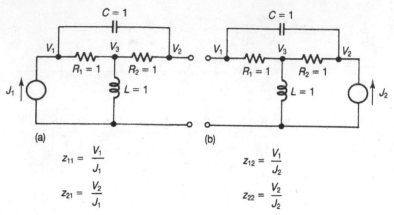

$$z_{11} = \frac{V_1}{J_1}$$

$$z_{21} = \frac{V_2}{J_1}$$

$$z_{12} = \frac{V_1}{J_2}$$

$$z_{22} = \frac{V_2}{J_2}$$

FIGURE 13.1.4 Z parameters for a bridged T network: (a) Connection to calculate z_{11} and z_{21}. (b) Connection to calculate z_{12} and z_{22}.

$$\begin{bmatrix} s+1 & -s & -1 \\ -s & s+1 & -1 \\ -1 & -1 & \dfrac{2s+1}{s} \end{bmatrix} \begin{bmatrix} V_1 \\ V_2 \\ V_3 \end{bmatrix} = \begin{bmatrix} J_1 \\ 0 \\ 0 \end{bmatrix}$$

The determinant is

$$D = \frac{2s+1}{s}$$

Replacing the first column by the right hand side we get

$$N_1 = \frac{2s^2 + 2s + 1}{s} J_1$$

and

$$z_{11} = \frac{V_1}{J_1} = \frac{2s^2 + 2s + 1}{2s + 1}$$

Replacing the second column by the right hand side we get

$$N_2 = (2s + 2)J_1$$

and

$$z_{21} = \frac{V_2}{J_1} = \frac{2s^2 + 2s}{2s + 1}$$

(b) The second solution differs from the first one in the right hand side vector only; J_2 will be in the second position and zeros in the first and third. The determinant is already known. Since the network is reciprocal, it must be true that $z_{12} = z_{21}$. In addition, since the network is symmetrical, it will also be true that $z_{22} = z_{11}$ (check!). The open circuit impedance matrix is

$$Z = \begin{bmatrix} \dfrac{2s^2 + 2s + 1}{2s + 1} & \dfrac{2s^2 + 2s}{2s + 1} \\[3mm] \dfrac{2s^2 + 2s}{2s + 1} & \dfrac{2s^2 + 2s + 1}{2s + 1} \end{bmatrix}$$

13.1.2. Admittance Parameters

The other obvious selection is to take the voltages as independent variables. Then the general two-port equations will be

$$\begin{aligned} I_1 &= y_{11}V_1 + y_{12}V_2 \\ I_2 &= y_{21}V_1 + y_{22}V_2 \end{aligned} \qquad (13.1.8)$$

or in matrix form,

$$\begin{bmatrix} I_1 \\ I_2 \end{bmatrix} = \begin{bmatrix} y_{11} & y_{12} \\ y_{21} & y_{22} \end{bmatrix} \begin{bmatrix} V_1 \\ V_2 \end{bmatrix} \qquad (13.1.9)$$

The equations can be clearly simplified if we select $V_2 = 0$, which is achieved by short-circuiting the second port. Attaching a source at the first port we get

$$\begin{aligned} I_1 &= y_{11}V_1 \\ I_2 &= y_{21}V_1 \end{aligned}$$

Dividing by V_1

$$y_{11} = \left. \frac{I_1}{V_1} \right|_{V_2 = 0} \qquad (13.1.10)$$

$$y_{21} = \left. \frac{I_2}{V_1} \right|_{V_2 = 0} \qquad (13.1.11)$$

Transferring the source to the second port and short circuiting the first port

$$\begin{aligned} I_1 &= y_{12}V_2 \\ I_2 &= y_{22}V_2 \end{aligned}$$

and dividing by V_2

$$y_{12} = \left. \frac{I_1}{V_2} \right|_{V_1 = 0} \qquad (13.1.12)$$

$$y_{22} = \left. \frac{I_2}{V_2} \right|_{V_1 = 0} \qquad (13.1.13)$$

The parameters y_{ij} have the dimension of an admittance and are sometimes referred to as *two-port admittance parameters*. Since they were obtained by short-circuiting the ports, they are also known as *short-circuit parameters*.

We will demonstrate derivation of the parameters on the example of a Π network, shown in Fig. 13.1.5. Short-circuit the second port, attach a current source at the first port and note the direction of I_2 in Fig. 13.1.5a. Because of the short-circuit, we have

$$I_1 = V_1(Y_1 + Y_3)$$

from which we get

$$y_{11} = \frac{I_1}{V_1} = Y_1 + Y_3$$

In determining I_2 we realize that this current is exactly equal to the current flowing through Y_3 but *in opposite direction:*

$$-I_2 = V_1 Y_3$$

and

$$y_{21} = \frac{I_2}{V_1} = -Y_3$$

In the next step we place the current source at the second port and shortcircuit the first port, as shown in Fig. 13.1.5b. Proceeding similarly as above, we obtain

$$y_{12} = \frac{I_1}{V_2} = -Y_3$$

$$y_{22} = \frac{I_2}{V_2} = Y_2 + Y_3$$

Collecting the terms

$$\mathbf{Y} = \begin{bmatrix} Y_1 + Y_3 & -Y_3 \\ -Y_3 & Y_2 + Y_3 \end{bmatrix} \tag{13.1.14}$$

By setting Y_1 or Y_2 or both equal to zero we get the short-circuit parameters for three additional networks. The networks and their matrices are shown in Fig. 13.1.6. We note again that the simplest network, consisting of only Y_3, has a singular admittance matrix. If the network is without dependent sources, then y_{12} is always equal to y_{21} (reciprocal network).

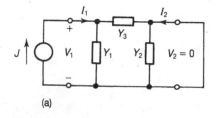

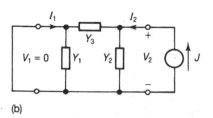

(a)   (b)

FIGURE 13.1.5 Y parameters for a Π network, (a) For y_{11} and y_{21} (b) For y_{12} and y_{22}.

Example 2.

Find the short-circuit parameters using the network in Fig. 13.1.7a and b. The network is the same as in Fig. 13.1.4, but is redrawn for this purpose.

We are free to select the independent sources (current or voltage), but in order to go via nodal formulation we select current sources. They are in the figure (a). Nodal equations for the network are

$$\begin{bmatrix} s+1 & -1 \\ -1 & \dfrac{2s+1}{s} \end{bmatrix} \begin{bmatrix} V_1 \\ V_2 \end{bmatrix} = \begin{bmatrix} J_1 \\ 0 \end{bmatrix}$$

The determinant is

$$D = \frac{2s^2 + 2s + 1}{s}$$

$$V_1 = \frac{2s+1}{2s^2 + 2s + 1} J_1$$

$$V_2 = \frac{s}{2s^2 + 2s + 1} J_1$$

and we have immediately

$$y_{11} = \frac{J_1}{V_1} = \frac{2s^2 + 2s + 1}{2s+1}$$

We also need the current I_2. It flows in opposite direction and therefore

$$-I_2 = sV_1 + V_2$$

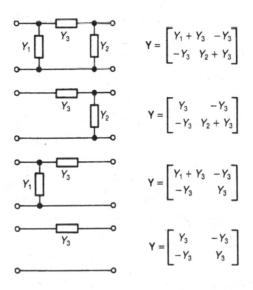

FIGURE 13.1.6 Π network and its simplifications.

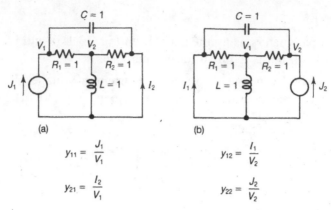

$$y_{11} = \frac{J_1}{V_1} \qquad\qquad y_{12} = \frac{I_1}{V_2}$$

$$y_{21} = \frac{I_2}{V_1} \qquad\qquad y_{22} = \frac{J_2}{V_2}$$

FIGURE 13.1.7 Y parameters for a bridged T network: (a) Connection to calculate y_{11} and y_{21}. (b) Connection to calculate y_{12} and y_{22}.

with the result

$$I_2 = -\frac{2s^2 + 2s}{2s^2 + 2s + 1}$$

Forming the ratio

$$y_{21} = \frac{I_2}{V_1} = -\frac{2s^2 + 2s}{2s + 1}$$

The network is reciprocal and symmetric and thus $y_{12} = y_{21}$ and also $y_{11} = y_{22}$ (check!). The short circuit admittance matrix is

$$\mathbf{Y} = \begin{bmatrix} \dfrac{2s^2 + 2s + 1}{2s + 1} & -\dfrac{2s^2 + 2s}{2s + 1} \\[2mm] -\dfrac{2s^2 + 2s}{2s + 1} & \dfrac{2s^2 + 2s + 1}{2s + 1} \end{bmatrix}$$

13.2.   TWO-PORTS WITH LOAD

The two-port currents I_1 and I_2 are defined as flowing into the two-port, see Fig. 13.1.1. This requires a certain caution when connecting some other elements. Here we derive a few useful relationships by using the impedance and admittance parameters.

Consider again the network in Fig. 13.1.1, but now assume that on the right is connected a resistor. Under normal conditions, the current through the resistor would flow from + to −, that is down. However, in this situation the two-port current, I_2, actually flows the other way. We take this into consideration by writing

$$I_2 = -\frac{V_2}{R} \tag{13.2.1}$$

Let us now find the input and transfer impedances of the combination starting with (13.1.1):

$$V_1 = z_{11} I_1 + z_{12} I_2$$
$$V_2 = z_{21} I_1 + z_{22} I_2$$
(13.2.2)

Inserting (13.2.1) into (13.2.2) we get

$$V_1 = z_{11} I_1 - z_{12} \frac{V_2}{R}$$
(13.2.3)

$$V_2 = z_{21} I_1 - z_{22} \frac{V_2}{R}$$
(13.2.4)

From (13.2.4) we get

$$V_2 = \frac{z_{21}}{1 + \frac{z_{22}}{R}} I_1$$
(13.2.5)

which already leads to the transfer impedance

$$Z_{TR} = \frac{V_2}{I_1} = \frac{z_{21}}{1 + \frac{z_{22}}{R}}$$
(13.2.6)

We can also substitute (13.2.5) into (13.2.3) and obtain the input impedance

$$Z_{in} = \frac{V_1}{I_1} = z_{11} - \frac{z_{12} z_{21}}{R\left(1 + \frac{z_{22}}{R}\right)}$$
(13.2.7)

When using the admittance parameters, we start with (13.1.8)

$$I_1 = y_{11} V_1 + y_{12} V_2$$
$$I_2 = y_{21} V_1 + y_{22} V_2$$
(13.2.8)

and modify (13.2.1) to the form

$$V_2 = -R I_2$$
(13.2.9)

Inserting into (13.2.8) we get

$$I_1 = y_{11} V_1 - y_{12} R I_2$$
(13.2.10)

$$I_2 = y_{21} V_1 - y_{22} R I_2$$
(13.2.11)

From (13.2.11) we get

$$I_2 = \frac{y_{21} V_1}{1 + y_{22} R}$$
(13.2.12)

which already represents the transfer admittance

$$Y_{TR} = \frac{I_2}{V_1} = \frac{y_{21}}{1 + y_{22} R}$$
(13.2.13)

Also inserting (13.2.12) into (13.2.10) we obtain

$$I_1 = \left(y_{11} - \frac{y_{12}y_{21}}{1 + y_{22}R} \right) V_1 \tag{13.2.14}$$

and the input admittance is

$$Y_{in} = \frac{I_1}{V_1} = y_{11} - \frac{y_{12}y_{21}R}{1 + y_{22}R} \tag{13.2.15}$$

PROBLEMS CHAPTER 13

P.13.1 Find the z-parameters for the bridged network we used in the text, Fig. 13.1.4, but now keeping the elements as variables.

P.13.2 Find the impedance parameters.

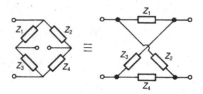

P.13.3 Find the impedance parameters.

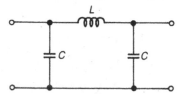

P.13.4 Find the impedance parameters.

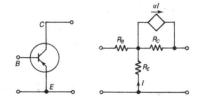

14 TRANSFORMERS

INTRODUCTION

Transformers are the last linear elements which we will consider. For application in low frequency regions, they have two or more coils on a magnetic core. In high frequency applications, transformers may be constructed with or without magnetic cores. This chapter will cover the important aspects of transformers and will explain how to analyze networks which use them.

14.1. PRINCIPLE OF TRANSFORMERS

A current flowing through a coil forms around it a magnetic field. If we place another coil sufficiently close so that the magnetic field reaches this second coil, then a voltage is induced in it. Such an arrangement is the simplest form of a transformer.

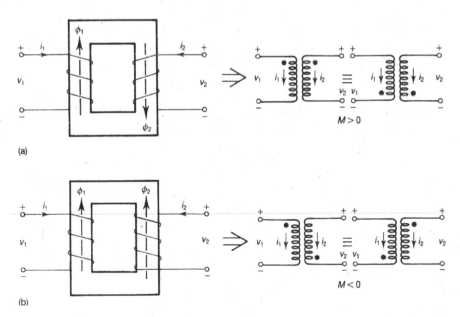

FIGURE 14.1.1 Transformer and symbols: (a) Fluxes in the same direction. (b) Fluxes in opposite directions.

The transformer usually has a magnetic core which concentrates the magnetic field for better coupling of the coils. Although technical construction may differ from case to case, two simple examples of transformers are shown in Fig. 14.1.1. One of the coils, for instance the left one, may be called the primary winding, the other is then the secondary winding. Assume that we let a current flow through the left coil as indicated, for instance by connecting to it an independent current source. The magnetic flux which is created by the current is determined by the right hand rule:

> Open the palm of your *right* hand, with the thumb pointing sidewise. Place the palm on the coil so that the fingers point in the direction of the current; your thumb indicates the direction of the magnetic flux through the core.

In both cases drawn in Fig. 14.1.1, the flux of the left coil, Φ_1, points upwards and goes clockwise around the core. If we move the current source to the right side so that the current flows into the upper terminal, then the right hand rule tells us that the magnetic flux of the right coil, Φ_2, will also point clockwise in Fig. 14.1.1a, in *the same* direction as the flux of the first coil. A different situation is in Fig. 14.1.1b, where the right hand rule indicates that the two fluxes act in opposite directions. We will return to these examples later.

In the case of a single inductor, the voltage across it is given by the derivative of the flux with respect to time. In the case of a transformer, the situation is more complicated; the fluxes are

$$\Phi_1 = \Phi_{11}(i_1) + \Phi_{12}(i_2)$$
$$\Phi_2 = \Phi_{21}(i_1) + \Phi_{22}(i_2)$$

$$(14.1.1)$$

Differentiation with respect to time gives

$$v_1 = \frac{d\Phi_1}{dt} = \frac{\partial\Phi_{11}}{\partial i_1}\frac{di_1}{dt} + \frac{\partial\Phi_{12}}{\partial i_2}\frac{di_2}{dt}$$

$$v_2 = \frac{d\Phi_2}{dt} = \frac{\partial\Phi_{21}}{\partial i_1}\frac{di_1}{dt} + \frac{\partial\Phi_{22}}{\partial i_2}\frac{di_2}{dt}$$

These expressions are still complicated but if we make the assumption that the partial derivatives are constants (the transformer is a linear device), then we can write

$$v_1 = L_{11}\frac{di_1}{dt} + L_{12}\frac{di_2}{dt}$$

$$(14.1.2)$$

$$v_2 = L_{21}\frac{di_1}{dt} + L_{22}\frac{di_2}{dt}$$

It is always true in all transformers that $L_{12} = L_{21} = M$ and we can simplify our notation by dropping the second subscript:

$$v_1 = L_1\frac{di_1}{dt} + M\frac{di_2}{dt}$$

$$(14.1.3)$$

$$v_2 = M\frac{di_1}{dt} + L_2\frac{di_2}{dt}$$

We could now take advantage of our knowledge of the Laplace transform, use

$$L[\frac{di_1}{dt}] = sI_1 - J_{10}$$

$$L[\frac{di_2}{dt}] = sI_2 - J_{20}$$

(14.1.4)

and take the Laplace transform of (14.1.3):

$$V_1 = sL_1 I_1 + sMI_2 - L_1 J_{10} - MJ_{20}$$

$$V_2 = sMI_1 + sL_2 I_2 - MJ_{10} - L_2 J_{20}$$

This would prepare our expressions for a time domain solution with initial conditions.

If the initial currents are zero, then

$$V_1 = sL_1 I_1 + sMI_2$$

$$V_2 = sMI_1 + sL_2 I_2$$

(14.1.5)

Here L_1 is the *primary inductance*, L_2 the *secondary inductance* and M is the *mutual inductance*. We can also write a matrix equation

$$\begin{bmatrix} V_1 \\ V_2 \end{bmatrix} = \begin{bmatrix} sL_1 & sM \\ sM & sL_2 \end{bmatrix} \begin{bmatrix} I_1 \\ I_2 \end{bmatrix}$$

(14.1.6)

We now return to Fig. 14.1.1. If we reverse the direction of the current flowing through the secondary coils, the direction of the secondary flux will change. We need some simple method to draw a transformer and simultaneously clearly indicate how the winding of the coil influences the flux. This is done with the dot convention. We draw the coils as simple inductors, as shown on the right of Fig. 14.1.1, and indicate the direction of the fluxes by two dots and by two currents flowing through the coils. If the directions of currents through the coils are such that they *both* point either away from the dots or to the dots, then the fluxes in the core flow in the same direction and mutual inductance, M, has a positive sign. If one of the currents points to the dot and the other away from the dot, as shown in Fig. 14.1.1b, then the mutual inductance, M, is negative.

The dot convention is useful only for two coupled coils but transformers may have several coils. An example with four coils is given in Fig. 14.1.2. Using the right hand rule we easily find the signs of the mutual inductances; they are written in the figure. The equations will be

$$V_1 = sL_{11} I_1 + sL_{12} I_2 - sL_{13} I_3 - sL_{14} I_4$$

$$V_2 = sL_{21} I_1 + sL_{22} I_3 - sL_{23} I_3 - sL_{24} I_4$$

(14.1.7)

$$V_3 = - sL_{31} I_1 - sL_{32} I_2 + sL_{33} I_3 + sL_{34} I_4$$

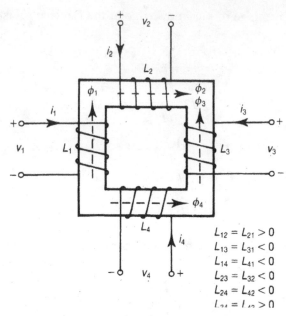

$$L_{12} = L_{21} > 0$$
$$L_{13} = L_{31} < 0$$
$$L_{14} = L_{41} < 0$$
$$L_{23} = L_{32} < 0$$
$$L_{24} = L_{42} < 0$$
$$L_{34} = L_{43} > 0$$

FIGURE 14.1.2 Transformer with four coils.

$$V_4 = -sL_{41} I_1 - sL_{42} I_2 + sL_{43} I_3 + sL_{44} I_4$$

or in matrix form

$$
\begin{bmatrix} V_1 \\ V_2 \\ V_3 \\ V_4 \end{bmatrix} =
\begin{bmatrix}
+sL_{11} & +sL_{12} & -sL_{13} & -sL_{14} \\
+sL_{21} & +sL_{22} & -sL_{23} & -sL_{24} \\
-sL_{31} & -sL_{32} & +sL_{33} & +sL_{34} \\
-sL_{41} & -sL_{42} & +sL_{43} & +sL_{44}
\end{bmatrix}
\begin{bmatrix} I_1 \\ I_2 \\ I_3 \\ I_4 \end{bmatrix}
\tag{14.1.8}
$$

We must always keep in mind that $L_{ij} = L_{ji}$. Another mathematical way to indicate that we have a matrix equation is to use bold letters

$$\mathbf{V} = \mathbf{L}\,\mathbf{I} \tag{14.1.9}$$

If we invert the matrix,

$$\mathbf{M} = \mathbf{L}^{-1} \tag{14.1.10}$$

we get the admittance form of the equations

$$\mathbf{I} = \mathbf{M}\,\mathbf{V} \tag{14.1.11}$$

The above examples have shown that it is easy to write the equations if we know directions of the windings and directions of the currents. In the following we will restrict our explanations to transformers having only two coils.

There still remains the question of how to find the value of the mutual inductance. This can be done experimentally. For the transformer in Fig. 14.1.1, attach a current source to the primary winding and open circuit the secondary winding, thus making $I_2 = 0$. If the source frequency is $s = j\omega$, then equations (14.1.5) reduce to

$$V_1 = j L_1 I_1$$

$$V_2 = j M I_1$$

Measurement of V_2 gives us the possibility of calculating M. The same steps can be applied to equations (14.1.7): attach a similar source to one of the windings and open circuit all the others. The equations simplify and from the measurement we calculate some of the mutual inductances. In the next step, transfer the source to another winding and do the same. Repeat until all mutual inductances have been found. We can reduce the number of such measurements by realizing that $L_{ij} = L_{ji}$.

14.2. COUPLING COEFFICIENT

When we apply a current to the first coil only, a voltage is induced in the other coil; if we do not attach any load, the current I_2 will be zero. Equations (14.1.5) simplify to

$$V_1 = sL_1 I_1$$

$$V_2 = \pm sMI_1$$

and their ratio is

$$A = \frac{V_1}{V_2}\bigg|_{I_2=0} = \frac{L_1}{\pm M} \tag{14.2.1}$$

Doing the same from the other side of the transformer without loading the first coil we get

$$V_1 = \pm sMI_2$$

$$V_2 = sL_2 I_2$$

and

$$B = \frac{V_1}{V_2}\bigg|_{I_1=0} = \pm \frac{M}{L_2} \tag{14.2.2}$$

The ratio of (14.2.2) and (14.2.1) is

$$\frac{B}{A} = \frac{M^2}{L_1 L_2} = k^2$$

The constant k is called the *coupling coefficient* and is always positive, regardless of which sign we took with the mutual inductance.

Using the concept of stored energy we will now establish that the coefficient of coupling always satisfies the inequality

$$0 \le k \le 1 \qquad (14.2.3)$$

The power delivered by a source connected to the first coil is

$$p_1 = v_1 i_1 = L_1 \frac{d i_1}{d t} i_1 \pm M \frac{d i_2}{d t} i_1$$

Power delivered by the source at the other coil is

$$p_2 = v_2 i_2 = \pm M \frac{d i_1}{d t} i_2 + L_2 \frac{d i_2}{d t} i_2$$

and the total power is

$$p = p_1 + p_2 = L_1 i_1 \frac{d i_1}{d t} \pm M \left(i_1 \frac{d i_2}{d t} + i_2 \frac{d i_1}{d t} \right) + L_2 i_2 \frac{d i_2}{d t}$$

This can be rewritten in the form

$$p = \frac{d}{d t} \left(\frac{L_1 i_1^2}{2} \pm M i_1 i_2 + \frac{L_2 i_2^2}{2} \right)$$

An integral over certain time will provide the energy stored in the transformer. The energy cannot be negative,

$$w = \frac{L_1 I_1^2}{2} + \frac{L_2 I_2^2}{2} \pm M I_1 I_2 \ge 0 \qquad (14.2.4)$$

We can modify (14.2.4) by simultaneously adding and subtracting the value $\sqrt{L_1 L_2 I_1 I_2}$:

$$w = \frac{1}{2} \left(\sqrt{L_1 I_1} - \sqrt{L_2 I_2} \right)^2 + \left(\sqrt{L_1 L_2} \pm M \right) I_1 I_2 \ge 0$$

The first term is always positive, but can be made zero, for instance by selecting $L_1 = L_2$ and $I_1 = I_2$. If we select a positive sign of M, the second term will be always positive. However, if we select a negative sign for M, then the limiting case will be

$$\sqrt{L_1 L_2} - M \ge 0$$

which is the same as writing

$$k = \frac{M}{\sqrt{L_1 L_2}}, \qquad 0 \le k \le 1. \qquad (14.2.5)$$

14.3. PERFECT AND IDEAL TRANSFORMERS

The name *perfect transformer* is given to a transformer with a coupling coefficient $k = 1$. In such case

$$L_1 L_2 = M^2 \tag{14.3.1}$$

Fig. 14.3.1 shows a transformer with a loading resistor. If the resistor is *not* connected, then $I_2 = 0$,

$$V_1 = sL_1 I_1$$

$$V_2 = sMI_1$$

and the voltage transfer is

$$\frac{V_2}{V_1} = \frac{M}{L_1} = \left[\frac{L_2}{L_1}\right]^{1/2} = \frac{1}{n} \tag{14.3.2}$$

For a perfect (or nearly perfect) transformer, the ratio $\frac{1}{n}$ is equal to the ratio of the turns of the primary and secondary coils. In some cases, this is even written as $\frac{n_1}{n_2}$. The choice of the dots indicates that the transformer equations are

$$V_1 = sL_1 I_1 + sMI_2$$

$$V_2 = sMI_1 + sL_2 I_2 \tag{14.3.3}$$

If we attach the loading resistor and select the loop currents, called now I_a and I_b, to circulate in clockwise directions, then

$$I_a = I_1$$

$$I_b = -I_2 \tag{14.3.4}$$

Substituting into the transformer equations (14.3.3)

$$V_1 = sL_1 I_a - sMI_b$$

$$V_2 = sMI_a - sL_2 I_b \tag{14.3.5}$$

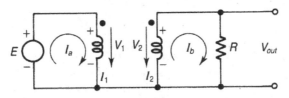

FIGURE 14.3.1 Transformer with a loading resistor.

We see from the figure that

$$V_1 = E$$

$$-V_2 + RI_b = 0$$

(14.3.6)

Combining with (14.3.5)

$$\begin{bmatrix} sL_1 & -sM \\ -sM & sL_2 + R \end{bmatrix} \begin{bmatrix} I_a \\ I_b \end{bmatrix} = \begin{bmatrix} E \\ 0 \end{bmatrix}$$

The determinant is

$$D = s^2(L_1L_2 - M^2) + sL_1R$$

and substituting condition (14.3.1)

$$D = sL_1R$$

(14.3.7)

The currents are

$$I_a = \frac{sL_2 + R}{sL_1R}E$$

$$I_b = \frac{sM}{sL_1R}E = \left[\frac{L_2}{L_1}\right]^{1/2}\frac{1}{R}E = \frac{1}{n}\frac{1}{R}E$$

(14.3.8)

The output voltage is

$$V_{out} = I_bR = \frac{1}{n}E,$$

(14.3.9)

the input impedance is

$$\frac{E}{I_a} = Z_{in} = \frac{sL_1R}{sL_2 + R} = \frac{sL_1R}{sL_2(1 + \frac{R}{sL_2})} = n^2\frac{1}{\frac{1}{R} + \frac{1}{sL_2}},$$

(14.3.10)

and the current ratio is

$$\frac{I_b}{I_a} = \frac{sM}{sL_2(1 + \frac{R}{sL_2})} = \left[\frac{L_1}{L_2}\right]^{1/2}\frac{1}{1 + \frac{R}{sL_2}} = n\frac{1}{1 + \frac{R}{sL_2}}$$

(14.3.11)

The voltage ratio is independent of the load and the input impedance is represented by parallel connection of the resistor and the secondary inductor, multiplied by n^2. If both inductances are large, we can simplify (14.3.10) and (14.3.11) to

$$Z_{in} = n^2 R,$$ (14.3.12)

$$I_b = nI_a$$ (14.3.13)

and still have

$$V_2 = \frac{1}{n} V_1$$ (14.3.14)

The product $V_2 I_b = V_1 I_a$ indicates that this simplified device does not consume any power. If we return back to currents I_1 and I_2 by using (14.3.4), we get

$$V_1 = nV_2$$
$$I_1 = -\frac{1}{n} I_2$$ (14.3.15)

These equations define an *ideal transformer*. Its symbol is in Fig. 14.3.2.

Ideal transformers have been used extensively in network theory but they cannot be realized technically. Their equations indicate that they transform the dc current, something we cannot achieve with coils.

Technical transformers which are used in power transfer are always realized with as high a coefficient of coupling as possible. Properties of such transformers are very similar to those of a perfect transformer, equations (14.3.9) to (14.3.11). If we need only approximate answers, we could use equations (14.3.12) to (14.3.14), but we must keep in mind that they do not apply for dc.

14.4.   NETWORKS WITH TRANSFORMERS

In this section we give a method to analyze networks with transformers. We assume that the current directions and winding directions have already been determined and that the networks have the necessary dots at the proper positions.

The first network is in Fig. 14.4.1; I_1, I_2, V_1 and V_2 refer to equations (14.1.5), repeated here for convenience:

$$V_1 = sL_1 I_1 + sMI_2$$
$$V_2 = sMI_1 + sL_2 I_2$$ (14.4.1)

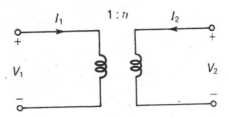

FIGURE 14.3.2   Ideal transformer.

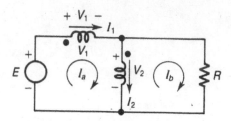

FIGURE 14.4.1   Network with a transformer.

Since we have chosen both currents I_1 and I_2 to point in the direction away from the dots, the mutual inductance in (14.4.1) has positive sign.

The loop currents are denoted by I_a and I_b and we always select clockwise direction for them. The two sets of currents are related:

$$I_1 = I_a$$

$$I_2 = I_a - I_b$$

(14.4.2)

Substituting (14.4.2) into (14.4.1) we obtain

$$V_1 = sL_1 I_a + sM(I_a - I_b)$$

$$V_2 = sMI_a + sL_2(I_a - I_b)$$

(14.4.3)

We are now ready to write the loop equations:

$$V_1 + V_2 = E$$

$$RI_b - V_2 = 0$$

(14.4.4)

Substituting for V_1 and V_2 from (14.4.3) we get

$$(sL_1 + 2sM + sL_2)I_a - (sL_2 + sM)I_b = E$$

$$- (sL_2 + sM)I_b + (sL_2 + R)I_b = 0$$

As always, it is convenient to put these equations into matrix form

$$\begin{bmatrix} s(L_1 + L_2 + 2M) & -s(L_2 + M) \\ -s(L_2 + M) & sL_2 + R \end{bmatrix} \begin{bmatrix} I_a \\ I_b \end{bmatrix} = \begin{bmatrix} E \\ 0 \end{bmatrix}$$

and solution can proceed along established lines.

Example 1.

Analyze the network in Fig. 14.4.2.

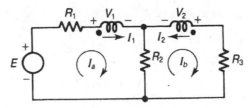

FIGURE 14.4.2 Network with a transformer.

The currents are

$$I_1 = I_a \tag{14.4.6}$$

$$I_2 = -I_b$$

Inserting into (14.4.1) results in

$$V_1 = sL_1 I_a - sMI_b \tag{14.4.7}$$

$$V_2 = sMI_a - sL_2 I_b$$

The loop equations are

$$R_1 I_a + V_1 + R_2(I_a - I_b) = E \tag{14.4.8}$$

$$R_2(I_b - I_a) - V_2 + R_3 I_b = 0$$

Substitution of V_1 and V_2 from (14.4.7) gives the final set, written here directly in matrix form:

$$\begin{bmatrix} R_1 + R_2 + sL_1 & -R_2 - sM \\ -R_2 - sM & R_1 + R_2 + sL_2 \end{bmatrix} \begin{bmatrix} I_a \\ I_b \end{bmatrix} = \begin{bmatrix} E \\ 0 \end{bmatrix}$$

Example 2.

Analyze the network in Figure 14.4.3.

The currents are

$$I_1 = I_a - I_b$$

$$I_2 = I_b$$

When substituted into the transformer equations, we get

$$V_1 = sL_1(I_a - I_b) + sMI_b \tag{14.4.9}$$

$$V_2 = sM(I_a - I_b) + sL_2 I_b$$

The loop equations for the network are

$$R_1 I_a + V_1 = E \tag{14.4.10}$$

$$-V_1 + V_2 + R_2 I_b = 0$$

When we now substitute V_1 and V_2 from (14.4.9), we get a system of equations which leads to the following matrix form

$$\begin{bmatrix} R_1 + sL_1 & s(M - L_1) \\ s(M - L_1) & s(L_1 + L_2 - 2M) + R_2 \end{bmatrix} \begin{bmatrix} I_a \\ I_b \end{bmatrix} = \begin{bmatrix} E \\ 0 \end{bmatrix}$$

Example 3.

Analyze the network in Fig. 14.4.4.

The currents are

$$I_1 = I_c - I_a \tag{14.4.11}$$

$$I_2 = I_b - I_c$$

When substituted into the transformer equations, we get

$$V_1 = sL_1(I_c - I_a) + sM(I_b - I_a)$$

$$V_2 = sM(I_c - I_a) + sL_2(I_b - I_c) \tag{14.4.12}$$

The loop equations for this network are

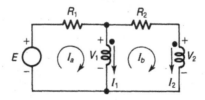

FIGURE 14.4.3   Network with a transformer.

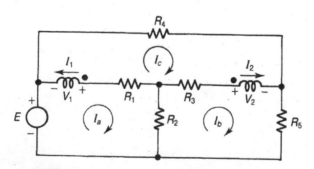

FIGURE 14.4.4   Network with three meshes.

$$-V_1 + R_1(I_a - I_c) + R_2(I_a - I_b) = E$$

$$R_2(I_b - I_a) + R_3(I_b - I_c) + V_2 + R_5 I_b = 0 \qquad (14.4.13)$$

$$R_4 I_c + R_3(I_c - I_b) + R_1(I_c - I_a) + V_1 - V_2 = 0$$

When we substitute from (14.4.13), we get fairly lengthy expressions; they can be simplified and written in matrix form. Because of the size, we give only the system matrix

$$\begin{bmatrix} R_1 + R_2 + sL_1 & -R_2 - sM & sM - sL_1 - R_1 \\ -R_2 - sM & R_2 + R_3 + R_5 + sL_2 & sM - sL_2 - R_3 \\ sM - sL_1 - R_1 & sM - sL_2 - R_3 & R_1 + R_3 + R_4 + s\left(L_1 + L_2 - 2M\right) \end{bmatrix}$$

The examples have given a reliable method how to handle networks with transformers. We write a bit more than absolutely necessary, but we avoid mistakes which usually happen when we try to do too many steps simultaneously.

14.5. EQUIVALENT TRANSFORMER NETWORKS

The transformer matrix equation

$$\begin{bmatrix} V_1 \\ V_1 \end{bmatrix} = \begin{bmatrix} sL_1 & sM \\ sM & sL_2 \end{bmatrix} \begin{bmatrix} I_1 \\ I_2 \end{bmatrix} \qquad (14.5.1)$$

is actually an open-circuit two-port description, compare with equation (13.1.2). If we succeed in drawing another network which will be described by the same matrix equation, then the two networks will behave in the same way. If any of them is placed into a box which cannot be opened, no amount of measurements can determine which of the networks is inside.

Consider the network to the left of Fig. 14.5.1a and derive its twoport impedance equations. We get exactly the expression (14.5.1) and thus the network is equivalent to the transformer on the right. Another network, in Fig. 15.5.1b is equivalent to the transformer with negative M. Note that the two windings of the transformer are connected at the bottom of the drawings.

We can also invert the matrix equation above and get

$$\begin{bmatrix} I_1 \\ I_2 \end{bmatrix} = \begin{bmatrix} y_{11} & y_{12} \\ y_{21} & y_{22} \end{bmatrix} \begin{bmatrix} V_1 \\ V_2 \end{bmatrix} \qquad (14.5.2)$$

Where

$$y_{11} = \frac{L_2}{s(L_1 L_2 - M^2)}$$

$$y_{12} = y_{21} = \frac{-M}{s(L_1 L_2 - M^2)} \qquad (14.5.3)$$

$$y_{22} = \frac{L_1}{s(L_1 L_2 - M^2)}$$

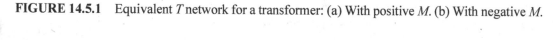

FIGURE 14.5.1   Equivalent T network for a transformer: (a) With positive M. (b) With negative M.

We try the network in Fig. 14.5.2 as an equivalent. Its admittance twoport parameters are

$$y_{11} = \frac{1}{sL_a} + \frac{1}{sL_c}$$

$$y_{12} = y_{21} = \frac{-1}{sL_c} \tag{14.5.4}$$

$$y_{22} = \frac{1}{sL_b} + \frac{1}{sL_c}$$

Comparing (14.5.4) with (14.5.3) we get

$$\frac{1}{sL_a} + \frac{1}{sL_c} = \frac{L_2}{s(L_1 L_2 - M^2)}$$

$$\frac{1}{sL_c} = \frac{M}{s(L_1 L_2 - M^2)}$$

$$\frac{1}{sL_b} + \frac{1}{sL_c} = \frac{L_1}{s(L_1 L_2 - M^2)}$$

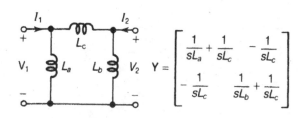

FIGURE 14.5.2   Network and its two-port Y matrix

and finally

$$L_a = \frac{L_1 L_2 - M^2}{L_2 - M}$$

$$L_b = \frac{L_1 L_2 - M^2}{L_1 - M}$$ (14.5.5)

$$L_c = \frac{L_1 L_2 - M^2}{M}$$

The equivalent network is drawn in Fig. 14.5.3a for positive M and in (b) for negative M. Again, note that the bottoms of the coils are connected.

Inversion of the matrix, used in the above steps, is not possible if the determinant $L_1 L_2 - M^2$ is zero. We showed in Section 14.3 that this condition describes a perfect transformer and we conclude that it is not possible to get the admittance form of two-port equations for a perfect transformer.

The above derivations can be applied to solve the following two examples. Fig. 14.5.4 shows a transformer with a series connection of its coils. We would like to know what the value of the resulting equivalent inductance is. We start with the transformer equations

$$V_1 = sL_1 I_1 + sMI_2$$

$$V_2 = sMI_1 + sL_2 I_2$$

and realize that $I_1 = I_2 = I$. Inserting into the equations:

$$V_1 = (sL_1 + sM)I$$

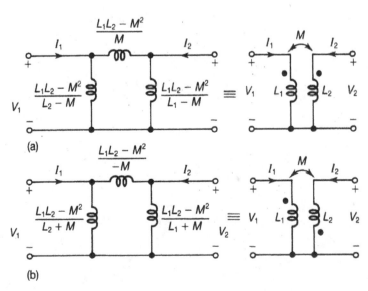

FIGURE 14.5.3 Equivalent Π network for a transformer: (a) With positive M. (b) With negative M.

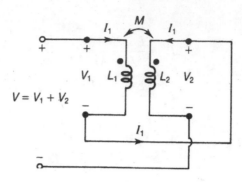

FIGURE 14.5.4   Transformer with series connection of coils.

$$V_2 = (sM + sL_2)I$$

The overall voltage is equal to $V = V_1 + V_2$. Adding the two equations we obtain

$$V = s(L_1 + L_2 + 2M)I$$

from which we conclude that the equivalent inductance is

$$L_{EQ} = L_1 + L_2 + 2M$$

connect the two coils in parallel, as shown in Fig. 14.5.5, then we must start with the admittance equations

$$I_1 = y_{11}V_1 + y_{12}V_2$$

$$I_2 = y_{21}V_1 + y_{22}V_2$$

Since both coils have the same voltage, $V = V_1 = V_2$, we substitute and get

$$I_1 = (y_{11} + y_{12})V$$

$$I_2 = (y_{21} + y_{22})V$$

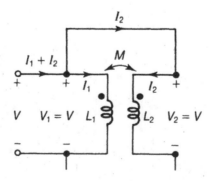

FIGURE 14.5.5   Transformer with parallel connection of coils.

The overall current is $I = I_1 + I_2$, as can be seen from the figure. We add the equations and obtain

$$I = (y_{11} + y_{12} + y_{21} + y_{22})V$$

The values were derived in (14.5.3). Inserting we get

$$I = \frac{L_1 + L_2 - 2M}{s(L_1 L_2 - M^2)} V$$

The equivalent input impedance is

$$Z_{in} = \frac{V}{I} = s\frac{L_1 L_2 - M^2}{L_1 + L_2 - 2M}$$

and the equivalent inductance is

$$L_{EQ} = \frac{L_1 L_2 - M^2}{L_1 + L_2 - 2M}$$

14.6. DOUBLE TUNED CIRCUITS

Double tuned circuits are formed by two parallel tuned circuits with a loose magnetic coupling of the coils. They are used in radios and have quite interesting properties. We give the derivations and results as a matter of interest

Consider the network in Fig. 14.6.1a. If we use the equivalent circuit of the transformer, we get the network shown in Fig. 14.6.1b. To simplify our work we select $C_1 = C_2 = C$, $L_1 = L_2 = L$ and $G_1 = G_2 = G$. We also substitute $L_e = L - M$ for easier writing. The nodal matrix equation is

$$\begin{bmatrix} sC + G + \dfrac{1}{sL_e} & -\dfrac{1}{sL_e} & 0 \\[3mm] -\dfrac{1}{sL_e} & \dfrac{2M + L_e}{sL_e M} & -\dfrac{1}{sL_e} \\[3mm] 0 & -\dfrac{1}{sL_e} & sC + G + \dfrac{1}{sL_e} \end{bmatrix} \begin{bmatrix} V_1 \\ V_2 \\ V_3 \end{bmatrix} = \begin{bmatrix} J(s) \\ 0 \\ 0 \end{bmatrix}$$

Because we have equal values for the elements, the determinant has terms which can be factored. After a few simple steps we first get

$$D = \frac{1}{s^3 L_e^3}(s^2 L_e C + s L_e G + 1)[(s^2 L_e C + s L_e G + 1)\frac{2M + L_e}{M} - 2]$$

and then

$$D = \frac{C^2}{s^3 L_e} \frac{2M + L_e}{M}(s^2 + s\frac{G}{C} + \frac{1}{L_e C})(s^2 + s\frac{G}{C} + \frac{1}{C(2M + L_e)})$$

It is now convenient to substitute back $L - M$ for L_e.

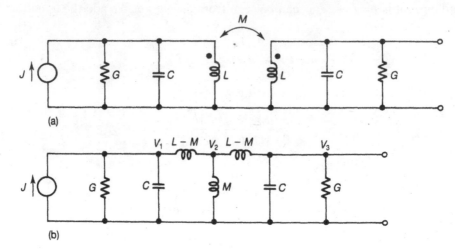

(a)

(b)

FIGURE 14.6.1 Double tuned circuit with magnetic coupling: (a) Original network. (b) Equivalent for analysis.

$$D = \frac{(L+M)C^2}{s^3 M(L-M)}(s^2 + s\frac{G}{C} + \frac{1}{C(L-M)})(s^2 + s\frac{G}{C} + \frac{1}{C(L+M)})$$

Since the numerator for V_3 is

$$N_3 = \frac{J}{s^2 L_e^2} = \frac{J}{s^2(L-M)^2}$$

the transfer impedance becomes

$$\frac{V_3}{J} = \frac{M}{(L^2 - M^2)C^2} \frac{s}{(s^2 + s\frac{G}{C} + \frac{1}{C(L-M)})(s^2 + s\frac{G}{C} + \frac{1}{C(L+M)})}$$

For the first quadratic term of the denominator we get

$$Q_1 = R\left[\frac{C}{L-M}\right]^{1/2}$$

$$\omega_1 = \frac{1}{[C(L-M)]^{1/2}}$$

and for the second term

$$Q_2 = R\left[\frac{C}{L+M}\right]^{1/2}$$

$$\omega_2 = \frac{1}{[C(L+M)]^{1/2}}$$

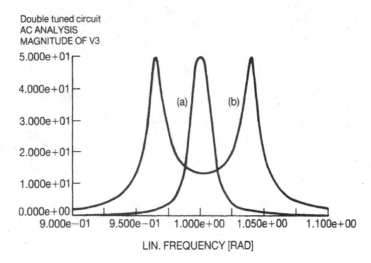

FIGURE 14.6.2 Amplitude responses for the network in Fig. 14.6.1. Element values are $C_1 = C_2 = 1$ F, $L_1 = L_2 = 1$ H, $G_1 = G_2 = 0.05$. Curve (a) is for M = 0.05 H. Curve (b) is for $M = 0.075$ H.

Linear networks can be scaled without losing any information. For our scaling we choose $C = 1\,F$ and $L = 1\,H$; this means each of the tuned circuits would separately resonate at $\omega = 1$. The transfer impedance simplifies to

$$\frac{V_3}{J} = \frac{M}{(1-M^2)} \frac{s}{(s^2 + sG + \frac{1}{1-M})(s^2 + sG + \frac{1}{1+M})}$$

where $0 < M < 1$. The transfer function has one zero at the origin and four poles. Resonant frequencies of the poles are above and below $\omega = 1$. If M is small, the two Q are approximately equal. If M grows, the two resonant frequencies pull apart and the Q will start to differ.

Amplitude response of this network is plotted in Fig. 14.6.2 for element values $C_1 = C_2 = 1\,F$, $L_1 = L_2 = 1\,H$ and $G_1 = G_2 = 0.05\,S$. Curve a is for $M = 0.05\,H$, curve b for $M = 0.075\,H$. If we select M between these two values, the peaks will get closer to each other.

PROBLEMS CHAPTER 14

P.14.1 Find the voltage transfer function using (a) the method of this chapter, (b) the equivalents in Fig. 14.5.1, (c) the equivalents in Fig. 14.5.3.

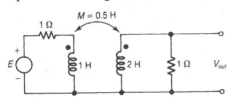

P.14 2 Do the same as in P.14.1.

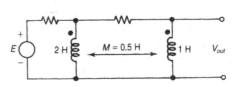

P.14.3   Do the same as in P.14.1.

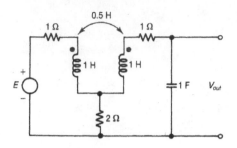

P.14.4   Use the transformation in Fig. 14.5.1 and the mesh method to find the voltage transfer function.

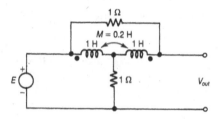

P.14.5   Find the voltage transfer function using the given element values.

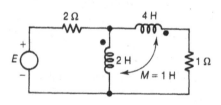

P.14.6   Do the same as in P.14.5.

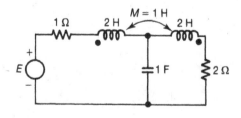

P.14.7   Do the same as in P.14.5.

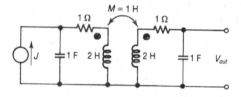

P.14.8   Use numerical values and the transformation in Fig. 14.5.1 to write the mesh equations. Find the transfer impedance.

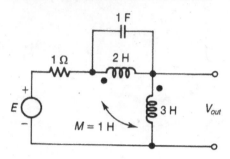

P.14.9   Calculate the expression for the output current. Use numerical values.

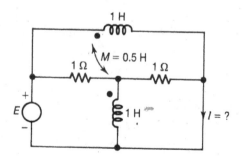

P.14.10   Find the voltage transfer function.

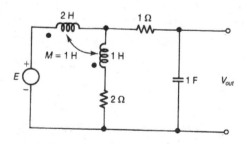

P.14.11   Find the transfer function.

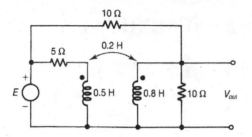

P.14.12   Find the transfer function.

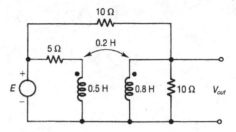

MODELING AND NUMERICAL METHODS

INTRODUCTION

Modeling of systems and devices, especially nonlinear devices like transistors, is an important area which requires considerable knowledge of physics, mathematics and other disciplines. A very small part of the modeling problem, modeling of linear devices, is suitable for this text and we will first derive linear models for amplifiers and transistors.

Afterwards we turn our attention to the consequences resulting from the presence of nonlinearities. Since the purpose of this chapter is to introduce concepts without going into too many details, we will only use two nonlinear devices, the bipolar diode and the field effect transistor. With these two elements we will show that writing equations for networks with nonlinear devices is not much different from what we have learned so far. Solutions of such equations must be done, in general, by iteration, and we will explain the Newton-Raphson method for the solution of systems of nonlinear equations. As a result, we will see that the steps are reduced to repeated solutions of linear systems written in matrix form. Linear elements appear in the matrix exactly the same way as they did in all our studies so far. Nonlinear elements also appear in the same positions as if the elements were linear, only their values are replaced by the values of their derivatives. In other words, everything we learned in linear network theory is re-used in nonlinear systems. In the last section we introduce the simplest formula for time domain calculations and show that this also uses material we explained earlier. Only one additional mathematical tool is necessary, the so called LU decomposition. It is a program available from many sources and we leave its explanation to the courses on numerical mathematics.

15.1. MODELING OF SOURCES

An ideal voltage source (independent or dependent) should, theoretically, deliver an arbitrary current and maintain the specified voltage, even if it is short-circuited. An ideal current source (independent or dependent), should deliver a given amount of current anywhere, theoretically even into an open circuit. Such sources do not exist in nature and the current or voltage are influenced by some internal property. We can model a nonideal voltage source as a combination of an ideal voltage source and an *internal resistance, R_S,* see Fig. 15.1.1. If we do not draw current by connecting some load at the output terminals, the voltage at these terminals is E. If we short circuit this model, then the current will be limited, $I = \dfrac{E}{R_S}$ and the power $W = \dfrac{E^2}{R_S}$ will be dissipated in the internal resistance.

There is also a limitation on the maximum power which a nonideal source can deliver into a load. To find the condition for maximum power transfer consider Fig. 15.1.2 with a variable load resistor R_L. The current flowing from the ideal voltage source is

$$I = \frac{E}{R_S + R_L},$$

the output voltage is

$$V = \frac{ER_L}{R_S + R_L},$$

and the power dissipated in the load is

$$W = E^2 \frac{R_L}{(R_S + R_L)^2}$$

We wish to find the condition under which the source with an internal resistance R_S delivers the maximum amount of power into the load resistor.

As is well know from mathematics, the maximum is found by first obtaining the derivative with respect to the variable and then setting the derivative equal to zero. Differentiating W with respect to R_L gets us the condition

$$\frac{dW}{dR_L} = E^2 \frac{R_S - R_L}{(R_S + R_L)^3} = 0$$

which is satisfied for

$$R_S = R_L \tag{15.1.1}$$

In this case the output voltage is

$$V = \frac{E}{2} \tag{15.1.2}$$

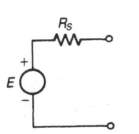

FIGURE 15.1.1   Nonideal voltage source.

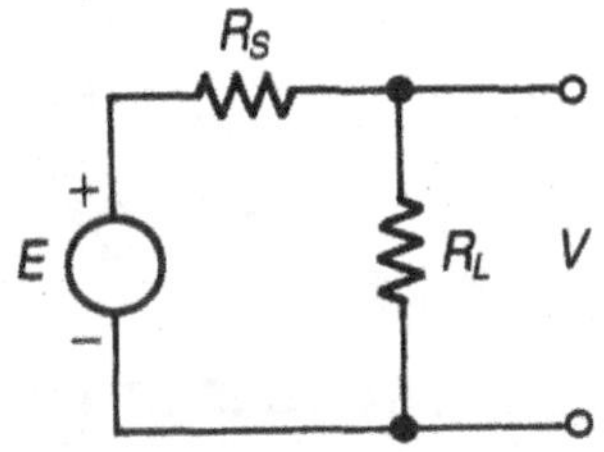

FIGURE 15.1.2   Nonideal voltage source loaded by R_L.

and the power dissipated in the load (as well as in the internal resistance) is

$$W_{max} = \frac{E^2}{4R_L} = \frac{E^2}{4R_S} \tag{15.1.3}$$

If the load is not a pure resistor but is, at some frequency, equal to

$$Z_L = R_L + jX_L$$

then it can be shown that the maximum power transfer occurs for complex source impedance

$$Z_S = R_S + jX_S$$

such that

$$R_S = R_L \tag{15.1.4}$$

$$X_S = -X_L \tag{15.1.5}$$

This situation applies when we are dealing with networks in steady state using phasors.

Ideal current sources are very convenient for theoretical considerations, especially when we are using nodal formulation, but there are certain peculiarities which we must be aware of. First of all, an ideal current source cannot be left on its own, unconnected to anything else, because, by definition, it must deliver a given current somewhere. For this reason, there should always be at least one element in parallel with the current source. A nonideal current source can be modeled by the network in Fig. 15.1.3. The Thevenin and Norton theorems show that this source can be equivalent to the voltage source in Fig. 15.1.1 *as far as the loading network is concerned.* Maximum power delivered into a load is subject to the same conditions as derived above. However, it is important to realize that the overall behavior of a nonideal current source model with a resistance in parallel differs from the model of the nonideal voltage source with a resistance in series. If the model in Fig. 15.1.3 is not connected to any other network, then the internal resistance is consuming power, which is not the case of the nonideal voltage source.

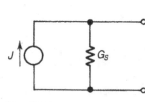

Fig. 15.1.3

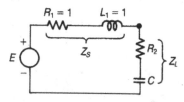

Fig. 15.1.4

FIGURE 15.1.3   Nonideal current source.

FIGURE 15.1.4   Load for maximum power transfer.

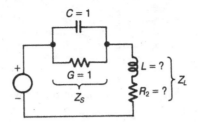

FIGURE 15.1.5   Load for maximum power transfer.

Example 1.

In the network Fig. 15.1.4 find such values R_2 and C to get maximum power transfer at the angular frequency $\omega = 2$.

The source impedance is $Z_S = R_1 + j\omega L = 1 + j2$. Impedance of the load is $Z_L = R_2 - j\dfrac{1}{2C}$. Condition for maximum power transfer will be $R_1 = R_2$ and $C = \dfrac{1}{4}$.

Example 2.

Find the values of R_2 and L to get maximum power transfer at $\omega = 2$ for the network in Fig. 15.1.3.

The source impedance is $Z_S = \dfrac{1}{1 + j2} = \dfrac{1 - j2}{5}$. The load impedance is $Z_L = R_2 + j2L$. Comparison of the real and imaginary parts gives $R_2 = \dfrac{1}{5}, L_2 = \dfrac{1}{5}$.

15.2.   LINEAR AMPLIFIERS

Ideal amplifiers with gain A are nothing but voltage controlled voltage sources. The usual symbol for a *differential* amplifier, already discussed in Chapter 12, is in Fig. 15.2.1.

In the case of an ideal amplifier the gain A is considered to be independent of frequency. This is not true in actual life. Due to various internal capacitances which are inherently present in any actual realization, the gain drops as we go to higher frequencies. Such a frequency dependence is usually simulated by the formula

$$A = \frac{A_0 \omega_0}{s + \omega_0} \tag{15.2.1}$$

Inserting $s = 0$ we see that A_0 is the gain of the amplifier at dc. Suppose that we now select signal frequency $s = j\omega_0$. In such case

$$A = \frac{A_0 \omega_0}{j\omega_0 + \omega_0} = \frac{A_0}{j + 1}$$

The absolute value becomes

$$|A| = 0.7071 A_0$$

and in decibels

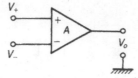

FIGURE 15.2.1   Differential amplifier with gain A.

$$\left|\frac{A}{A_0}\right| dB = 20\log 0.7071 \sim -3db$$

Thus ω_0 is a signal frequency (in radians per second) at which the gain of the amplifier drops 3 dB as compared to the gain at dc. The product in the numerator of (15.2.1) was given the name *gain-bandwidth product*,

$$GB_\omega = A_0\omega_0. \tag{15.2.2}$$

Specification sheets of manufacturers usually give the gain-bandwidth product in Hz,

$$GB_f = \frac{GB_\omega}{2\pi}. \tag{15.2.3}$$

The gain-bandwidth product is sometimes called the unity-gain frequency. The name becomes clear when we substitute $A_0 = 1$ into the definition (15.2.2). As a rule of thumb we will remember that commercial operational amplifiers may have a dc gain A_0 approximately equal to 100,000 and f_0 may be as low as $10Hz$.

Work with operational amplifiers is simpler if we define the inverted gain

$$B = -\frac{1}{A} \tag{15.2.4}$$

which modifies the mutual relationship of the voltages to

$$V_+ - V_- + BV_0 = 0 \tag{15.2.5}$$

If the OPAMP is ideal, with infinitely large gain, then $B = 0$ is substituted into this equation, making $V_+ = V_-$. If we must consider the frequency dependent properties of the amplifier, we substitute (15.2.1) into (15.2.4) and get

$$B = -\frac{s + \omega_0}{GB_\omega} \tag{15.2.6}$$

The expression can be simplified if we realize that just about all practical networks work at frequencies which are considerably higher than ω_0. In such case $s = j\omega$ is in absolute value much larger than ω_0 and we simplify the above expression to

$$B = -\frac{s}{GB_\omega} \tag{15.2.7}$$

This is a practical approximation of a linear nonideal operational amplifier.

If we wish to simulate the property (15.2.1) by a linear network, possibly taking into account an output resistance, we can use Fig. 15.2.2. It has two voltage controlled voltage sources. The input voltage, $V_{in} = V_+ - V_-$, controls the first VV with the gain μ_1. The voltage V_1 is obtained from the RC voltage divider,

$$V_1 = \frac{\mu_1 V_{in}}{CR(s + \frac{1}{CR})} = \frac{\omega_0 \mu_1 V_{in}}{s + \omega_0}$$

where we defined

$$\omega_0 = \frac{1}{RC}$$

The overall output voltage without loading is

$$V_{out} = \frac{\mu_1 \mu_2 \omega_0 V_{in}}{s + \omega_0}$$

If we now select

$$\mu_1 = 1$$
$$\mu_2 = A_0$$

then

$$\frac{V_{out}}{V_{in}} = \frac{A_0 \omega_0}{s + \omega_0},$$

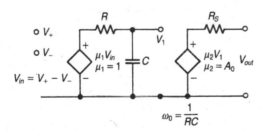

FIGURE 15.2.2 Equivalent network for amplifier with one pole.

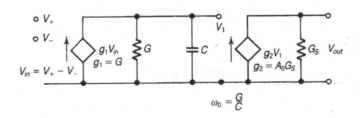

FIGURE 15.2.3 Equivalent network for amplifier with one pole.

exactly the expression above. The network in Fig. 15.2.2 is suitable for mesh formulation. For nodal formulation we can transform the controlled voltage sources into controlled current sources by the Thevenin-Norton transformation. The resulting equivalent network is in Fig. 15.2.3.

15.3. LINEAR TRANSISTOR MODELS

Transistors are widely used elements and a huge amount of literature deals with their modeling. They are nonlinear devices and we will consider their nonlinear models later in this chapter; here we derive their linear models from their most basic properties.

15.3.1. Field Effect Transistor

The symbol for a field effect transistor (FET) is in Fig. 15.3.1. It is a three terminal device with a current flowing between terminals D, drain, and S, source. The amount of the current is controlled by the voltage between the terminals G (gate) and S but no current is drawn at G. To simplify notation, it is common to denote $v_{GS} = v_G - v_S$ and $v_{DS} = v_D - v_S$. Since no current flows from the gate into the transistor, the properties are described by two general equations

$$i_G = 0$$
$$i_D = f(v_{GS}, v_{DS}) \tag{15.3.1}$$

If we keep the voltages constant, a constant current

$$i_{D,0} = f(v_{GS,0}, v_{DS,0})$$

will flow through the device. This is the operating point of the transistor. Increments about the operating point can be obtained by differentiation. Since i_D is a function of two voltages, we must use the mathematical formula for total differential

$$d i_D = \frac{\partial f(v_{GS}, v_{DS})}{\partial v_{GS}} d v_{GS} + \frac{\partial f(v_{GS}, v_{DS})}{\partial v_{DS}} d v_{DS} \tag{15.3.2}$$

The first partial derivative is the derivative of the current flowing from D to S with respect to the voltage between G and S. This indicates that the term is a transconductance of a voltage controlled current source; we denote it by g_m. The second partial derivative is with respect to the voltage between D and S and is thus a simple conductance, G_{DS}. We can rewrite (15.3.2) in terms of increments

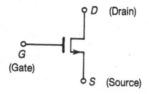

FIGURE 15.3.1   Field-effect transistor (FET) symbol.

$$I_D = g_m V_{GS} + G_{DS} V_{DS} \tag{15.3.3}$$

This equation corresponds to the network in Fig. 15.3.2. Since there are invariably some parasitic capacitances between the three terminals, a more complete linear model of the field effect transistor is in Fig. 15.3.3.

15.3.2. Bipolar Transistor

A bipolar transistor symbol is shown in Fig. 15.3.4. The terminals are given the names collector (C), base (B), and emitter (E). To simplify notation, we use $v_{BE} = v_B - v_E$ and $v_{CE} = v_C - v_E$. The device is more complicated than the field effect transistor since currents flow at all terminals; this can be expressed by two general equations

$$i_B = f_1(v_{BE}, v_{CE})$$
$$i_C = f_2(v_{BE}, v_{CE}) \tag{15.3.4}$$

Following the same steps as above we find total differentials

$$d i_B = \frac{\partial f_1(v_{BE}, v_{CE})}{\partial v_{BE}} d v_{BE} + \frac{\partial f_1(v_{BE}, v_{CE})}{\partial v_{CE}} d v_{CE}$$
$$d i_C = \frac{\partial f_2(v_{BE}, v_{CE})}{\partial v_{BE}} d v_{BE} + \frac{\partial f_2(v_{BE}, v_{CE})}{\partial v_{CE}} d v_{CE} \tag{15.3.5}$$

Looking at (15.3.5) we recognize that this is a two-port admittance equation. Using capital letters for the increments

$$I_B = y_{11} V_{BE} + y_{12} V_{CE}$$
$$I_C = y_{12} V_{BE} + y_{22} V_{CE} \tag{15.3.6}$$

These equations have the equivalent network in Fig. 15.3.5, but certain modifications can remove one of the dependent sources. Rewrite (15.3.6) by simultaneously adding and subtracting $y_{12} V_{BE}$ in the second equation,

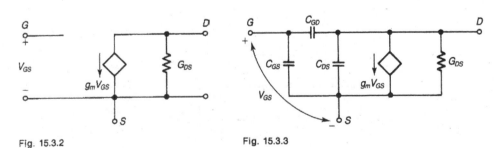

Fig. 15.3.2

Fig. 15.3.3

FIGURE 15.3.2 Small-signal equivalent for a FET.

FIGURE 15.3.3 Small-signal equivalent for a FET with capacitances.

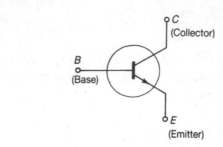

FIGURE 15.3.4   Bipolar transistor symbol.

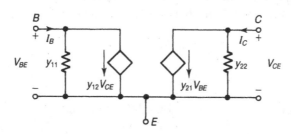

FIGURE 15.3.5   Two-port equivalent for a bipolar transistor.

$$I_B = y_{11}V_{BE} + y_{12}V_{CE}$$
$$I_C = y_{12}V_{BE} + y_{22}V_{CE} + (y_{12} - y_{12})V_{BE}$$

(15.3.7)

The term in the brackets represents a VC at the right hand port and the rest describes a reciprocal network with $y_{12} = y_{21}$. We can thus use the equivalent network shown in Fig. 15.3.6. Capacitances, which are invariably present, were also drawn into the equivalent network. More complicated models exist as well but we will not need them.

Fig. 15.3.7a shows a simple FET amplifier. On the left the voltage source delivers the signal, the output can be taken either at the point D or S, with respect to ground. In order to make the steps similar to our previous developments, we have marked the voltages by V_1 and V_2 and the resistors by G_1 and G_2. V_{DD} is commonly used to denote the dc voltage supply; it is not normally drawn in the figures, but we have done so for clarification. Every

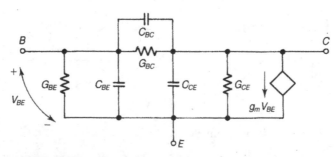

FIGURE 15.3.6   Small-signal network for a bipolar transistor.

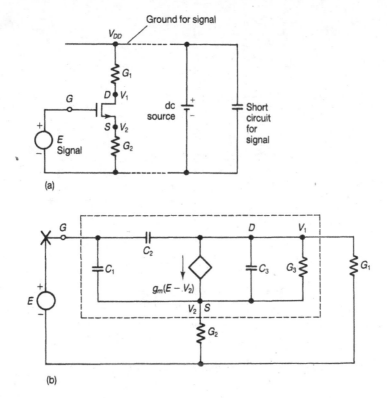

15.1
15.2
15.3
15.4
15.5
15.6
15.7
15.8

FIGURE 15.3.7 FET amplifier (a) Original network with a dc source. (b) Small-signal equivalent.

dc source is always put in parallel with a capacitor large enough to effectively ground the point V_{DD} for all nonzero signal frequencies. When we redraw the network for analysis, Fig. 15.3.7b, the resistor G_1 is connected to ground. We also copy the transistor model between the nodes marked by G, D and S, in this case taken from Fig. 15.3.3. The model is inside the dashed box. The network is prepared for nodal analysis and we can use all the steps we learned earlier. First, we take advantage of the fact that the signal voltage source is grounded, mark the node with a cross and use the method we learned in the study of active networks. We do not write the KCL equations at this node. Nodal equations for the other nodes are:

$$(G_1 + G_3 + sC_2 + sC_3)V_1 - (G_3 + sC_3)V_2 + g_m(E - V_2) - sC_2E = 0$$
$$-(G_3 + sC_3)V_1 + (G_2 + G_3 + sC_1 + sC_3)V_2 - g_m(E - V_2) - sC_1E = 0$$

Expressions multiplied by E are transferred to the right side of the equations, with the result

$$\begin{bmatrix} (G_1 + G_3 + sC_2 + sC_3) & -(G_3 + sC_3 + g_m) \\ -(G_3 + sC_3) & (G_2 + G_3 + sC_1 + sC_3 + g_m) \end{bmatrix} \begin{bmatrix} V_1 \\ V_2 \end{bmatrix} = \begin{bmatrix} (sC_2 - g_m)E \\ (sC_1 + g_m)E \end{bmatrix}$$

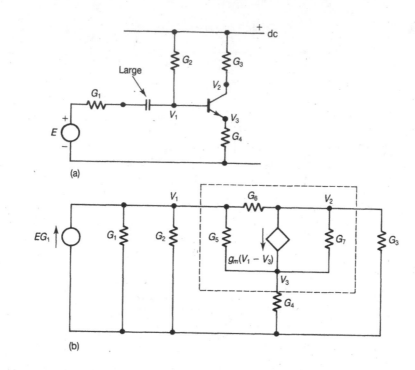

FIGURE 15.3.8   Bipolar amplifier: (a) Original network. (b) Small-signal equivalent.

In this example we used all the components of the FET model. At low frequencies we may be able to neglect the capacitors; this is achieved by setting their values equal to zero. The same applies for any missing resistors whose conductance values are also set equal to zero.

Example 1.

Draw the small signal equivalent circuit for the bipolar amplifier in Fig. 15.3.8a.

The only function of the capacitor is to separate the dc source from the signal source; it is large and acts as a short circuit for signal frequencies. For simplicity, neglect all capacitances in the transistor model.

In Fig. 15.3.8b, the voltage source with G_1 in series is transformed into a current source by the Thevenin transformation. Conductances G_2 and G_3 are grounded for signal frequencies. The model of the transistor, taken from Fig. 15.3.6, is in the dashed box and voltages are indicated at all nodes. Nodal formulation will lead to a 3×3 matrix equation.

15.4.   NONLINEAR ELEMENTS

Until now we assumed that the properties of all network elements did not change with changes in voltage, current or environment. Real life devices do not behave as nicely. Even a simple resistor changes its value with time: The current flowing through it influences its temperature and consequently also influences its resistance. Some kind of nonlinear behavior will exist in any element we can think of.

Nonlinear properties are sometimes undesirable but sometimes we need them. To name some useful applications of nonlinearities let us mention modulation, frequency multiplication or frequency division, and class C or E amplifiers. Applications where we need linearity are high fidelity amplifiers and all kinds of filters.

Although every element is to some extent nonlinear, proper choice of conditions can make it behave linearly. If the current through the resistor is small, the temperature and thus the resistance will not change noticeably. If a small signal is applied to a nonlinear device, it operates over a small portion of the device characteristic and we can approximate the nonlinearity by a tangent straight line. This is called *linearization* of the characteristic and linear models are sometimes called *small signal* models. The transistor models in Section 3 are such small signal models.

The moment we can consider the network as a linear one, all the methods explained in this book can be applied: time domain analysis, frequency domain analysis, poles and zeros, symbolic functions, the superposition principle and frequency domain sensitivities. In linearized networks we do not even consider the sources of dc power. All we need are the signal sources.

The most widely used nonlinear elements, transistors, need a source of dc power (a battery) to make them work. If this dc source has been applied, the network first adjusts the voltages at its various nodes according to the nonlinear properties of the devices. We say that the network reaches its *operating point*. Afterwards we can apply small signals and consider the network as linear, or apply large signals, and then take the nonlinear properties fully into account.

Nonlinearities of elements may be quite complicated but fundamental principles can be explained with the help of simple concepts. Let the current flowing through a nonlinear two-terminal element be a continuous and infinitely differentiable function of voltage,

$$i = f(v) \tag{15.4.1}$$

Such a function can be expanded into a Taylor series around $v = 0$

$$f(v) = f(0) + \frac{f^{(1)}(0)}{1!} v + \frac{f^{(2)}(0)}{2!} v^2 + \frac{f^{(3)}(0)}{3!} v^3 + ... \tag{15.4.2}$$

We could consider any number of such terms but for simplicity take only the quadratic term and let the multiplying constant be one,

$$i = v^2 \tag{15.4.3}$$

If the voltage is a composed signal

$$v = 1 + \alpha_1 \cos \omega_1 t + \alpha_2 \cos \omega_2 t \tag{15.4.4}$$

then the current flowing through such element will be the square of this expression,

$$i = 1 + \alpha_1^2 \cos^2 \omega_1 t + \alpha_2^2 \cos^2 \omega_2 t + 2\alpha_1 \cos \omega_1 t + 2\alpha_2 \cos \omega_2 t +$$
$$2\alpha_1\alpha_2 \cos \omega_2 t \cos \omega_2 t \cos \omega_2 t$$

Using Table 8.1.1 we can replace the second powers and the product of the two cosines and get, after a few steps,

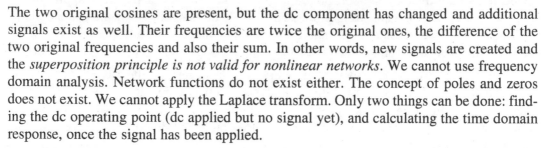

$$i(t) = 1 + \frac{\alpha_1^2}{2} + \frac{\alpha_2^2}{2} + 2\alpha_2 \cos\omega_1 t + 2\alpha_2 \cos\omega_2 t + \frac{\alpha_1^2}{2}\cos 2\omega_1 t$$

$$+ \frac{\alpha_2^2}{2}\cos 2\omega_2 t + \alpha_1\alpha_2\cos(\omega_1 - \omega_2)t + \alpha_1\alpha_2\cos(\omega_1 - \omega_2)t \qquad (15.4.5)$$

The two original cosines are present, but the dc component has changed and additional signals exist as well. Their frequencies are twice the original ones, the difference of the two original frequencies and also their sum. In other words, new signals are created and the *superposition principle is not valid for nonlinear networks*. We cannot use frequency domain analysis. Network functions do not exist either. The concept of poles and zeros does not exist. We cannot apply the Laplace transform. Only two things can be done: finding the dc operating point (dc applied but no signal yet), and calculating the time domain response, once the signal has been applied.

We will not go into time domain analysis of nonlinear networks because this requires considerable knowledge which is beyond the scope of this text. However, we will present a mathematical method to obtain the operating dc point. It gives us at least the possibility of using small signal analysis and thus applying all the theory covered in this book. To establish the principles we will use two nonlinear elements: a bipolar diode and a field effect transistor. We will not go into physical background or details.

15.4.1 Diode

A bipolar diode is shown in Fig. 15.4.1a. Its *dc* characteristic is expressed by the equation

$$i = I_s(e^{\frac{v_D}{V_T}} - 1) \qquad (15.4.6)$$

and is sketched in Fig. 15.4.1a. In (15.4.6) I_s is the *saturation* current, a constant which depends on the size of the diode, v_D is the voltage across the diode and V_T is a constant, approximately equal to 25 mV. In order to simplify the expressions we will insert the number for V_T and use

$$i = I_s(e^{40v_D} - 1). \qquad (15.4.7)$$

The current is large for $v_D > 0$ and very small, flowing in opposite direction, for $v_D < 0$. For any selected fixed voltage, $v_{D,0}$, the current will be fixed, i_0. If we increment $v_{D,0}$ by a small amount, Δv_D, the current will also increment by an amount Δi. Instead of increments we can use differentiation and obtain

$$\frac{di}{dv_D} = g(v_{D,0}) = 40I_s e^{40v_D} \qquad (15.4.8)$$

This is an incremental conductance of the diode at the operating point $v_{D,0}$. For small variations of the voltage and current around the operating point we will use upper case letters and

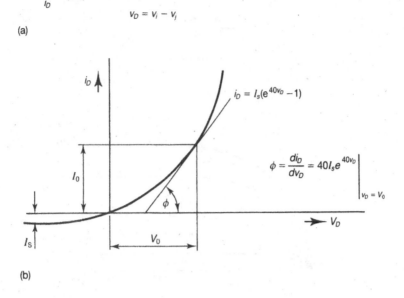

FIGURE 15.4.1   Bipolar diode: (a) Symbol. (b) Characteristic.

$$G = g(v_{D,0}) = \frac{1}{V} \qquad (15.4.9)$$

The value of G is the conductance we would use in small signal analysis. The characteristic of the diode and the way to determine G (the tangent) are shown in Fig. 15.4.1b.

15.4.2. Field Effect Transistor

The element symbol is Fig. 15.4.2a and a sketch of its characteristics is in (b). It is described by two sets of equations: In the triode region (see figure) the equations are

$$i_D = \beta[(v_{GS} - V_T)v_{DS} - \frac{1}{2}v_{DS}^2], \qquad (15.4.10)$$

valid for

$$\begin{aligned} v_{GS} &\geq V_T \\ v_{DS} &\leq v_{GS} - V_T \end{aligned} \qquad (15.4.11)$$

In the pinch-off, or saturation, region

$$i_D = \frac{1}{2}\beta(v_{GS} - V_T)^2, \qquad (15.4.12)$$

valid for

$$\begin{aligned} v_{GS} &\geq V_T \\ v_{DS} &\leq v_{GS} - V_T \end{aligned} \qquad (15.4.13)$$

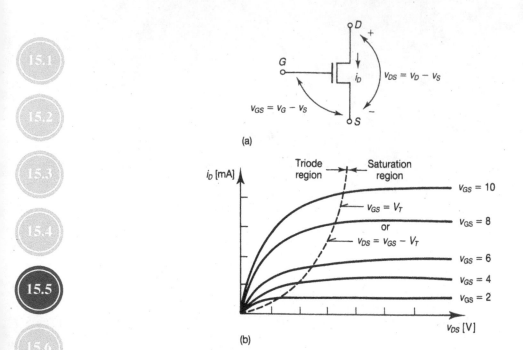

FIGURE 15.4.2   FET transistor: (a) Symbol with voltages. (b) Characteristics.

In the equations V_T is a constant which depends on technology. We will assume that $V_{T\_} = 2$. The boundary between the two regions is reached when we use the equal sign in the inequalities. Inserting $v_{DS} = v_{GS} - V_T$ into (15.4.10) results in

$$i_D = \beta[(v_{GS} - V_T)(v_{GS} - V_T) - \frac{1}{2}(v_{GS} - V_T)^2] = \frac{\beta}{2}(v_{GS} - V_T)^2$$

which is (15.4.12). There is thus smooth transition from one set of the equations to the other set.

The first question which comes into mind is how does one know which of these equations should be used? In simple networks, and with experience, one can make a reasonable guess. Without experience, or when one deals with a larger network, only one thing remains: take one set, solve the network and check whether the inequalities are satisfied. If they are not, take the other set and solve the equations again. Another way involves graphical solutions which use characteristics like those in Fig. 15.4.2b. The methods are reasonably simple for single transistor networks and are the subject of courses on electronics.

15.5. NETWORKS WITH NONLINEAR ELEMENTS

In this section we set up equations for several networks with nonlinear elements. First we consider networks with diodes, each of the diodes described by

$$i_D = I_s(e^{40v_D} - 1) \tag{15.5.1}$$

The first network is in Fig. 15.5.1; we identify

$$v_D = v_1 - v_2 \tag{15.5.2}$$

and write KCL for both nodes:

$$
\begin{aligned}
f_1 &= G_1 v_1 + i_D - J = 0 \\
f_2 &= -i_D + G_2 v_2 = 0
\end{aligned} \tag{15.5.3}
$$

The steps do not differ, in principle, from our previous ways of writing the equations. Nevertheless, there are two differences:

(a) nonlinear equations *can not* be put into a matrix form and *must* be left as separate equations.

(b) The independent sources, E or J, are kept on the left side because our solution method, to be explained in Section 6, will be based on this form.

The second network, Fig. 15.5.2, is slightly more complicated. For the diode we identify

$$v_D = v_1 - v_2 \tag{15.5.4}$$

The KCL are

$$
\begin{aligned}
f_1 &= (G_1 + G_2)v_1 - G_2 v_2 + i_D - J_1 = 0 \\
f_2 &= -G_2 v_1 + (G_2 + G_3)v_2 - i_D - J_2 = 0
\end{aligned} \tag{15.5.5}
$$

The third network, Fig. 15.5.3, has two voltage sources and three diodes. One terminal of the dc sources is grounded and we can write KCL similarly as we have done in active networks by marking them with crosses and not taking them into KCL. The nodal voltages are indicated in the figure and we identify

$$
\begin{aligned}
v_{D1} &= v_1 \\
v_{D2} &= v_1 - v_3 \\
v_{D3} &= v_3 - v_2
\end{aligned} \tag{15.5.6}
$$

The three nodal equations are:

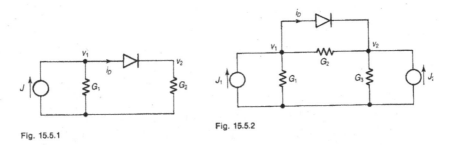

Fig. 15.5.1

Fig. 15.5.2

FIGURE 15.5.1   Network with one diode.

FIGURE 15.5.2   Network with one diode and two current sources.

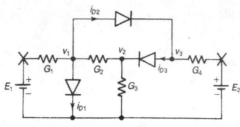

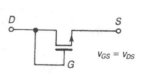

Fig. 15.5.3

Fig. 15.5.4

FIGURE 15.5.3   Network with three diodes.

FIGURE 15.5.4   FET connected as a diode.

$$f_1 = (G_1 + G_2)v_1 - G_2 v_2 - E_1 G_1 + i_{D1} + i_{D2} = 0$$
$$f_2 = -G_2 v_1 + (G_2 + G_3)v_2 - i_{D3} = 0$$
$$f_3 = G_4 v_3 - G_4 E_2 - i_{D2} + i_{D3} = 0$$

(15.5.7)

A transistor can also be connected as a diode, as shown in Fig. 15.5.4. The connection makes the voltage between gate and source, v_{GS}, qual to the voltage between drain and source, v_{DS} and thus (15.4.12) simplifies to

$$i_D = \frac{\beta}{2}(v_{DS} - V_T)^2$$

(15.5.8)

Considering the network in Fig. 15.5.5 we first identify

$$v_{DS} = v_1 - v_2$$

(15.5.9)

and then the KCL equations are

$$f_1 = 4v_1 - 5 - 2v_2 + i_D = 0$$
$$f_2 = -2v_1 + 3v_2 - i_D = 0$$

In this case we will substitute for i_D. To simplify writing, we set $V_T = 2$ and $\frac{\beta}{2} = 1$. This makes $i_D = (v_1 - v_2 - 2)^2$ and the equations reduce to

$$f_1 = v_1^2 + v_2^2 + 2v_2 - 2v_1 v_2 - 1 = 0$$
$$f_2 = -v_1^2 - v_2^2 + 2v_1 - v_2 + 2v_1 v_2 - 4 = 0$$

(15.5.10)

Although the FET transistor is described by a quadratic function, even in this simple case we cannot find the unknown voltages by algebraic methods. We will derive a numerical method in the next section, but first let us consider a network where the transistor acts as an amplifier. We are asked to set up the equations for the amplifier in Fig. 15.5.6. We use (15.4.12)

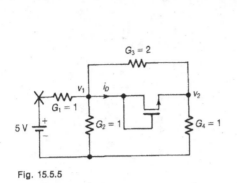

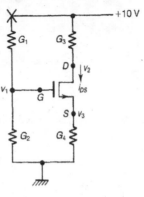

Fig. 15.5.5

Fig. 15.5.6

FIGURE 15.5.5 Network with FET connected as a diode.

FIGURE 15.5.6 FET amplifier.

$$i_D = \frac{\beta}{2}(v_{GS} - V_T)^2,$$

and identify

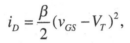

$$v_{GS} = v_1 - v_3.$$

The three KCL equations are:

$$f_1 = (G_1 + G_2)v_1 - 10G_1 = 0$$
$$f_2 = G_3 V_2 - 10G_3 + i_{DS} = 0$$
$$f_3 = G_4 v_3 - i_{DS} = 0$$

We could substitute for i_{DS} but it would not make the result simpler and we leave the set in the above form.

15.6.   NEWTON-RAPHSON ITERATIVE METHOD

We will derive a general method which can be used to solve systems of n nonlinear equations in n unknowns, but as an introduction we start with just one nonlinear equation with one variable,

$$f(x) = 0 \qquad\qquad (15.6.1)$$

We know that at a point x we can expand such function into a Taylor series about a point which is Δx away from x. The expansion is

$$f(x + \Delta x) = f(x) + \frac{f^{(1)}(x)}{1!}\Delta x + \frac{f^{(2)}(x)}{2!}(\Delta x)^2 + ...$$

Usually we terminate the expansion after the second term and get the condition

$$f(x) + f^{(1)}(x)\Delta x = 0$$

or

$$f^{(1)}(x)\Delta x = -f(x)$$

In the case of one equation we obtain the unknown Δx by division,

$$\Delta x = -\frac{f(x)}{f^{(1)}(x)}$$

Usually we indicate the starting point and the starting increment with a superscript zero, x^0 and Δx^0; then the new x is given a superscript one,

$$x^1 = x^0 + \Delta x^0$$

The process is repeated with the new value, x^1. We can thus write an iterative scheme

$$\Delta x^k = -\frac{f(x^k)}{f^{(1)}(x)^k}$$

$$x^{k+1} = x^k + \Delta x^k \qquad (15.6.2)$$

In many books, the first expression above is inserted into the second to obtain

$$x^{k+1} = x^k - \frac{f(x^k)}{f^{(1)}(x^k)}$$

but it is not advantageous to do so. The iterations on (15.6.2) have the advantage that we can see the sequence of Δx^k, which must be decreasing.

As an example we take the simple case

$$f(x) = x^2 - 1 = 0$$

which clearly has the solution $x = \pm 1$. The derivative is $f^{(1)} = 2x$ and $\Delta x = \dfrac{1 - x^2}{2x}$. To start iteration by means of (15.6.2) we need an initial point. If we have an idea what the root could be, then we should use it as an initial value. If we know nothing, then we simply select some arbitrary number. We will start with $x^0 = 2$; the sequence of iterations is shown in Table 15.6.1.

Solution of *systems* of nonlinear equations proceeds similarly; we will derive the method by using a system of two equations:

$$f_1(x_1, x_2) = 0$$
$$f_2(x_1, x_2) = 0 \qquad (15.6.3)$$

Each of these equations can expanded into a Taylor series about points $x_1 + \Delta x_1$ and $x_2 + \Delta x_2$, with the assumption that numerical values are available for x_1 and x_2.

Table 15.6.1

Iterations on $f(x) = x^2 - 1 = 0$

k	x^k	Δx^k
0	2	-0.75
1	1.25	-0.225
2	1.025	-0.0246951
3	1.003049	-0.0003048
4	1.000000	-4.598E-8

$$f_1(x_1 + \Delta x_1, x_2 + \Delta x_2) = f_1(x_1, x_2) + \frac{\partial f_1}{\partial x_1} \Delta x_1 + \frac{\partial f_1}{\partial x_2} \Delta x_2 + higher\ terms = 0$$

$$f_2(x_1 + \Delta x_1, x_2 + \Delta x_2) = f_2(x_1, x_2) + \frac{\partial f_2}{\partial x_1} \Delta x_1 + \frac{\partial f_2}{\partial x_2} \Delta x_2 + higher\ terms = 0$$

We neglect all higher-order terms and rewrite

$$f_1(x_1, x_2) + \frac{\partial f_1}{\partial x_1} \Delta x_1 + \frac{\partial f_1}{\partial x_2} \Delta x_2 = 0$$

$$f_2(x_1, x_2) + \frac{\partial f_2}{\partial x_1} \Delta x_1 + \frac{\partial f_2}{\partial x_2} \Delta x_2 = 0$$

Since both x_1 and x_2 are known as numbers, we can insert them into the equations and get numerical values for the functions and the partial derivatives. Transfer the function values to the right hand side:

$$\frac{\partial f_1}{\partial x_1} \Delta x_1 + \frac{\partial f_1}{\partial x_2} \Delta x_2 = -f_1(x_1, x_2)$$

$$\frac{\partial f_2}{\partial x_1} \Delta x_1 + \frac{\partial f_2}{\partial x_2} \Delta x_2 = -f_2(x_1, x_2)$$

(15.6.4)

This is a set of 2 linear equations in two unknowns,

$$\begin{bmatrix} \frac{\partial f_1}{\partial x_1} & \frac{\partial f_1}{\partial x_2} \\ \frac{\partial f_2}{\partial x_1} & \frac{\partial f_2}{\partial x_2} \end{bmatrix} \begin{bmatrix} \Delta x_1 \\ \Delta x_2 \end{bmatrix} = \begin{bmatrix} -f_1(x_1, x_2) \\ -f_2(x_1, x_2) \end{bmatrix}$$

(15.6.5)

The matrix of the partial derivatives is called the *Jacobian* and is usually denoted by the letter J. Since in this book J has been used to denote independent current sources, we will use the letter M. Thus in matrix notation we have

$$M = \begin{bmatrix} \dfrac{\partial f_1}{\partial x_1} & \dfrac{\partial f_1}{\partial x_2} \\[2mm] \dfrac{\partial f_2}{\partial x_1} & \dfrac{\partial f_2}{\partial x_2} \end{bmatrix} \tag{15.6.6}$$

$$\Delta x = \begin{bmatrix} \Delta x_1 \\ \Delta x_2 \end{bmatrix} \tag{15.6.7}$$

$$f = \begin{bmatrix} f_1 \\ f_2 \end{bmatrix} \tag{15.6.8}$$

After solving the system we obtain new values by adding the increments to the original values of the variables. Since this must be repeated, we distinguish the steps by using a superscript. The starting values are given the superscript zero, the next value the superscript one:

$$\begin{aligned} x_1^1 &= x_1^0 + \Delta x_1^0 \\ x_2^1 &= x_2^0 + \Delta x_2^0 \end{aligned} \tag{15.6.9}$$

Equations (15.6.5) and (15.6.9) form the Newton-Raphson iterative algorithm. We write them usually in the form

$$M^k \Delta x^k = -f^k \tag{15.6.10}$$

$$x^{k+1} = x^k + \Delta x^k \tag{15.6.11}$$

where the superscript k starts at zero and is incremented with every iteration. We start with x_1^0 and x_2^0. If we know anything about their possible values, we use them as an estimate; it will reduce the number of iterations. If we do not know anything, then we make an arbitrary guess (and hope for the best). The iterations are stopped when the increments Δx_1 and Δx_2 are sufficiently small. The method applies for any number of equations; for n equations the matrix **M** will have the size $n \times n$ and the vectors Δx and **f** will have length n.

We will use a simple example to show how the method is used. Consider two equations

$$\begin{aligned} f_1 &= x_1^2 + x_2^2 - 4 = 0 \\ f_2 &= x_1 - x_2 = 0 \end{aligned} \tag{15.6.12}$$

The first one describes a circle with radius $r = 2$ centered at the origin, the second is a straight line passing through the origin. In this simple example we can substitute the second equation into the first one and get $x_1 = x_2 = \pm\sqrt{2} = \pm 1.4142$.

The derivatives for the Jacobian are:

$$\frac{\partial f_1}{\partial x_1} = 2x_1$$

$$\frac{\partial f_1}{\partial x_2} = 2x_2$$

$$\frac{\partial f_2}{\partial x_1} = 1$$

$$\frac{\partial f_2}{\partial x_2} = -1$$

Equation (15.6.10) will have the form

$$\begin{bmatrix} 2x_1 & 2x_2 \\ 1 & -1 \end{bmatrix} \begin{bmatrix} \Delta x_1 \\ \Delta x_2 \end{bmatrix} = \begin{bmatrix} -x_1^2 - x_2^2 + 4 \\ -x_1 + x_2 \end{bmatrix}$$

Since this is only a 2×2 system, it is easiest to invert the matrix

$$\begin{bmatrix} \Delta x_1 \\ \Delta x_2 \end{bmatrix} = \frac{1}{-2(x_1 + x_2)} \begin{bmatrix} -1 & -2x_2 \\ -1 & 2x_1 \end{bmatrix} \begin{bmatrix} -x_1^2 - x_2^2 + 4 \\ -x_1 + x_2 \end{bmatrix} \qquad (15.6.13)$$

We will use (15.6.13) repeatedly. To start we select $x_1^0 = x_2^0 = 1$:

$$\begin{bmatrix} \Delta x_1^0 \\ \Delta x_2^0 \end{bmatrix} = \frac{1}{-4} \begin{bmatrix} -1 & -2 \\ -1 & 2 \end{bmatrix} \begin{bmatrix} 2 \\ 0 \end{bmatrix} = \begin{bmatrix} 0.5 \\ 0.5 \end{bmatrix}$$

using (15.6.11) we get the next approximation to the solution

$$\begin{bmatrix} x_1^1 \\ x_2^2 \end{bmatrix} = \begin{bmatrix} x_1^0 \\ x_2^0 \end{bmatrix} + \begin{bmatrix} \Delta x_1^0 \\ \Delta x_2^0 \end{bmatrix} = \begin{bmatrix} 1.5 \\ 1.5 \end{bmatrix}$$

In the next iteration we substitute these results into (15.6.13) to obtain

$$\begin{bmatrix} \Delta x_1^1 \\ \Delta x_2^1 \end{bmatrix} = \frac{1}{-6} \begin{bmatrix} -1 & -3 \\ -1 & 3 \end{bmatrix} \begin{bmatrix} -0.5 \\ 0 \end{bmatrix} = \begin{bmatrix} -1/12 \\ -1/12 \end{bmatrix}$$

The next approximation is

$$\begin{bmatrix} x_1^2 \\ x_2^2 \end{bmatrix} = \begin{bmatrix} 1.5 - 1/12 \\ 1.5 - 1/12 \end{bmatrix} = \begin{bmatrix} 17/12 \\ 12/12 \end{bmatrix} = \begin{bmatrix} 1.4166667 \\ 1.4166667 \end{bmatrix}$$

If we make one more iteration, we will get $\Delta x_1^2 = \Delta x_2^2 = -\dfrac{1}{408} = 0.00245098$ and $x_1^3 = x_2^3 = \dfrac{577}{408} = 1.4142157$, almost equal to $\sqrt{2} = 1.41421356$.

Example 1

Apply the Newton-Raphson formula to the linear set of equations

$$f_1 = 3x_1 + 2x_2 - 7 = 0$$
$$f_2 = 2x_1 + 4x_2 - 10 = 0$$

and show that the result is obtained in one iteration starting from any initial estimate. The solution is easy to find directly: $x_1 = 1$, $x_2 = 2$.

We take an arbitrary initial guess $x_1^0 = x_2^0 = 10$. Equation (15.6.10) gets the form

$$\begin{bmatrix} 3 & 2 \\ 2 & 4 \end{bmatrix} \begin{bmatrix} \Delta x_1^0 \\ \Delta x_2^0 \end{bmatrix} = \begin{bmatrix} -30 - 20 + 7 \\ -20 - 40 + 10 \end{bmatrix}$$

The solution is

$$\begin{bmatrix} \Delta x_1^0 \\ \Delta x_2^0 \end{bmatrix} \begin{bmatrix} -9 \\ -8 \end{bmatrix}$$

and inserting into (15.6.11)

$$\begin{bmatrix} x_1^1 \\ x_2^1 \end{bmatrix} = \begin{bmatrix} 10 \\ 10 \end{bmatrix} + \begin{bmatrix} -9 \\ -8 \end{bmatrix} = \begin{bmatrix} 1 \\ 2 \end{bmatrix}$$

which is the final result.

There are many more mathematical details which should be known. The Newton-Raphson method converges rapidly if the initial estimates for the variables are close to the final values. It may not converge if they are far away. It is not the purpose of this chapter to give a comprehensive treatment. What we need are the consequences of its use in the solution of nonlinear networks. They are discussed in the next section.

15.7. SOLUTIONS OF NONLINEAR NETWORKS

Let us now turn our attention to the solution of network equations and take for this purpose several examples. As a first one we consider Fig. 15.7.1 where the diode is described by $i_D = e^{40v} - 1$. Writing the nodal equation for the single node

$$Gv + i_D = 2$$

or, in a form suitable for Newton-Raphson solution,

$$f(v) = v + (e^{40v} - 1) - 2 = 0$$

The derivative is

$$f^{(1)}(v) = 40e^{40v} + 1$$

and thus

$$\Delta v^k = \frac{3 - v - e^{40v}}{40e^{40v} + 1}$$

The iterations are shown in Table 15.7.1

The second case will consider the network in Fig. 15.5.2, equations (15.5.4) and (15.5.5), repeated here:

$$i_D = i(v_D) = I_s(e^{40v_D} - 1), \hspace{2cm} (15.7.1)$$

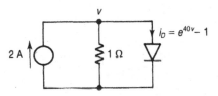

FIGURE 15.7.1   Network with one diode.

	Table 15.7.1	
	Iterations for Fig. 15.7.1	
k	x^k	Δx^k
0	0.2	-0.0249763
1	0.1750237	-0.0249351
2	0.1500886	-0.0248225
3	0.1252661	-0.0245168
4	0.1007493	-0.0237011
5	0.0770482	-0.0216233
6	0.0554249	-0.0169346
7	0.0384903	-0.0090729
8	0.0294174	-0.0020885
9	0.0273289	-0.0000915
10	0.0272375	-0.000002
11	0.0272373	-8.3396E-12

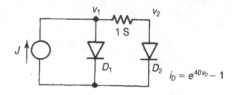

FIGURE 15.7.2   Network with two diodes.

$$v_D = v_1 - v_2 \tag{15.7.2}$$

and

$$
\begin{aligned}
f_1 &= (G_1 + G_2)v_1 - G_2v_2 + i_D - J_1 = 0 \\
f_2 &= -G_2v_1 + (G_2 + G_3)v_2 - i_D - J_2 = 0
\end{aligned}
\tag{15.7.3}
$$

The partial derivatives are

$$\frac{\partial f_1}{\partial v_1} = G_1 + G_2 + \frac{\partial i_D}{\partial v_D}\frac{\partial v_D}{\partial v_1}$$

$$\frac{\partial f_1}{\partial v_2} = -G_2 + \frac{\partial i_D}{\partial v_D}\frac{\partial v_D}{\partial v_2}$$

$$\frac{\partial f_2}{\partial v_1} = -G_2 - \frac{\partial i_D}{\partial v_D}\frac{\partial v_D}{\partial v_1}$$

$$\frac{\partial f_2}{\partial v_2} = G_1 + G_3 - \frac{\partial i_D}{\partial v_D}\frac{\partial v_D}{\partial v_2}$$

We have used the differentiation chain rule to obtain the derivatives of the current with respect to the nodal voltages. However, since we know (15.7.2), the derivatives are

$$\frac{\partial v_D}{\partial v_1} = 1$$

$$\frac{\partial v_D}{\partial v_2} = -1$$

Substituting these partial derivatives into the Jacobian

$$
M = \begin{bmatrix}
G_1 + G_2 + \dfrac{\partial i_D}{\partial v_D} & -G_2 - \dfrac{\partial i_D}{\partial v_D} \\[4mm]
-G_2 - \dfrac{\partial i_D}{\partial v_D} & G_2 + G_3 + \dfrac{\partial i_D}{\partial v_D}
\end{bmatrix}
$$

The last step is to substitute into this matrix the derivative of the current. From (15.7.1)

$$\frac{\partial i_D}{\partial v_D} = 40I_s e^{40v_D} = 40I_s e^{40(v_1 - v_2)}$$

From the example we conclude that the Jacobian matrix has the same form as the nodal matrix. All linear elements in the matrix are in the same positions as they were before. Every nonlinear element will also occupy the same positions as if it were linear. The number will change from iteration to iteration and will be equal to the derivative, evaluated at the voltages of that particular iteration. When the iterations finish with sufficiently small Δv_i, then this derivative represents the *small signal* equivalent of the nonlinear element. These conclusions are valid for any other formulation method as well.

Example 1.

Write the Jacobian matrix for the network Fig. 15.5.4. Use equations (15.5.10), repeated here:

$$f_1 = v_1^2 + v_2^2 + 2v_2 - 2v_1v_2 - 1 = 0$$
$$f_2 = -v_1^2 - v_2^2 + 2v_1 - v_2 + 2v_1v_2 - 4 = 0$$

The partial derivatives are:

$$\frac{\partial f_1}{\partial v_1} = 2v_1 - 2v_2$$

$$\frac{\partial f_1}{\partial v_2} = 2v_2 + 2 - 2v_1$$

$$\frac{\partial f_2}{\partial v_1} = -2v_1 + 2 + 2v_2$$

$$\frac{\partial f_2}{\partial v_2} = -2v_2 - 1 + 2v_1$$

The Jacobian matrix is

$$M = \begin{bmatrix} 2v_1 - 2v_2 & 2v_2 + 2 - 2v_1 \\ -2v_1 + 2 + 2v_2 & -2v_2 - 1 + 2v_1 \end{bmatrix}$$

Example 2.

Consider the network in Fig. 15.7.2, apply Newton-Raphson iteration and find the voltages v_1 and v_2. Use the initial estimate $v_1^0 = v_2^0 = 0.1$ and let $I_s = 1$. The diodes are described by the equation (15.7.1).

The nodal equations are

$$f_1 = i_{D1} + Gv_1 - Gv_2 - J = 0$$
$$f_2 = -Gv_1 + Gv_2 + i_{D2} = 0$$

Inserting numerical values

$$f_1 = e^{40v_1} + v_1 - v_2 - 2 = 0$$
$$-v_1 + v_2 + e^{40v_2} + 1 = 0$$

The Jacobian is

$$\mathbf{M} = \begin{bmatrix} 40e^{40v_1} + 1 & -1 \\ -1 & 40e^{40v_2} + 1 \end{bmatrix}$$

Evaluation at the initial estimates provides

$$f_1 = 52.59815$$
$$f_2 = 53.59815$$

Inserting the values we must solve

$$\begin{bmatrix} 2184.926 & -1 \\ -1 & 2184.926 \end{bmatrix} \begin{bmatrix} \Delta v_1^0 \\ \Delta_2^0 \end{bmatrix} = \begin{bmatrix} -52.59815 \\ -53.59815 \end{bmatrix}$$

The result is

$$\Delta v_1^0 = -0.0240844$$
$$\Delta v_2^0 = -0.0245419$$

and the first approximation to the nodal voltages is

$$v_1^1 = 0.0759156$$
$$v_2^1 = 0.07545810$$

Table 15.7.2			
Iterations for network Fig. 15.7.2			
k	Δv_1^k	Δv_2^k	v_1^{k+1}	v_2^{k+1}
0	-0.02408	-0.02454	0.07592	0.07546
1	-0.02260	-0.02378	0.05331	0.05168
2	-0.01909	-0.02182	0.03423	0.02986
3	-0.01234	-0.01736	0.02188	0.01250
4	-0.00432	-0.00962	0.01757	0.00289
5	-0.00044	-0.00236	0.01712	0.00053
6	-0.00001	-0.00012	0.01712	0.00041
7	-0.00000	-0.00000	0.01712	0.00041

The iterations were run on a computer with the sequence shown in Table 15.7.2. The final nodal voltages are $v_1 = 0.01712$ and $v_2 = 0.00041$.

15.8. NUMERICAL TIME DOMAIN RESPONSES

In Chapter 9 we used Laplace transformation to obtain mathematical formulas for time domain responses of small *linear* networks. In practical situations we must use numerical methods and computers. Our explanation will be limited to linear networks and will use the simplest possible formula. To start, we recall from Chapter 7, that multiplication by s is equivalent to the differentiation. We introduced there, for instance, equation (7.1.1)

$$i = C \frac{dv}{dt}$$

and replaced it by equation (7.1.2)

$$I = sCV$$

Our problem is to find a numerical formula which will approximate this differentiation. Consider Fig. 15.8.1. where at the point t_0 we know both the value x_0 and the derivative x_0,

$$x_0' = \frac{x_1 - x_0}{h} \tag{15.8.1}$$

This expression is called the *forward Euler formula*. For reasons we will not go into, this is not the best formula. A better one is given by

$$x_1' = \frac{x_1 - x_0}{h} \tag{15.8.2}$$

Its name is the *backward Euler formula*. It looks somewhat strange because we are using the derivative at a point which we actually do not know yet, but for very small h the values of the derivatives will not differ very much. For further explanations we will use Fig. 15.8.2. In the frequency domain, the system equation is

$$\begin{bmatrix} sC_1 + G & -G \\ -G & sC_2 + G \end{bmatrix} \begin{bmatrix} V_1 \\ V_2 \end{bmatrix} = \begin{bmatrix} J \\ 0 \end{bmatrix} \tag{15.8.3}$$

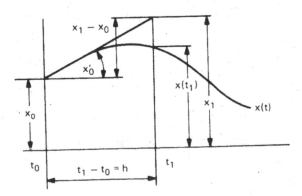

FIGURE 15.8.1 Approximating the derivative.

Let us separate the terms multiplied by s from the other terms. This leads to

$$\begin{bmatrix} G & -G \\ -G & G \end{bmatrix}\begin{bmatrix} V_1 \\ V_2 \end{bmatrix} + \begin{bmatrix} C_1 & 0 \\ 0 & C_2 \end{bmatrix}\begin{bmatrix} sV_1 \\ sV_2 \end{bmatrix} = \begin{bmatrix} J \\ 0 \end{bmatrix}$$
(15.8.4)

We now replace the differentiation indicated by the variable s by the formula (15.8.2). It is a common practice to use lower case letters for time domain variables and we will adhere to this practice. In addition, the lower subscript will refer to the nodes of the network and the upper subscript to the time variable

$$\begin{bmatrix} G & -G \\ -G & G \end{bmatrix}\begin{bmatrix} v_1^1 \\ v_2^1 \end{bmatrix} + \frac{1}{h}\begin{bmatrix} C_1 & 0 \\ 0 & C_2 \end{bmatrix}\begin{bmatrix} v_1^1 - v_1^0 \\ v_2^1 - v_2^0 \end{bmatrix} = \begin{bmatrix} j^1 \\ 0 \end{bmatrix}$$

This can be re-written in the form

$$\begin{bmatrix} G & -G \\ -G & G \end{bmatrix}\begin{bmatrix} v_1^1 \\ v_2^1 \end{bmatrix} + \frac{1}{h}\begin{bmatrix} C_1 & 0 \\ 0 & C_2 \end{bmatrix}\begin{bmatrix} v_1^1 \\ v_2^1 \end{bmatrix} - \frac{1}{h}\begin{bmatrix} C_1 & 0 \\ 0 & C_2 \end{bmatrix}\begin{bmatrix} v_1^0 \\ v_2^0 \end{bmatrix} = \begin{bmatrix} j^1 \\ 0 \end{bmatrix}$$
(15.8.5)

The first two matrices are multiplied by the same voltage vectors and can be added. The third term is multiplied by known initial voltages and we transfer it to the right. This leads to the final expression

$$\begin{bmatrix} G+\frac{1}{h}C_1 & -G \\ -G & G+\frac{1}{h}C_2 \end{bmatrix}\begin{bmatrix} v_1^1 \\ v_2^1 \end{bmatrix} = \frac{1}{h}\begin{bmatrix} C_1 & 0 \\ 0 & C_2 \end{bmatrix}\begin{bmatrix} v_1^0 \\ v_2^0 \end{bmatrix} + \begin{bmatrix} j^1 \\ 0 \end{bmatrix}$$
(15.8.6)

Compare the upper matrix with the system matrix (15.8.3). The only difference is the replacement of s by $1/h$. On the right is the matrix containing capacitors, multiplied by the vector of known initial voltages, and divided by h. Finally, the source vector uses known value at $t = h$. Once we have solved the equation, we know the voltages v_1^1 and v_2^1. They are now used on the right of (15.8.5) for the next time step.

Let us now introduce the usual notation of matrices and vectors by bold letters. Equation (15.9.5) can be re-written in the form

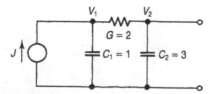

FIGURE 15.8.2　Network with two capacitors.

$$(G + \frac{1}{h}C)v^{k+1} = \frac{1}{h}Cv^k + j^k \qquad (15.8.7)$$

where the superscript indicates the sequence of steps starting from $k = 0$. To derive equation (15.8.7), we needed the network elements in symbolic form, using G and C. In actual solutions we must use numerical values and a small step size h. For the sake of simplicity, let us select $h = 0.1$ and also let the source of current be a unit step (which will not change as we proceed with the steps). Equation (15.8.6) will get the numerical form

$$\begin{bmatrix} 12 & -2 \\ -2 & 32 \end{bmatrix} \begin{bmatrix} v_1^1 \\ v_2^2 \end{bmatrix} = \begin{bmatrix} 10 & 0 \\ 0 & 30 \end{bmatrix} \begin{bmatrix} v_1^0 \\ v_2^0 \end{bmatrix} + \begin{bmatrix} 1 \\ 0 \end{bmatrix}$$

Notice that if we do not change the step size, both matrices will not change as we proceed with time steps. Theoretically, we could invert the matrix on the left one and get the solutions. Let us stress here that inversions of matrices are avoided in numerical calculations. The method to use is called LU *decomposition*. Programs for it are widely available; we will leave its explanation to courses on numerical mathematics.

All networks with resistors and capacitors, as well as all active networks used in Chapter 12, could be handled by the above steps. If the network contains capacitors as well as inductors, then the formulation methods we have learned so far (nodal and mesh) will not work, because we will not be able to avoid terms with $1/s$. Fortunately, such a division can be avoided if we use *modified nodal formulation*, described in detail in Chapter 17. In other words, the above steps, in connection with the new formulation method, can be used for time domain analysis of any linear network.

If resistors are nonlinear, but capacitors and inductors are linear, we would have to use iteration. What if the capacitors and inductors are nonlinear? The correct steps will use charges and fluxes, but such steps are beyond the scope of this book. The reader can find all necessary explanations in the book written by J. Vlach and K. Singhal: "Computer Methods for Circuit Analysis and Design," Van Nostrand Reinhold, New York (second edition 1994).

PROBLEMS CHAPTER 15

P.15.1. Redraw the network for small signal analysis by using the model on the right. Consider the capacitors as short circuits. Find the overall gain of the amplifier.

P.15.2. Do the same as in P.15.1.

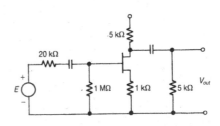

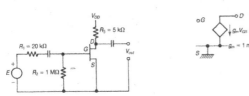

P.15.3. Redraw the network (a) for small signal analysis using the model in (b).

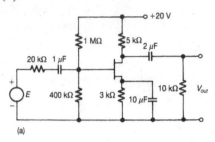

(a)

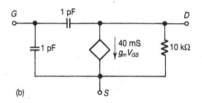

(b)

P.15.4. The nonlinear element is defined by $i = v^2$ for $v \geq 0$ and $i = 0$ for $v < 0$. (a) find the voltage v as a solution of one quadratic equation. (b) Apply the Newton-Raphson algorithm starting with $v^0 = 1.2$

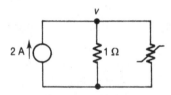

P.15.5. Use mesh and nodal formulation (after transforming the voltage source into a current source). Find the dc voltage across the diode.

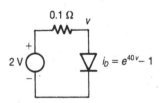

P.15.6. Set up nodal equations and make two Newton-Raphson iterations starting with $v_1^0 = 1$ and $v_2^0 = 0.5$.

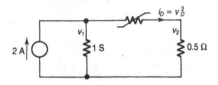

P.15.7. Given are two equations:

$$f_1 = x_1^2 - x_2 - 2 = 0$$
$$f_2 = 2x_1 - x_2 + 3 = 0$$

(a) Find the result by eliminating one variable (b) Start with $x_1^0 = 3.45$ and $x_2^0 = 9.9$ and make two steps in the Newton-Raphson iteration.

P.15.8. Do the same as in P.15.7 with the equations

$$f_1 = x_1 - x_2^2 = 0$$
$$f_2 = x_1 + 3x_2 - 5 = 0$$

Start the iteration with $x_1^0 = x_2^0 = 1$.

16

SENSITIVITIES

INTRODUCTION

All our previous studies assumed that the element values are accurate and do not change. This may be an ideal situation but is not true in applications. Any system which is supposed to perform a certain function is built from components which always differ somewhat from the nominal value. The differences may not be large but if all components are not what they are supposed to be, it may happen that the system does not meet its specifications. It would certainly be nice to have an idea which components are critical and which are not. In electrical networks, this is achieved by the theory of sensitivities, the subject of this chapter.

Calculating sensitivities for larger networks is normally done by a computer, but in the case of small, second order networks, we can get them relatively easily. We will restrict our explanations to such small networks.

16.1. MOTIVATION

We will use a small example to show how the change of an element influences performance of a network. The voltage divider in Fig. 16.1.1 has the transfer function

$$F = \frac{V_{out}}{E} = \frac{1}{sCR + 1} \tag{16.1.1}$$

In this chapter we will use the letter F for any function which will be subject to sensitivity studies; they need not be network functions only.

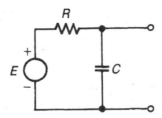

FIGURE 16.1.1   Sensitivity principle on RC voltage divider.

Suppose that we change the resistor by a small amount ΔR. This will influence the function and we will have

$$F + \Delta F = \frac{1}{sC(R + \Delta R) + 1} \qquad (16.1.2)$$

Subtracting (16.1.1) from (16.1.2)

$$\Delta F = (F + \Delta F) - F = \frac{1}{sC(R + \Delta R) + 1} - \frac{1}{sCR + 1}$$

or

$$\Delta F = \frac{-sC\Delta R}{\left[sC(R + \Delta R) + 1\right]\left[sCR + 1\right]}$$

Opening the brackets in the denominator we arrive at

$$\frac{\Delta F}{\Delta R} = \frac{-sC}{s^2 R^2 C^2 (1 + \frac{\Delta R}{R}) + sCR(2 + \frac{\Delta R}{R}) + 1} \qquad (16.1.3)$$

For comparisons insert numerical values: $C = 1, R = 1, s = j1$ and $\dfrac{\Delta R}{R} = 0.001$. This leads to

$$\frac{\Delta F}{\Delta R} = \frac{-j}{-0.001 + j2.001}$$

and the absolute value is

$$\left|\frac{\Delta F}{\Delta R}\right| = 0.49975 \qquad (16.1.4)$$

The above steps required fair amount of algebra even though the network is so simple. If the changes are small (in fact infinitesimally small), then we get the result much faster by differentiating (16.1.1) with respect to R:

$$\frac{dF}{dR} = \frac{-sC}{(sCR + 1)^2} \qquad (16.1.5)$$

Substituting the same numerical values as above we get

$$\left|\frac{dF}{dR}\right| = 0.5, \qquad (16.1.6)$$

a result which is almost the same as above, but obtained with much less work. We differentiated with respect to R but could have done so with respect to C or s, or any other variable which may appear in the function F. Such derivatives are the simplest sensitivity functions. They indicate how a small change in any variable influences the function under consideration.

16.2. DEFINITION OF SENSITIVITIES

Having established the reason why we use the derivatives to check on the influence of element changes, we now turn to the various sensitivity definitions. The simple derivative we used above is an obvious choice but it has problems. If we change the capacitor to some other value, for instance $C = 2$, and simultaneously change the resistor to $R = 0.5$ so that their product RC does not change, we get the *same* transfer function but *another* sensitivity at the same frequency,

$$\frac{dF}{dR} = \frac{j2}{(j+1)^2}.$$

This is quite unpleasant. In order to eliminate this problem, a normalized sensitivity has been defined:

$$S_h^F = \frac{dF}{dh}\frac{h}{F} \tag{16.2.1}$$

Here S_h^F is the symbol for this definition of sensitivity, F is the function under consideration, and h is any variable in F, for instance any network element. Let us calculate this sensitivity for the network in Fig. 16.1.1. Using (16.1.1) and (16.1.5)

$$S_R^F = \frac{-sC}{(sCR+1)^2}\frac{R}{\dfrac{1}{sCR+1}} = -\frac{sCR}{sCR+1}.$$

This normalized sensitivity is the same for any choice of C and R, as long as their product remains the same.

The normalized sensitivity definition (16.2.1) can be calculated by first finding the derivative and then multiplying by the ratio $\dfrac{h}{F}$. In some applications, it is more convenient to apply an *equivalent* definition

$$S_h^F = \frac{d\ln F}{d\ln h} \tag{16.2.2}$$

where *ln* denotes natural logarithm. We will see later that this definition can save us lots of work in certain situations.

The normalized sensitivity is clearly advantageous but becomes meaningless in two situations: if either F or h become zero. In such cases we apply instead two alternate normalizing definitions. If $h = 0$, then we use

$$\Sigma_1 = \frac{dF}{dh}\frac{1}{F}, \tag{16.2.3}$$

simply skipping multiplication by zero. If $F = 0$, we do not divide by F and use instead

$$\Sigma_2 = \frac{dF}{dh}h \tag{16.2.4}$$

If both happen to be zero, then all we can use is the derivative itself.

It is sometimes difficult to visualize that the derivative with respect to an element can be nonzero although the element itself has zero value. To clarify this possibility we take a small example, Fig. 16.2.1, in which the capacitor C_1 actually does not exist and has a value $C_1 = 0$. Using all elements we first derive the voltage transfer function:

$$F = \frac{V_{out}}{E} = \frac{sC_1 + G}{s(C_1 + C_2) + G}$$

The derivative with respect to C_1 is

$$\frac{dF}{dC_1} = \frac{s^2 C_2}{\left[s(C_1 + C_2) + G\right]^2}$$

If we now set $C_1 = 0$, we get

$$F = \frac{G}{sC_2 + G}$$

(the same as if C_1 did not exist at all), but

$$\frac{dF}{dC_1} = \frac{s^2 C_2}{\left[sC_2 + G\right]^2}$$

which is nonzero. In this situation, the use of the fully normalized sensitivity definition (16.2.1) would clearly be wrong and we must apply Σ_1.

Sensitivity with respect to a zero-valued element is a useful concept because the element does not change the network function but we can get information on how a small, actually existing, parasitic element would influence our results.

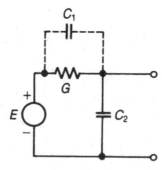

FIGURE 16.2.1   Sensitivity to a prasitic $C_1 = 0$.

16.3. SENSITIVITIES OF TUNED CIRCUITS

A tuned circuit in Fig. 16.3.1 is the best starting example for the study of sensitivities. Its impedance is

$$Z = \frac{1}{C} \frac{s}{s^2 + s\frac{G}{C} + \frac{1}{LC}} \qquad (16.3.1)$$

In most applications the value of the conductance is chosen such that the tuned circuit has two complex conjugate poles. In such case

$$Z = \frac{1}{C} \frac{s}{s^2 + s\frac{\omega_0}{Q} + \omega_0^2} \qquad (16.3.2)$$

Comparing the above two expressions we identify

$$\omega_0 = \frac{1}{\sqrt{LC}} \qquad (16.3.3)$$

and

$$Q = \frac{\omega_0 C}{G} = \frac{1}{\omega_0 GL} = \frac{1}{G}\left[\frac{C}{L}\right]^{1/2} \qquad (16.3.4)$$

We now have three functions, Z, ω_0 and Q, all of which describe, in some way, the properties of the tuned circuit. We can calculate sensitivities for each of them.

For instance, differentiating (16.3.1) with respect to G results in

$$\frac{dZ}{dG} = \frac{1}{C^2} \frac{-s^2}{(s^2 + s\frac{G}{C} + \frac{1}{LC})^2}$$

and the normalized sensitivity is

$$S_G^Z = \frac{dZ}{dG}\frac{G}{Z} = \frac{-s\frac{G}{C}}{s^2 + s\frac{G}{C} + \frac{1}{LC}}$$

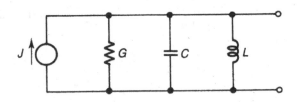

FIGURE 16.3.1 Parallel tuned circuit.

This sensitivity is a function of frequency and when we substitute $s = j\omega$, we get a complex number which is difficult to interpret. To get a meaningful result we can do two things: Either consider the absolute value of Z and find its sensitivity (we will do this later), or use Q and ω_0 because they are real variables, independent of frequency.

Suppose that we want to derive sensitivities of Q. We can differentiate with respect to, say, L, get $\dfrac{dQ}{dL}$ and then multiply by $\dfrac{L}{Q}$, as required by (16.2.1). However, since (16.3.4) does not contain any additions or subtractions, we can take advantage of the normalized sensitivity definition (16.2.2) and proceed as follows: Take first the logarithm of (16.3.4)

$$\ln Q = -\ln G + \frac{1}{2}\ln C - \frac{1}{2}\ln L$$

and then its differential

$$d\ln Q = -d\ln G + \frac{1}{2}d\ln C - \frac{1}{2}d\ln L \qquad (16.3.5)$$

Since we want sensitivity with respect to L, we simply divide the equation by $d\ln L$

$$\frac{d\ln Q}{d\ln L} = \frac{d\ln G}{d\ln L} + \frac{1}{2}\frac{d\ln C}{d\ln L} - \frac{1}{2}\frac{d\ln L}{d\ln L}.$$

Comparing the left side of this equation with the definition (16.2.2) we see that it is in fact the desired S_L^Q. The elements G and C are independent of L, the ratios $\dfrac{d\ln G}{d\ln L}$ and $\dfrac{d\ln C}{d\ln L}$ are zero and $\dfrac{d\ln L}{d\ln L} = 1$. The result is

$$S_L^Q = \frac{d\ln Q}{d\ln L} = -\frac{1}{2}\frac{d\ln L}{d\ln L} = -\frac{1}{2}$$

Dividing (16.3.6) by $d\ln G$ or $d\ln C$ will result in

$$S_G^Q = -1$$

$$S_C^Q = \frac{1}{2}$$

Calculation of normalized sensitivities of ω_0 proceeds the same way. First we take the logarithm

$$\ln \omega_0 = -\frac{1}{2}\ln L - \frac{1}{2}\ln C,$$

apply differentiation by writing d in front of every term and then divide by $d\ln L$ or $d\ln C$. The sensitivities are

$$S_L^{\omega_0} = \frac{d\ln \omega_0}{d\ln L} = -\frac{1}{2}$$

$$S_C^{\omega_0} = \frac{d\ln \omega_0}{d\ln C} = -\frac{1}{2}$$

Finally, since ω_0 does not depend on G, we also have

$$S_G^{\omega_0} = 0$$

The above derivations lead to one important conclusion: In the case of a tuned circuit, all $\left|S_h^Q\right|$ and $\left|S_h^{\omega_0}\right|$ are less than or at most equal to one. As we will discover later, these values are extremely small and no other circuit, described by a second order network function, has sensitivities as low as the tuned circuit.

How can we get some practical conclusions from the sensitivity calculation? We will indicate the steps on the result $S_L^Q = -\dfrac{1}{2}$. If we take a small increment instead of the derivative, then we can write

$$S_L^Q \sim \frac{\Delta Q}{\Delta L}\frac{L}{Q} = -\frac{1}{2}$$

or

$$\frac{\Delta Q}{Q} \sim -0.5\frac{\Delta L}{L}$$

Suppose that the value of L increases by 1%; then $\dfrac{\Delta L}{L} = 0.01$ and

$$\frac{\Delta Q}{Q} \sim -0.5 \times 0.01 = -0.005$$

If, in addition, our Q is 200, then

$$\Delta Q \sim -0.005 \times 200 = -1$$

These simple steps indicate that a 1% increase in inductor value results in an approximate 0.5% drop of Q. If the original Q was 200, the change in L will cause a reduction of Q to approximately 199. We can also check how good this approximation is. To get $Q = 200$, let us select $C = 1$, $L = 1$ and $R = 1/G = 200$. If we increase the inductor to the value $L = 1.01$, then $Q = \dfrac{200}{\sqrt{1.01}} = 199.007438$. The approximation is very good.

16.4.   NETWORK FUNCTION SENSITIVITY

All network functions are rational functions but even if the network is not large, it is often difficult to find the function in terms of variable elements. Finding the derivatives with respect to the elements is also lengthy and any help from the theory is welcome.

Our study will start with the transfer function sensitivity of the Sallen-Key network in Fig. 16.4.1. Using the methods of Chapter 12 we set up the system equation

$$\begin{bmatrix} G_1 + G_2 + sC_1 & -G_2 - sC_1 A \\ -G_2 & G_2 + sC_2 \end{bmatrix}\begin{bmatrix} V_1 \\ V_2 \end{bmatrix} = \begin{bmatrix} EG_1 \\ 0 \end{bmatrix} \tag{16.4.1}$$

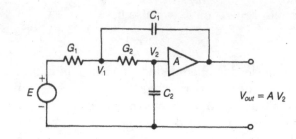

FIGURE 16.4.1   Sallen–Key Second-order network.

and get the transfer function

$$F = \frac{N}{D} = \frac{AG_1G_2}{s^2C_1C_2 + s\left[C_1G_2(1-A) + C_2G_1 + C_2G_2\right] + G_1G_2} \qquad (16.4.2)$$

If we need the normalized sensitivity with respect to, say, the gain of the amplifier, A, then we must first calculate $\dfrac{dF}{dA}$ and then multiply this derivative by $\dfrac{A}{F}$. It is not possible to use the definition (16.2.2) because the logarithm cannot be applied to additions or subtractions.

We can simplify our task by first deriving a formula for the sensitivity of a rational function

$$F = \frac{N}{D}$$

Since it is only a ratio of two terms, we can use the sensitivity formula based on the natural logarithm

$$\ln F = \ln N - \ln D,$$

differentiate

$$d \ln F = d \ln N - d \ln D$$

and divide by $d \ln h$

$$\frac{d\ln F}{d\ln h} = \frac{d\ln N}{d\ln h} - \frac{d\ln D}{d\ln h}$$

This is equivalent to

$$S_h^F = S_h^N - S_h^D \qquad (16.4.3)$$

The formula says that we first obtain separately normalized sensitivities of the numerator and denominator and then merely subtract them.

We will apply the steps for the above transfer function. Its numerator is

$$N = AG_1G_2$$

and its normalized sensitivity with respect to the gain is

$$S_A^N = 1.$$

Differentiating the denominator

$$\frac{dD}{dA} = -sC_1G_2$$

and normalizing

$$S_A^D = \frac{A}{D}\frac{dD}{dA} = -\frac{sC_1G_2A}{D}.$$

Using (16.4.3)

$$S_A^F = 1 + \frac{sC_1G_2A}{s^2C_1C_2 + s\left[C_1G_2(1-A) + C_2G_1 + C_2G_2\right] + G_1G_2} \tag{16.4.4}$$

The steps were easy but this result is rarely needed because (16.4.4) is complex for $s = j\omega$, and thus difficult to interpret. What we really need is the sensitivity of the absolute value of the function.

In order to derive the formula for $S_h^{|F|}$ we resort to a trick and start with the equivalence

$$F = |F|e^{j\phi} \tag{16.4.5}$$

Taking the logarithm and differentiating

$$d\ln F = d\ln|F| + jd\phi$$

Dividing by $d\ln h$

$$\frac{d\ln F}{d\ln h} = \frac{d\ln|F|}{d\ln h} + j\frac{d\phi}{d\ln h}$$

which is equivalent to

$$S_h^F = S_h^{|F|} + j\frac{d\phi}{\frac{dh}{h}}$$

On the left side is a complex number which we can write as $\operatorname{Re}S_h^F + \operatorname{Im}S_h^F$ and compare with the right hand. The real parts and the imaginary parts must be equal

$$S_h^{|F|} = \operatorname{Re}S_h^F \tag{16.4.6}$$

and

$$\operatorname{Im} S_h^F = \frac{d\phi}{dh} h$$

If we wish to obtain the last expression as a normalized sensitivity, we divide both sides by ϕ and get

$$\frac{1}{\phi} \operatorname{Im} S_h^F = \frac{h}{\phi} \frac{d\phi}{dh}$$

which is equivalent to

$$S_h^\phi = \frac{1}{\phi} \operatorname{Im} S_h^F \tag{16.4.7}$$

We will demonstrate application of (16.4.6) on the Sallen-Key network, but simplify our task by inserting numerical values: $C_1 = C_2 = 1$, $G_1 = G_2 = 1$ and $A = 2$. Using (16.4.4)

$$S_A^F = 1 + \frac{2s}{s^2 + s + 1}$$

If we select, for instance, $s = j2$, this reduces to

$$S_A^F = 1 + \frac{j4}{-3 + j2} = \frac{21}{13} - j\frac{12}{13}$$

and, as follows from (16.4.6), sensitivity of the absolute value is

$$S_A^{|F|} = \frac{21}{13}$$

Let us summarize the steps of this section. For any frequency $s = j\omega$, the sensitivity of the transfer function, S_h^F, is obtained as an intermediate step using (16.4.3). The result is a complex number and its real part is the desired $S_h^{|F|}$.

Example 1.

For the network in Fig. 16.4.2 calculate the voltage transfer function $F = \dfrac{V_2}{E}$ by keeping the transconductance g of the VC as a variable. For the frequency $s = j2$ calculate the absolute value of F and also the sensitivity $S_g^{|F|}$.

Using nodal formulation explained in Chapter 12 and taking numerical values for all elements except g we arrive at the matrix equation

$$\begin{bmatrix} s+2 & -1 \\ -1+g & s+1 \end{bmatrix} \begin{bmatrix} V_1 \\ V_2 \end{bmatrix} = \begin{bmatrix} E \\ 0 \end{bmatrix}$$

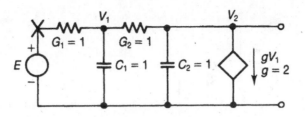

FIGURE 16.4.2   Network with a *CV.*

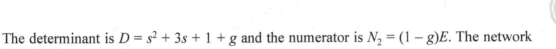

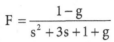

The determinant is $D = s^2 + 3s + 1 + g$ and the numerator is $N_2 = (1 - g)E$. The network function is

$$F = \frac{1-g}{s^2 + 3s + 1 + g}$$

The derivative of the numerator is $\dfrac{dN}{dg} = -1$ and the normalized sensitivity is

$$S_g^N = \frac{dN}{dg}\frac{g}{N} = \frac{-g}{1-g}$$

The derivative of D is $\dfrac{dD}{dg} = 1$ and the normalized sensitivity is

$$S_g^D = \frac{g}{s^2 + 3s + 1 + g}$$

Since we now have all the derivatives we can substitute the numerical value for the VC, $g = 2$.

$$S_g^F = S_g^N - S_g^D = 2 - \frac{2}{s^2 + 3s + 3} = \frac{2s^2 + 6s + 4}{s^2 + 3s + 3}$$

Let us now select $s = j2$. The absolute value of F will be

$$|F| = \left|\frac{-1}{-1 + j6}\right| = \frac{1}{\sqrt{37}}$$

and the sensitivity will be

$$S_g^F = \frac{-4 + j12}{-1 + j6} = \frac{76}{37} + j\frac{12}{37}$$

According to formula (16.4.6) we must take only the real part to get the sensitivity of the absolute value:

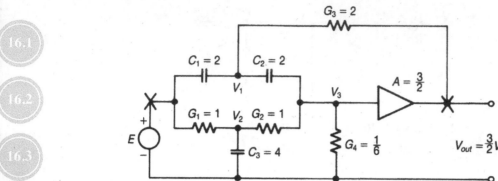

FIGURE 16.4.3 Network with a Twin-T.

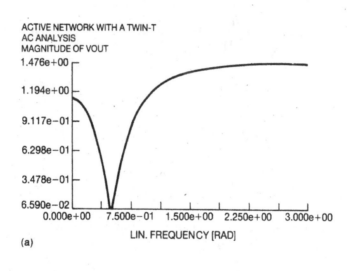

(a)

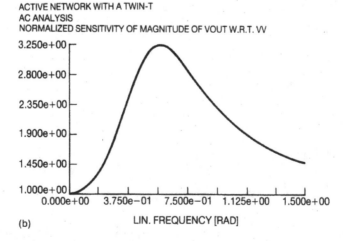

(b)

FIGURE 16.4.4 Responses of the network in Fig. 16.4.3: (a) Amplitude response, (b) Normalized sensitivity $S_A^{|F|}$.

$$S_g^{|F|} = \operatorname{Re} S_g^F = \frac{76}{37}$$

Calculation of network function sensitivities by hand is an extremely laborious process. Fortunately, many specialized programs offer this option and we will show computer results in the next example.

Example 2.

Consider the network in Fig. 16.4.3 (redrawn Fig. 12.3.6). Calculate the amplitude response and the sensitivity with respect to the gain of the amplifier.

The amplitude response is given in Fig. 16.4.4a, the sensitivity in Fig. 16.4.4b. From Fig. 16.4.4b we see that changes in the gain of the amplifier will have their largest influence in the approximate frequency range of 0.38 to 0.75 rad/s.

16.5.   SENSITIVITIES OF Q AND ω_0

A second order network function is a rational function whose denominator is a polynomial of second degree and whose numerator has a degree which may be less than or equal to the degree of the denominator.

If both the numerator and denominator are polynomials of second degree and if their roots are complex conjugate, we can distinguish them by using for the denominator the notation Q_p, ω_p (p standing for the pole) and for the numerator Q_z, ω_z (z standing for the zero). We will use the notation Q and ω_0 and derive formulas for their evaluation and also for their sensitivities.

Two complex conjugate poles (or zeros) form a second-order polynomial

$$P(s) = s^2 + s\frac{\omega_0}{Q} + \omega_0^2 \tag{16.5.1}$$

but this polynomial may still be multiplied by some element values. As an example, consider the Sallen-Key network in Fig. 16.4.1 which we have analyzed and for which we found

$$D = s^2 C_1 C_2 + s[C_1 G_2(1 - A) + C_2 G_1 + C_2 G_2] + G_1 G_2 \tag{16.5.2}$$

In order to save us lots of unnecessary work, we will derive formulas for Q and ω_0 and for their sensitivities by assuming that the polynomial has the general form

$$P = Us^2 + Vs + W \tag{16.5.3}$$

which we rewrite as

$$P = U\left[s^2 + \frac{V}{U}s + \frac{W}{U}\right]$$

Now we can compare the expression in the square bracket with (16.5.1) and get the following identities

$$\frac{V}{U} = \frac{\omega_0}{Q}$$

$$\frac{W}{U} = \omega_0^2$$

or

$$\omega_0 = \left[\frac{W}{U}\right]^{1/2}$$

$$Q = \frac{\sqrt{UW}}{V} \tag{16.5.4}$$

Applying these formulas to the polynomial (16.5.2) we identify

$$U = C_1 C_2$$
$$V = C_1 G_2 (1 - A) + C_2 G_1 + C_2 G_2 \tag{16.5.5}$$
$$W = G_1 G_2.$$

and obtain

$$\omega_0 = \left[\frac{W}{U}\right]^{1/2} = \left[\frac{G_1 G_2}{C_1 C_2}\right]^{1/2}$$

$$Q = \frac{\sqrt{G_1 G_2 C_1 C_2}}{C_1 G_2 (1 - A) + C_2 G_1 + C_2 G_2} \tag{16.5.6}$$

More important is the fact that we can use (16.5.4) to derive general rules for sensitivities of these variables. Since (16.5.4) does not contain any additions or subtractions, we resort to the already known trick by taking the logarithm

$$In \, \omega_0 = \frac{1}{2} In \, W - \frac{1}{2} In \, U,$$

differentiating and dividing by $d \ln h$:

$$\frac{d \ln \omega_0}{d \ln h} = \frac{1}{2}\frac{d \ln W}{d \ln h} - \frac{1}{2}\frac{d \ln U}{d \ln h}.$$

This is equivalent to

$$S_h^{\omega_0} = \frac{1}{2} S_h^W - \frac{1}{2} S_h^U \tag{16.5.7}$$

Doing the same for Q:

$$\ln Q = \frac{1}{2}\ln U + \frac{1}{2}\ln W - \ln V$$

and

$$S_h^Q = \frac{1}{2}S_h^U + \frac{1}{2}S_h^W - S_h^V \tag{16.5.8}$$

If we want in our example the sensitivity with respect to the gain of the amplifier, we calculate first the auxiliary sensitivities

$$S_A^U = 0$$

$$S_A^V = -\frac{C_1 C_2 A}{V}$$

$$S_A^W = 0$$

and then use (16.5.7) and (16.5.8) to get

$$S_A^{\omega_0} = 0$$

$$S_A^Q = \frac{C_1 G_2 A}{V} \tag{16.5.9}$$

At this point we could insert for V and leave the result. However, it is very instructive to take one more theoretical step using (16.5.4), writing $\frac{1}{V} = \frac{Q}{\sqrt{UW}}$ and inserting into S_A^Q. We get

$$S_A^Q = AQ\left[\frac{C_1 G_2}{C_2 G_1}\right]^{1/2} \tag{16.5.10}$$

This is a surprising result because the sensitivity is proportional to the product AQ! Without sensitivities we might have expected that the active network we just analyzed could serve as a tuned circuit. After all, like the tuned circuit, it is described by a second order network function. We derived earlier that all sensitivities of the tuned circuit, S_h^Q and $S_h^{\omega_0}$, are in absolute value less than or equal to one. Our active network has a small $S_A^{\omega_0}$, but the sensitivity of Q is proportional to AQ. A small change in the gain of the amplifier will cause considerable change of Q and the situation will be aggravated if we need high Q.

In order to fully understand properties of the Sallen-Key (or any other) active second-order network, we must derive sensitivities with respect to all elements. We do so by first deriving the sensitivities S_h^U, S_h^V and S_h^W; they are summarized in Table 16.5.1. Next we apply formulas (16.5.7) and (16.5.8) and obtain the sensitivities of Q and ω_0. These are summarized in Table 16.5.2. We see that *all* Q sensitivities are proportional to Q, a fairly disappointing result. Without going into any more details, we should perhaps say that second-order active networks tend to have this property.

Example 1.

In the network Fig. 16.5.1 keep the two amplifiers as variables. Obtain the transfer function and select such gains that $Q = 5$. Calculate the Q sensitivities for both amplifier gains.

Table 16.5.1.

Auxiliary sensitivities of the Sallen-Key network

$S_{C_1}^U = 1$	$S_{C_1}^V = \dfrac{C_1 G_2 (1-A)}{V}$	$S_{C_1}^W = 0$
$S_{C_2}^U = 1$	$S_{C_2}^V = \dfrac{C_2 (G_1 + G_2)}{V}$	$S_{C_2}^W = 0$
$S_{G_1}^U = 0$	$S_{G_1}^V = \dfrac{G_1 C_2}{V}$	$S_{G_1}^W = 1$
$S_{G_2}^U = 0$	$S_{G_2}^V = \dfrac{G_2 (C_2 + C_1 - C_1 A)}{V}$	$S_{G_2}^W = 1$
$S_A^U = 0$	$S_A^V = -\dfrac{C_1 G_2 A}{V}$	$S_A^W = 0$

Table 16.5.2.

Sensitivities of the Sallen-Key network

$S_{C_1}^{\omega_0} = -\dfrac{1}{2}$	$S_{C_1}^Q = \dfrac{1}{2} - C_1 G_2 (1-A)\,\dfrac{Q}{\sqrt{C_1 C_2 G_1 G_2}}$
$S_{C_2}^{\omega_0} = -\dfrac{1}{2}$	$S_{C_2}^Q = \dfrac{1}{2} - C_2 (G_1 + G_2)\,\dfrac{Q}{\sqrt{C_1 C_2 G_1 G_2}}$
$S_{G_1}^{\omega_0} = +\dfrac{1}{2}$	$S_{G_1}^Q = \dfrac{1}{2} - G_1 C_2\,\dfrac{Q}{\sqrt{C_1 C_2 G_1 G_2}}$
$S_{G_2}^{\omega_0} = +\dfrac{1}{2}$	$S_{G_2}^Q = \dfrac{1}{2} - G_2 (C_2 + C_1 - C_1 A)\,\dfrac{Q}{\sqrt{C_1 C_2 G_1 G_2}}$
$S_A^{\omega_0} = 0$	$S_A^Q = C_1 G_2\, A\,\dfrac{Q}{\sqrt{C_1 C_2 G_1 G_2}}$

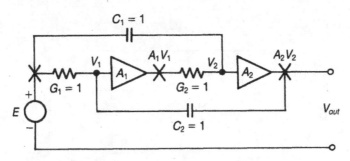

FIGURE 16.5.1   Network with two amplifiers.

Using element values the system matrix becomes

$$\begin{bmatrix} s+1 & -sA_2 \\ -A_1 & s+1 \end{bmatrix}\begin{bmatrix} V_1 \\ V_2 \end{bmatrix} = \begin{bmatrix} E \\ sE \end{bmatrix}$$

The determinant is $D = s^2 + s(2 - A_1A_2) + 1$, $N_2 = A_2(s^2 + s + A_1)$ and the network function is

$$F = \frac{A_2(s^2 + s + A_1)}{s^2 + s(2 - A_1A_2) + 1}$$

Comparing with (16.5.3) we identify $U = 1$, $V = 2 - A_1A_2$ and $W = 1$. Inserting into (16.5.4) we have $\omega_0 = 1$ and $Q = \dfrac{1}{2 - A_1A_2}$. Using the steps explained above we derive that

$$S_{A_1}^Q = S_{A_2}^Q = \frac{A_1A_2}{2 - A_1A_2}$$

In order to get $Q = 5$ the product A_1A_2 must be equal to $\dfrac{9}{5}$ and the Q sensitivity with respect to each amplifier will be 9.

Example 2.

In the network Fig. 16.5.2 keep the two capacitors as variables and find the transfer function. Calculate the Q and ω_0 sensitivities for changes in these two capacitors.

The matrix equation is

$$\begin{bmatrix} -sC_1 & 0 & -10/6 \\ -3 & sC_2+4 & -1 \\ 0 & 2 & -1 \end{bmatrix}\begin{bmatrix} V_1 \\ V_2 \\ V_3 \end{bmatrix} = \begin{bmatrix} 3E \\ 0 \\ 0 \end{bmatrix}$$

The determinant is $D = s^2C_1C_2 + 2sC_1 + 10$ and the transfer function is

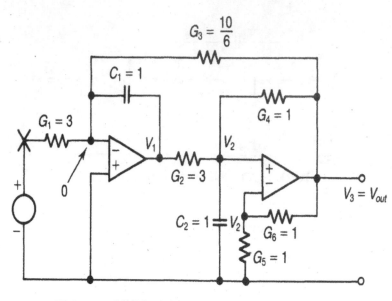

FIGURE 16.5.2   Network with two OPAMPs.

$$F = -\frac{18}{s^2 C_1 C_2 + 2 s C_1 + 10}$$

Following (16.5.3) we identify $U = C_1 C_2$, $V = 2C_1$ and $W = 1$, The sensitivities are $S_{C_1}^U = S_{C_2}^U = 1, S_{C_1}^V = 1, S_{C_2}^V = 0$ and W is a constant and thus does not contribute to sensitivities. Inserting into (16.5.7) and (16.5.8) we get $S_{C_1}^{\omega_0} = S_{C_2}^{\omega_0} = \frac{1}{2}$ and $S_{C_1}^Q = -\frac{1}{2}, S_{C_2}^Q = \frac{1}{2}$.

16.6.   SENSITIVITIES OF POLES AND ZEROS

The subject of this section is the sensitivity of a pole of a second-order network to any element. The derivations are also valid for the numerator of a network function with two complex conjugate zeros. For demonstration we will use the Sallen-Key network but simplify the steps by keeping only one variable, A, and by setting $C_1 = C_2 = 1$ and $G_1 = G_2 = 1$. After the necessary differentiations we will set the gain $A = 2$.

The denominator of the transfer function was derived in (16.4.2) and for the above values

$$D = s^2 + s(3 - A) + 1 \tag{16.6.1}$$

It has two complex conjugate roots. We will consider only the one in the upper half plane because results for the other one are complex conjugate. The root, denoted by r, is obtained after several steps in the form

$$r = \frac{1}{2}\left[A - 3 + j\sqrt{6A - A^2 - 5} \right] \tag{16.6.2}$$

Differentiation with respect to A results in

$$\frac{dr}{dA} = \frac{1}{2}\left[1 + j\frac{3-A}{\sqrt{6A-A^2-5}}\right]$$

If we now substitute $A = 2$, we obtain

$$\frac{dr}{dA} = \frac{1}{2}\left[1 + \frac{j}{\sqrt{3}}\right] \tag{16.6.3}$$

Differentiation of expressions containing square roots are not all that difficult but they are lengthy and it is easy to make a mistake. We would appreciate a method which could get us the result with less effort. This is possible if we proceed as follows. The polynomial, P, is a function of the element, h, and its root is also a function of h. The condition for finding the root can be written as

$$P[h, s(h)] = 0 \,,$$

Since we have here two variables we take partial derivatives

$$\frac{\partial P(h,s)}{\partial h} + \frac{\partial P(h,s)}{\partial s}\frac{\partial s}{\partial h}\bigg|_{s=r} = 0$$

to get

$$\frac{\partial r}{\partial h} = \frac{\partial s}{\partial h}\bigg|_{s=r} = -\frac{\dfrac{\partial P(h,s)}{\partial h}\bigg|}{\dfrac{\partial P(h,s)}{\partial s}\bigg|_{s=r}} \tag{16.6.4}$$

Finding the sensitivity by means of (16.6.4) is easy. We first calculate the root by substituting *all* values. This leads to the polynomial

$$P = s^2 + s + 1 \tag{16.6.5}$$

whose upper half plane root is

$$r = \frac{1}{2}\left[-1 + j\sqrt{3}\right]$$

Next we differentiate (16.6.1) with respect to A

$$\frac{\partial P(h,s)}{\partial A} = -s$$

and substitute the root for s

$$\frac{\partial P(h,s)}{\partial A} = \frac{1}{2}\left[1 - j\sqrt{3}\right]$$

Finally, we differentiate (16.6.5) with respect to s

$$\frac{\partial P(h,s)}{\partial s} = 2s + 1$$

and substitute the root for s again

$$\frac{\partial P(h,s)}{\partial s} = j\sqrt{3}$$

All entries needed for (16.6.4) are now available and

$$\frac{dr}{dA} = \frac{1}{2}(1 + \frac{j}{\sqrt{3}})$$

We can, of course, continue to get the normalized sensitivity as well. In this case it is

$$S_A^r = \frac{dr}{dA}\frac{A}{r} = -\frac{j}{2}\left[\sqrt{3} + \frac{1}{\sqrt{3}}\right]$$

The above steps are a mixture of polynomial differentiations and substitutions of numerical values but there is no differentiation of terms with square roots. If all element values are available numerically we can use (16.6.4) for higher-order polynomials as well but for this we invariably need a computer and a polynomial root-finding routine.

Example 1.

Find the pole sensitivity with respect to A_1 for the network in Fig. 16.5.1. Use the transfer function derived there and the values $A_1 = 1$, $A_2 = 1.5$.

The denominator of the transfer function is

$$D = s^2 + s(2 - A_1A_2) + 1$$

We need the derivatives

$$\frac{dD}{dA_1} = -sA_2$$

$$\frac{dD}{ds} = 2s + 2 - A_1A_2$$

Inserting the gains, the denominator simplifies to $D = s^2 + 0.5s + 1$ and the root (in the upper half plane) is

$$s = 0.25(-1 + j\sqrt{16}$$

To evaluate (16.6.4) substitute the pole coordinate for s in the above two derivatives and obtain

$$\frac{d \text{ pole}}{d A_1} = 1.5 \frac{s}{2s + 0.5} = 0.75(1 + j\frac{1}{\sqrt{15}})$$

As can be seen even from the simple example, sensitivities of poles are rarely needed and the section was added for completeness.

16.7.   SENSITIVITIES OF OPERATIONAL AMPLIFIERS

Ideal operational amplifiers have gains which are theoretically infinitely large. Dealing with infinity is unpleasant in theoretical work and computers cannot handle infinity at all. For these reasons we redefined the operational amplifier as a device with an inverted gain

$$B = -\frac{1}{A} \tag{16.7.1}$$

We also derived in Chapter 12, equation (12.1.4), that properties of nonideal operational amplifiers are expressed by

$$V_+ - V_- + BV_0 = 0 \tag{16.7.2}$$

If the amplifier is ideal, then $B = 0$ and we have an element with zero value. If we wish to find sensitivity with respect to this ideal element, we must use (16.7.2) in all our derivations, obtain the network function in terms of B, differentiate with respect to B and only afterwards insert $B = 0$.

We will demonstrate this case on the simple integrator shown in Fig. 16.7.1. Using the methods of Chapter 12 we write

$$V_1(G + sC) - sCV_{out} = EG$$

$$-V_1 + BV_{out} = 0.$$

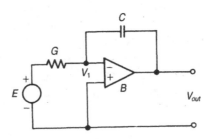

FIGURE 16.7.1   Network with one nonideal OPAMP.

The second equation is (16.7.2) where we identified $V_+ = 0$, $V_- = V_1$ and $V_0 = V_{out}$. The solution is

$$F = \frac{V_{out}}{E} = \frac{G}{-sC + B(G + sC)}$$

Differentiating with respect to B

$$\frac{dF}{dB} = \frac{-G(sC + G)}{\left[-sC + B(sC + G)\right]^2}$$

At this point we can insert $B = 0$ and obtain

$$F = -\frac{G}{sC}$$

$$\frac{dF}{dB} = \frac{G(sC + G)}{s^2C^2}$$

Since the element has zero value, we cannot use the usual normalized sensitivity S and must apply

$$\Sigma_1 = \frac{dF}{dB}\frac{1}{F} = \frac{sC + G}{sC}$$

Hand calculation of sensitivities to changes of operational amplifiers is always time consuming because we must keep the inverted gain B in the equations and cannot assume that the input terminals are at the same potential. This increases the system by one row and column per OPAMP.

PROBLEMS CHAPTER 16

Sensitivity analysis is fairly difficult even in small networks, but is crucial for the design and for checking. We give a few problems to indicate how to proceed in sensitivity analysis; every earlier problem in which some (or all) elements were kept as variables can be used for sensitivity studies.

P.16.1. Find the admittance of the network in terms of variable elements. Obtain expressions for Q, ω_0, S_h^Q and $S_h^{\omega_0}$ where h stands for all elements of the network. Select unit values for all elements and get numerical values for the the above expressions.

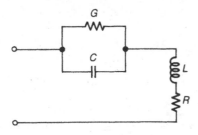

P.16.2. Find the transfer function, F, in terms of variables G and g. Calculate S_h^Q and $S_h^{\omega_0}$ where h is either g or G. Use $G = 1S$ and such g to get $Q = 4$. Insert numerical values into the sensitivity expressions.

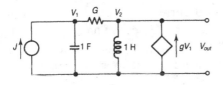

P.16.3. Find the transfer function. Select $G = 1S$ and select g such that $Q = 1$. What is the numerical value for ω_0? Obtain S_h^Q and $S_h^{\omega_0}$ where h is either G or g and insert numerical values.

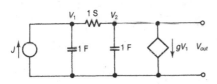

P.16.4. For the problem P.16.3 find where S_g^F is the transfer function. Select $G = 1$, $\omega = \omega_0 = 2$ and obtain $S_g^{|F|}$ at this frequency.

P.16.5. Find the transfer function, F, in terms of variables G_1, G_2 and A. Use $G_1 = G_2$ and design for $Q = 4$ and $\omega_0 = 3$. Find the sensitivity S_A^F. Obtain $S_A^{|F|}$ at $s = j2$.

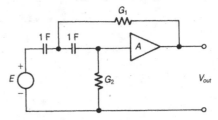

P.16.6. Derive the transfer function in terms of all G and A. Design to realize the denominator polynomial $s^2 + 2s + 2$. Find the sensitivity of S_A^F and $S_A^{|F|}$ at $s = j3$. Find the sensitivity of the pole in the upper half plane to the changes of the gain A.

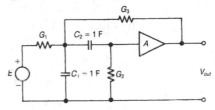

P.16.7. Let the operational amplifier be ideal. Derive the transfer function, F, in terms of all elements. Set afterwards $G_2 = G_3$; you should get a firstorder all-pass function. Find $S_{C_1}^{|F|}, S_{G_1}^{|F|}, S_{G_2}^{|F|}$ and $S_{G_3}^{|F|}$ for $s = j3$. Find the zero and sensitivity with respect to the elements.

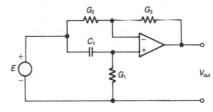

17

MODIFIED NODAL FORMULATION

INTRODUCTION

Until now we solved all our problems by writing either the nodal or the mesh equations. If the network had independent current sources and voltage controlled current sources, we used nodal formulation. If it had voltage sources and current controlled voltage sources, we used mesh equations without modifications. In all other cases we had to resort to pre-processing which sometimes lead to redrawing of the figures. We also discovered that if the network had capacitors as well as inductors, we were unable to avoid fractions of the type $1/s$, no matter which formulation we used.

The various transformations and re-drawings may be acceptable if we are dealing with small problems and hand calculations, but such steps are always sources of possible errors. For larger networks, which we cannot solve by hand anyway, we need a method which avoids all transformations and can be automated. Several possibilities exist and the most practical one is the modified nodal formulation. We pay a modest price for the convenience: the matrices are larger than we could obtain with the various transformations, but are not excessively large. Solving larger systems means more work, but if we leave it to a computer, it need not worry us too much. Here we explain the modified nodal formulation. It is the method used in most modern programs.

17.1. PASSIVE ELEMENTS

Modified nodal formulation starts as nodal formulation by first creating a $n \times n$ matrix where n is the number of ungrounded nodes. This portion handles the current sources, J, voltage controlled current sources, VC, conductances, G and admittances of capacitances, sC, exactly the same way as we have learned in the nodal formulation. We will repeat in this section how to handle the conductances and the capacitances and will also introduce the concept of the *stamp*.

If a conductance, G, is connected between nodes i and j, then the value G is added to the matrix in the position (i,i) and (j,j), and is subtracted from the entries in the positions (i,j) and (j,i). This can be symbolically described by the Fig. 17.1.1, the *stamp* of the conductance.

If the capacitor is connected between nodes i and j, then the product sC is added to the entries in the positions (i,i) and (j,j) and is subtracted from the entries in the positions (i,j) and (j,i). If the capacitor has an initial voltage, E_0, it can be taken into account by an impulsive current source in parallel with the capacitor; this was explained in the text related to Fig. 10.1.10. For the initial voltage, positive at node i and negative at node j, we enter $+CE_0$ in the ith row of the right hand side vector and $-CE_0$ in its jth row. This can be expressed schematically by the stamp of the capacitor as shown in Fig. 17.1.2. The signs of the entries of the right hand side vector will be interchanged if E_0 is negative at node i and positive at node j.

The first difference is encountered when we use inductors. We would like to avoid terms of the form $\dfrac{1}{sL}$ and this can be achieved by considering inductors in their impedance form. To clarify the concept we start with the example in Fig. 17.1.3 where the current through the inductor is indicated as leaving node 1 and entering node 2. Suppose that we remove the inductor and replace it by its current, I_L. This current must be added to the KCL equation of the first node and subtracted from the KCL equation of the second node:

$$V_1 sC_1 + I_L = J$$
$$V_2(sC_2 + G) - I_L = 0 \tag{17.1.1}$$

In addition, we know that

$$sLI_L = V_1 - V_2,$$

and since the voltages as well as the current are unknown, we transfer all to one side of the equation

$$V_1 - V_2 - sLI_L = 0 \tag{17.1.2}$$

Additional equations of this type are given the name *constitutive* equations. We will use this name throughout the text. Collecting (17.1.1) and (17.1.2) into one matrix equation

$$\begin{bmatrix} sC_1 & 0 & 1 \\ 0 & sC_2 + G & -1 \\ 1 & -1 & -sL \end{bmatrix} \begin{bmatrix} V_1 \\ V_2 \\ I_L \end{bmatrix} = \begin{bmatrix} J \\ 0 \\ 0 \end{bmatrix} \tag{17.1.3}$$

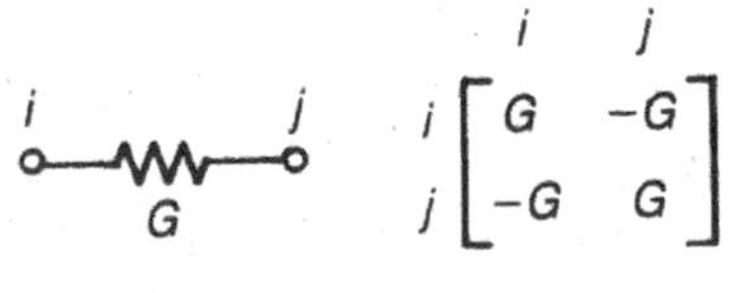

Fig. 17.1.1

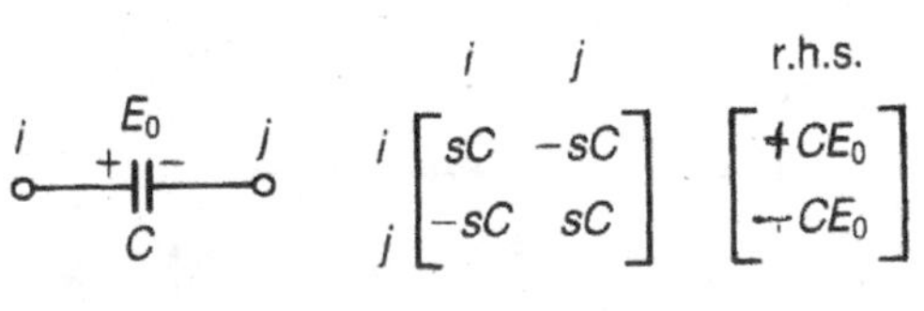

Fig. 17.1.2

FIGURE 17.1.1 Conductance and its stamp.

FIGURE 17.2.2 Capacitor and its stamp.

Inductors are elements which can have initial currents. As we explained in detail in Chapter 10 and shown in Fig. 10.2.4, this initial current can be taken into account by a Dirac impulse voltage source in series with the inductor.

Before deriving the stamp of an inductor, we take another example, Fig.17.1.4a. The initial voltage across the capacitor is represented in Fig. 17.1.4b by the current source in parallel with C. The initial current through the inductor is taken into account by the equivalent impulsive voltage source in series with L. The set of nodal equations is written as before, now taking into account the current source due to the initial voltage across the capacitor:

$$(G_1 + sC)V_1 - sCV_2 + I_L = J - CE_0$$
$$- sCV_1 + (sC + G_2)V_2 - I_L = CE_0$$

The constitutive equation for the inductor with an initial current flowing through it is

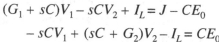

FIGURE 17.1.3   Network with one inductor.

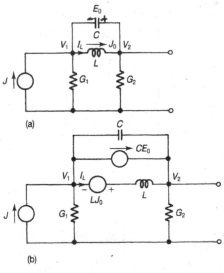

(a)

(b)

Fig. 17.1.4

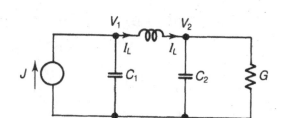

Fig. 17.1.5

FIGURE 17.1.4   Network with initial condition: (a) Original. (b) With equivalent impulse sources.

FIGURE 17.1.5   Inductor and its stamp.

$$V_1 - V_2 - sLI_L = -LJ_0 \qquad (17.1.4)$$

and the matrix equation is

$$\begin{bmatrix} G_1 + sC & -sC & 1 \\ -sC & sC + G_2 & -1 \\ 1 & -1 & -sL \end{bmatrix} \begin{bmatrix} V_1 \\ V_2 \\ I_L \end{bmatrix} = \begin{bmatrix} J - CE_0 \\ CE_0 \\ -LJ_0 \end{bmatrix}$$

We can now draw conclusions from the results of the two examples. Every inductor *increases* the size of the system matrix by one row and one column. If the inductor is connected between nodes i and j, as shown in Fig. 17.1.5, a new column for the current I_L is added to the matrix with +1 in the ith row and -1 in the jth row. The constitutive equation itself is added as a new row. If the inductor has an initial current, J_0, flowing from node i to node j, $-LJ_0$ is entered in the right hand side column in the same row as the constitutive equation. The stamp is in Fig. 17.1.5. Here the empty matrix portion, denoted by $m \times m$, is the *previous matrix* in which the inductor makes no modifications. The additional row and column are marked symbolically by $m + 1$.

The next element we consider is a short circuit. It must be taken into account by marking an *additional* node in the network. If the short circuit is between nodes i and j and we assume, as always, that the current flows from the first node, i, to the second node, j, then I is added in row i and subtracted from row j. In addition, we know that $V_i = V_j$. Since none of these voltages is known, we write this constitutive equation as

$$V_i - V_j = 0 \qquad (17.1.5)$$

and add it to the previous set of equations. This leads to the stamp in Fig. 17.1.6

An example will show how we set up the modified nodal equations if we have more than one element which increases the size of the matrix. The network in Fig. 17.1.7 has three ungrounded nodes; its nodal portion will have the size 3×3. We write the nodal equations by removing the inductors and replacing them with their currents. This leads to

$$(G_1 + sC_1)V_1 + I_{L1} = J$$

$$-I_{L1} + sC_2V_2 + I_{L2} = 0$$

$$-I_{L2} + (G_2 + sC_3)V_3 = 0$$

Next we add the two constitutive equations

$$V_1 - V_2 - sL_1I_{L1} = 0$$
$$V_2 - V_3 - sL_2I_{L2} = 0$$

to complete the set. In matrix form

$$\begin{bmatrix} G_1 + sC_1 & 0 & 0 & 1 & 0 \\ 0 & sC_2 & 0 & -1 & 1 \\ 0 & 0 & sC_3 + G_2 & 0 & -1 \\ 1 & -1 & 0 & -sL_1 & 0 \\ 0 & 1 & -1 & 0 & -sL_2 \end{bmatrix} \begin{bmatrix} V_1 \\ V_2 \\ V_3 \\ I_{L1} \\ I_{L2} \end{bmatrix} = \begin{bmatrix} J \\ 0 \\ 0 \\ 0 \\ 0 \end{bmatrix}$$

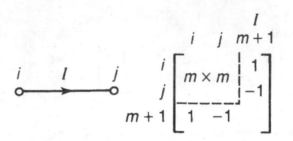

FIGURE 17.1.6   Short circuit and its stamp.

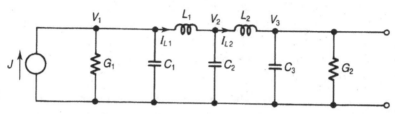

FIGURE 17.1.7   Network with two inductors.

In the above example we first wrote the KLC equations one by one and then collected them as a matrix. This is not necessary and we can write the matrix equation directly by using the stamps. Disregard first the presence of the inductors and write the nodal 3×3 matrix. Next use the stamp for the first inductor and add one row and one column. For second inductor use the same stamp and add *again* a new row and column to the previous matrix. Now we see why we have denoted the new row and column by $m + 1$: it simply indicates an additional row and column to whatever matrix we had before.

The last element which *should not* normally be used is a linear resistor, taken in impedance form. If it is connected between nodes i and j, with the current assumed to flow from i to j, then the constitutive equation is

$$V_i - V_j - RI = 0 \qquad (17.1.6)$$

In addition, current I must be added in row i and subtracted in row j. The stamp is similar to the stamp of the inductor and is in Fig. 17.1.8. It can be seen that a resistor, taken as a resistor, adds unnecessarily one row and one column to the system matrix. This is the reason why we *always* use conductances in linear networks. Even if the user gives the

FIGURE 17.1.8   Resistor and its stamp (not normally used).

resistor as a resistor, we convert the element internally into G and return to the given element only in the result.

It is advisable to compare the stamps for the inductor, resistor and short circuit; they are quite similar and this similarity should be remembered.

17.2.   INDEPENDENT SOURCES

The independent current source has not presented any problems in nodal formulation and we use it exactly as before. The voltage source was causing problems in nodal formulation and we always transformed it into a current source. In modified nodal formulation we take its current into account and do not perform any network transformations.

To introduce the concept we start with a simple network having one voltage source and one current source, see Fig. 17.2.1. Assign voltages to all three nodes. At the beginning disregard the presence of the voltage source but take into account its current I_E (in the direction from + to −). The three nodal equations are

$$I_E + sC_1V_1 - sC_1V_2 = 0$$

$$-sC_1V_1 + (sC_1 + G_1 + G_2)V_2 - G_2V_3 = 0$$

$$-G_2V_2 + (G_2 + sC_2)V_3 = J$$

At this point we have three equations and four unknowns: V_1, V_2, V_3 and I_E. However, we still did not use the fact that the voltage V_1 is known and equal to E,

$$V_1 = E$$

Now we have the necessary number of equations; in the matrix we first write the nodal portion and account for the voltage source in the fourth row and column:

$$
\begin{bmatrix}
sC_1 & -sC_1 & 0 & 1 \\
-sC_1 & sC_1 + G_1 + G_2 & -G_2 & 0 \\
0 & -G_2 & G_2 + sG_2 & 0 \\
1 & 0 & 0 & 0
\end{bmatrix}
\begin{bmatrix}
V_1 \\
V_2 \\
V_3 \\
I_E
\end{bmatrix}
=
\begin{bmatrix}
0 \\
0 \\
J \\
E
\end{bmatrix}
$$

The steps do not change if the voltage source is floating (not connected to ground), as in Fig. 17.2.2. First we disregard the voltage source, keep its current and write the nodal equations:

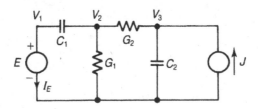

FIGURE 17.2.1   Network with two independent sources.

$$(G_1 + G_2)V_1 - G_2V_2 + I_E = 0$$

$$- G_2V_1 + (G_2 + G_3 + G_4)V_2 - G_4V_3 = 0$$

$$- G_4V_2 + (G_4 + G_5)V_3 - I_E = 0$$

In addition, we know the constitutive equation

$$V_1 - V_3 = E$$

which is the last equation needed. In matrix form

$$
\begin{bmatrix}
G_1 + G_2 & -G_2 & 0 & 1 \\
-G_2 & G_2 + G_3 + G_4 & -G_4 & 0 \\
0 & -G_4 & G_4 + G_5 & -1 \\
1 & 0 & -1 & 0
\end{bmatrix}
\begin{bmatrix}
V_1 \\
V_2 \\
V_3 \\
I_E
\end{bmatrix}
=
\begin{bmatrix}
0 \\
0 \\
0 \\
E
\end{bmatrix}
$$

These equations give us enough information to write the stamps for both independent sources. Let the current source be as in Fig. 17.2.3, connected to nodes i and j with current flowing away from node i. The stamp is in the same figure and follows the rules established for nodal formulation.

If the voltage source is connected as in Fig. 17.2.4, with the + sign at node i and the − sign at j, then the current I is added in row i and subtracted in row j. The constitutive equation is

$$V_i - V_j = E \tag{17.2.1}$$

and the stamp is as shown; it requires addition of one row and one column to the previously defined matrix.

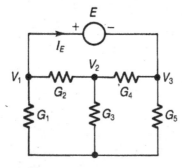

Fig. 17.2.2

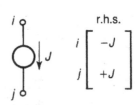

Fig. 17.2.3

FIGURE 17.2.2   Network with a floating voltage source.

FIGURE.17.2.3   Independent current source and its stamp.

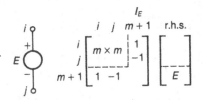

Fig. 17.2.4

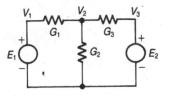

Fig. 17.2.5

FIGURE 17.2.4   Independent voltage sourece and its stamp.

FIGURE 17.2.5   Network with two voltage sources.

If the network has several independent voltage sources, then we must add one row and one column for each. For instance, for the network in Fig. 17.2.5 we get the matrix equation

$$
\begin{bmatrix}
G_1 & -G_1 & 0 & 1 & 0 \\
-G_1 & (G_1 + G_2 + G_3) & -G_3 & 0 & 0 \\
0 & -G_3 & G_3 & 0 & 1 \\
1 & 0 & 0 & 0 & 0 \\
0 & 0 & 1 & 0 & 0
\end{bmatrix}
\begin{bmatrix}
V_1 \\
V_2 \\
V_3 \\
I_{E1} \\
I_{E2}
\end{bmatrix}
=
\begin{bmatrix}
0 \\
0 \\
0 \\
E_1 \\
E_2
\end{bmatrix}
$$

17.3.   CONTROLLED SOURCES

The four dependent sources in Chapter 4 were already presented in a suitable form. There is no change in the use of the voltage controlled current source, *VC*. If the controlling voltage is taken between nodes i (noninverting) and node j (inverting) and the output is connected to nodes k and l with the current flowing from node k to l, then the currents

$$I_k = gV_i - gV_j$$

$$I_l = -gV_i + gV_j$$

(17.3.1)

and are added to the kth and lth nodal equation, respectively. This means that the stamp of the *VC* is as shown in Fig. 17.3.1. As examples use Fig. 4.1.2 through 4.1.6 in Chapter 4 without any modification.

The voltage controlled voltage source, *VV* is in Fig. 17.3.2. The output voltage is

$$V_k - V_l = \mu(V_i - V_j)$$

but since none of these voltages is known, we transfer all terms to the left and get the constitutive equation

$$-\mu V_i + \mu V_j + V_k - V_l = 0$$

(17.3.2)

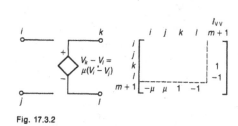

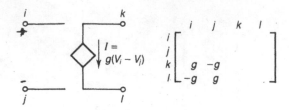

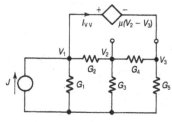

FIGURE 17.3.1   *VC and its stamp.*

Fig. 17.3.2

Fig. 17.3.3

FIGURE 17.3.2   *VV and its stamp.*

FIGURE 17.3.3   Network with a floating *VV.*

A current also flows through the voltage source, in the positive direction at node k and in the negative direction at node l. The stamp is in the same figure. As another example consider the network in Fig. 17.3.3. Similarly as in the case of the independent source, we first "remove" the voltage source but consider its current, I_{VV}. The nodal equations are

$$(G_1 + G_2)V_1 - G_2V_2 + I_{VV} = J$$

$$- G_2V_1 + (G_2 + G_3 + G_4)V_2 - G_4V_3 = 0$$

$$- G_4V_2 + (G_4 + G_5)V_3 - I_{VV} = 0$$

We also use the constitutive equation (17.3.2) in which we identify $V_i = V_2$, $V_j = V_3$, $V_k = V_1$ and $V_l = V_3$:

$$- \mu V_2 + \mu V_3 + V_1 - V_3 = 0$$

This results in

$$\begin{bmatrix} (G_1 + G_2) & -G_2 & 0 & 1 \\ -G_2 & (G_2 + G_3 + G_4) & -G_4 & 0 \\ 0 & -G_4 & (G_4 + G_5) & -1 \\ 1 & -\mu & \mu - 1 & 0 \end{bmatrix} \begin{bmatrix} V_1 \\ V_2 \\ V_3 \\ I_{VV} \end{bmatrix} = \begin{bmatrix} J \\ 0 \\ 0 \\ 0 \end{bmatrix}$$

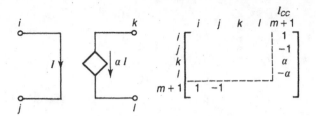

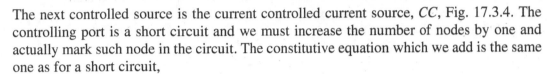

FIGURE 17.3.4   *CC* and Its stamp.

The next controlled source is the current controlled current source, *CC*, Fig. 17.3.4. The controlling port is a short circuit and we must increase the number of nodes by one and actually mark such node in the circuit. The constitutive equation which we add is the same one as for a short circuit,

$$V_i - V_j = 0. \tag{17.3.3}$$

We must also add the current I to the ith nodal equation and subtract it from the jth equation. In addition, the output of the *CC* adds the current αI at node k and subtracts the same current at node l. These explanations should now be sufficient to derive the stamp; it is shown in Fig. 17.3.4. For better understanding we will explain application on the network in Fig. 17.3.5 for which we write the nodal equations and attach to them (17.3.3):

$$G_1 V_1 + I + \alpha I = J$$

$$-I + V_2 G_2 - V_3 G_2 = 0$$

$$-V_2 G_2 + (G_2 + G_3 + G_4) V_3 - G_4 V_4 = 0$$

$$-G_4 V_3 + (G_4 + G_5) V_5 - \alpha I = 0$$

$$V_1 - V_2 = 0$$

In matrix form

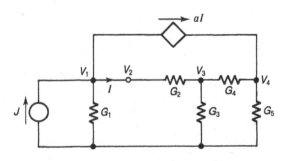

FIGURE 17.3.5   Network with a floating *CC*.

$$\begin{bmatrix} G_1 & 0 & 0 & 0 & (1+\alpha) \\ 0 & G_2 & -G_2 & 0 & -1 \\ 0 & -G_2 & (G_2+G_3+G_4) & -G_4 & 0 \\ 0 & 0 & -G_4 & (G_4+G_5) & -\alpha \\ 1 & -1 & 0 & 0 & 0 \end{bmatrix} \begin{bmatrix} V_1 \\ V_2 \\ V_3 \\ V_4 \\ I \end{bmatrix} = \begin{bmatrix} J \\ 0 \\ 0 \\ 0 \\ 0 \end{bmatrix}$$

The last element of this section is a current controlled voltage source, CV, shown in Fig. 17.3.6. The short circuit at its input port requires one additional row and column. The voltage source at its output port also requires one more row and column. Thus the set of equations must be increased by two and we must add two variables and two constitutive equations. The constitutive equation of the short circuit is

$$V_i - V_j = 0 \qquad (17.3.4)$$

For the voltage source we have $V_k = V_l = rI_1$. Since neither the voltages nor the current are known, we transfer all terms to one side and the constitutive equation becomes

$$V_k - V_l - rI_1 = 0 \qquad (17.3.5)$$

The current I_1 modifies the ith and jth nodal equations, the current of the voltage source, I_2, modifies also the kth and lth nodal equations. This leads to the stamp in Fig. 17.3.6.

To clarify the stamp, we also take the example in Fig. 17.3.7. The nodal equations are

$$G_1V_1 + I_1 + I_2 = J$$

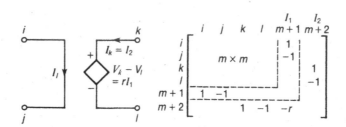

FIGURE 17.3.6 *CV* and its stamp.

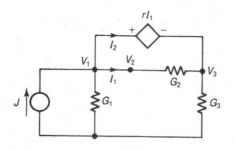

FIGURE 17.3.7 Network with a floating *CV*.

$$-I_1 + G_2V_2 - G_2V_3 = 0$$

$$-G_2V_2 + (G_2 + G_3)V_3 - I_2 = 0$$

To these equations we add the constitutive equations

$$V_1 - V_2 = 0$$

$$V_1 - V_3 - rI_1 = 0$$

and in matrix form

$$
\begin{bmatrix}
G_1 & 0 & 0 & 1 & 1 \\
0 & G_2 & -G_2 & -1 & 0 \\
0 & -G_2 & (G_2 + G_3) & 0 & -1 \\
1 & -1 & 0 & 0 & 0 \\
1 & 0 & -1 & -r & 0
\end{bmatrix}
\begin{bmatrix}
V_1 \\
V_2 \\
V_3 \\
I_1 \\
I_2
\end{bmatrix}
=
\begin{bmatrix}
J \\
0 \\
0 \\
0 \\
0
\end{bmatrix}
$$

17.4. SPECIAL ELEMENTS

In this section we will consider stamps of opertional amplifiers and two coil transformers. However, before doing so, it is useful to show how nodal formulation relates to the two-port definitions. The boxes in Fig. 17.4.1a and b denote the *same* network. In *a* is the two-port definition, in *b* is the nodal definition with voltages measured with respect to ground. Comparison immediately reveals the following equivalences:

$$V_i - V_j = V_1$$

$$V_k - V_l = V_2$$

$$I_i = I_1$$

$$I_j = -I_1$$

$$I_k = I_2$$

$$I_l = -I_2$$

(17.4.1)

We will use them to derive the stamps.

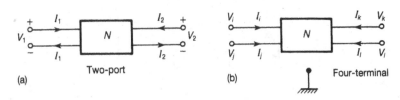

(a) Two-port (b) Four-terminal

FIGURE 17.4.1 A network and its: (a) Two-port description. (b) Four-terminal description.

7.4.1. Ideal Operational Amplifier

The ideal operational amplifier was already used in Chapter 12 but here we will make it more general and allow both output terminals to be floating, see Fig. 17.4.2. The constitutive equation is

$$(V_i - V_j)A = V_k - V_l \tag{17.4.2}$$

Since we wish to have one stamp for both ideal as well as nonideal OPAMPs, we cannot keep A; instead we again introduce the inverted gain

$$B = -\frac{1}{A} \tag{17.4.3}$$

and modify (17.4.2)

$$V_i - V_j + BV_k - BV_l = 0 \tag{17.4.4}$$

This is the constitutive equation to be used in the stamp. The ideal operational amplifier does not draw any current at its input terminals; only the kth and lth nodes will be influenced by the output current. The stamp is in Fig. 17.4.2. If terminal l is grounded (the usual case), then the floating operational amplifier symbol reduces to Fig. 17.4.3 and V_l becomes zero.

We will use the network in Fig. 17.4.4 to write modified nodal equations and also the matrix equation. For the three ungrounded nodes we have

$$GV_1 - GV_2 + I_E = 0$$

$$-GV_1 + (sC + G)V_2 - sCV_3 = 0$$

$$-sCV_2 + sCV_3 + I_{OP} = 0$$

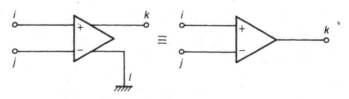

FIGURE 17.4.2 Operational amplifier and its stamp.

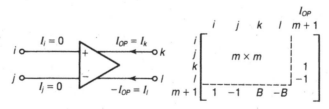

FIGURE 17.4.3 Equivalent symbols for an OPAMP.

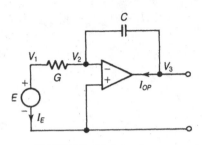

FIGURE 17.4.4   Integrator.

and the constitutive equations for the voltage source and operational amplifier are, respectively,

$$V_1 = E$$

$$-V_2 + BV_3 = 0$$

Collected together

$$\begin{bmatrix} G & -G & 0 & 1 & 0 \\ -G & G+sC & -sC & 0 & 0 \\ 0 & -sC & sC & 0 & 1 \\ 1 & 0 & 0 & 0 & 0 \\ 0 & -1 & B & 0 & 0 \end{bmatrix} \begin{bmatrix} V_1 \\ V_2 \\ V_3 \\ I_E \\ I_{OP} \end{bmatrix} = \begin{bmatrix} 0 \\ 0 \\ 0 \\ E \\ 0 \end{bmatrix}$$

This matrix equation can also be written directly by inspection. Set up first the nodal portion; it has the dimension 3×3 because there are three ungrounded nodes. Next attach the constitutive equation for the voltage source and also modify the first row by adding one column with unit entry in the first row. This takes care of the independent voltage source. Next attach the constitutive equation of the operational amplifier and add one column with a unit in row three, because the current of the OPAMP output influences the current balance of the third node.

As a reminder, we repeat that the gain of a practical OPAMP changes with frequency and we often assume that

$$A = \frac{A_0 \omega_0}{s + \omega_0} \tag{17.4.5}$$

17.4.2.   Transformer With Two Coils

We derived, in section 14.1, that a transformer with initial currents through the two coils is described by the equations

$$V_1 = sL_1 I_1 + sMI_2 - L_1 J_{10} - MJ_{20}$$

$$V_2 = sMI_1 + sL_2 I_2 - MJ_{10} - L_2 J_{20}$$

If we substitute from (17.4.1) for V_1 and V_2, we get two constitutive equations

$$V_i - V_j - SL_1 I_1 - sMI_2 = -L_1 J_{10} - MJ_{20}$$

$$V_k - V_l - sMI_1 - sL_2 I_2 = -MJ_{10} - L_2 J_{20}$$

(17.4.6)

Nodes i and j have their currents modified by I_1, nodes k and l by I_2; the stamp is in Fig. 17.4.5. The right hand side vector is also influenced if the initial currents are nonzero.

As an example consider the network in Fig. 17.4.6. Although we will set up the equations, the reader should attempt to write the matrix equation directly. The equations are

$$sC_1 V_1 - sC_1 V_2 + I_1 = J$$

$$-sC_1 V_1 + sC_1 V_2 + I_2 = 0$$

$$sC_2 V_3 - I_1 + I_2 = 0$$

$$V_1 - V_3 - sL_1 I_1 - sMI_2 = 0$$

$$V_2 - V_3 - sMI_1 - sL_2 I_2 = 0$$

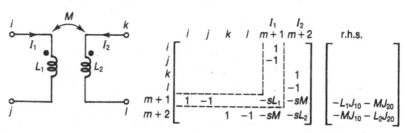

FIGURE 17.4.5   Transformer and its stamp.

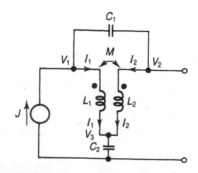

FIGURE 17.4.6   Network with a transformer.

and the matrix equation is

$$\begin{bmatrix} sC_1 & -sC_1 & 0 & 1 & 0 \\ -sC_1 & sC_1 & 0 & 0 & 1 \\ 0 & 0 & sC_2 & -1 & -1 \\ 1 & 0 & -1 & -sL_1 & -sM \\ 0 & 1 & -1 & -sM & -sL_2 \end{bmatrix} \begin{bmatrix} V_1 \\ V_2 \\ V_3 \\ I_1 \\ I_2 \end{bmatrix} = \begin{bmatrix} J \\ 0 \\ 0 \\ 0 \\ 0 \end{bmatrix}$$

Note how simple the handling of the transformer in this formulation really is.

17.5. COMPUTER APPLICATIONS

We are adding this section for readers interested in understanding how the theories presented in this chapter can be applied to computer applications. For additional information see: J. Vlach and K. Singhal: Computer Methods for Circuit Analysis and Design, Chapman & Hall, 1994, Chapter 4.

In all examples presented in this Chapter we always wrote one system matrix with all its components in it. For instance, for the network in Fig. 17.1.7 we obtained the system equation

$$\begin{bmatrix} G_1 + sC_1 & 0 & 0 & 1 & 0 \\ 0 & sC_2 & 0 & -1 & 1 \\ 0 & 0 & sC_3 + G_2 & 0 & -1 \\ 1 & -1 & 0 & -sL_1 & 0 \\ 0 & 0 & -1 & 0 & -sL_2 \end{bmatrix} \begin{bmatrix} V_1 \\ V_2 \\ V_3 \\ I_{L1} \\ I_{L2} \end{bmatrix} = \begin{bmatrix} J \\ 0 \\ 0 \\ 0 \\ 0 \end{bmatrix}$$

Denote the matrix by **T**, the vector of unknowns by **X** and the right hand side by **W**,

$$\mathbf{TX = W}$$

For computer application, it is convenient to decompose the matrix **T** into two real matrices which we will call **GM** and **CM**:

$$\mathbf{T = GM} + s\mathbf{CM}$$

The two matrices are:

$$\mathbf{GM} = \begin{bmatrix} G_1 & 0 & 0 & 1 & 0 \\ 0 & 0 & 0 & -1 & 1 \\ 0 & 0 & G_2 & 0 & -1 \\ 1 & -1 & 0 & 0 & 0 \\ 0 & 1 & -1 & 0 & 0 \end{bmatrix}$$

$$\mathbf{CM} = \begin{bmatrix} C_1 & 0 & 0 & 0 & 0 \\ 0 & C_2 & 0 & 0 & 0 \\ 0 & 0 & C_3 & 0 & 0 \\ 0 & 0 & 0 & -L_1 & 0 \\ 0 & 0 & 0 & 0 & -L_2 \end{bmatrix}$$

The right hand side vector is unchanged,

$$\mathbf{W}^T = [\, J\, 0\, 0\, 0\, 0\,].$$

To save space we have written it here in transposed form.

Let us take another network, Fig. 17.5.1. When writing the **GM** matrix we simply disregard all capacitors and inductors, as if they were erased from the figure. The network has 5 ungrounded nodes and the voltage source with the operational amplifier will increase the size of the overall matrix to 7×7. We can fill the nodal portion of the matrix by the methods we learned earlier, or element by element: for instance, G_1 will be positive in the position $(1,1)$ and $(2,2)$ and negative in positions $(1,2)$ and $(2,1)$. G_5 will be positive in positions $(1,1)$ and $(4,4)$, negative in $(1,4)$ and $(4,1)$. Any grounded conductance will appear only once: G_2 is in only one place, the entry $(2,2)$.

When the nodal portion has been prepared, we add entries for the other two elements. The voltage source is connected to node 1 and its current will influence row 1; as a result, 1 comes into the $(1,6)$ position. Constitutive equation for the source is $V_1 = E$ and 1 is entered into the position $(6,1)$. The operational amplifier influences the balance of currents at node 5; this is taken care of by an additional, seventh, column with 1 in the $(5,7)$ position. Finally we attach the constitutive equation for the operational amplifier, in this case

$$- V_3 + V_4 + BV_5 = 0$$

and the entries are -1 in position $(7,3)$, $+1$ in the position $(7,4)$ and B in the position $(7,5)$. This completes writing of the **GM** matrix:

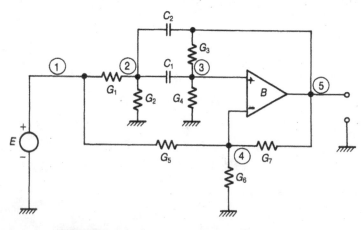

FIGURE 17.5.1   Second-order active network.

$$
\mathbf{GM} =
\begin{bmatrix}
G_1+G_5 & -G_1 & 0 & -G_5 & 0 & 1 & 0 \\
-G_1 & G_1+G_2 & 0 & 0 & 0 & 0 & 0 \\
0 & 0 & G_3+G_4 & 0 & -G_3 & 0 & 0 \\
-G_5 & 0 & 0 & G_5+G_6+G_7 & -G_7 & 0 & 0 \\
0 & 0 & -G_3 & -G_7 & G_3+G_7 & 0 & 1 \\
1 & 0 & 0 & 0 & 0 & 0 & 0 \\
0 & 0 & -1 & 1 & B & 0 & 0
\end{bmatrix}
$$

The **CM** matrix must have the same size as **GM**, 7×7. C_1 will be positive in positions (2,2) and (3,3), negative in (2,3) and (3,2). C_2 will be taken positive in (1,1) and (5,5), negative in (1,5) and (5,1). All other entries are zero,

$$
\mathbf{CM} =
\begin{bmatrix}
0 & 0 & 0 & 0 & 0 & 0 & 0 \\
0 & C_1+C_2 & -C_1 & 0 & -C_2 & 0 & 0 \\
0 & -C_1 & C_1 & 0 & 0 & 0 & 0 \\
0 & 0 & 0 & 0 & 0 & 0 & 0 \\
0 & -C_2 & 0 & 0 & C_2 & 0 & 0 \\
0 & 0 & 0 & 0 & 0 & 0 & 0 \\
0 & 0 & 0 & 0 & 0 & 0 & 0
\end{bmatrix}
$$

The right hand side vector length must be the same as the size of the matrices, 7, and only the voltage source will have an entry in this vector. Since the constitutive equation for the source was taken care of by the sixth row of the **GM** matrix, the only nonzero entry will be the sixth one. The vector will be

$$
\mathbf{W}^T = [\, 0\ 0\ 0\ 0\ 0\ E\ 0\,]
$$

written horizontally in transposed form to save space.

In programming, the matrices **GM** and **CM** and the vector **W** can be prepared by first setting zeros into all entries and then writing the nonzero entries as they are given above. These steps are done only once and the matrices are re-used. For instance, if we wish to solve the network in frequency domain, we must substitute $s = j\omega$ and create a complex matrix **T**

$$
\mathbf{T} = \mathbf{GM} + s\mathbf{CM} \tag{17.5.1}
$$

Whenever we change the frequency, all we have to do is change ω, define a new s and repeat. For each value of s we must solve the equation

$$
\mathbf{TX} = \mathbf{W}
$$

which means that we must have routines which work with complex numbers.

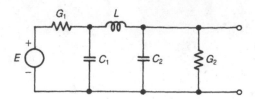

FIGURE 17.5.2   LCG network.

Example 1.

Write the **GM, CM** matrices and the **W** vector for the network in Fig. 17.5.2. The network has three ungrounded nodes, one inductor and one voltage source and thus the overall size of the matrices will be 5×5. The matrices are

$$
\mathbf{GM} = \begin{bmatrix}
G_1 & -G_1 & 0 & 0 & 1 \\
-G_1 & G_1 & 0 & 1 & 0 \\
0 & 0 & G_2 & -1 & 0 \\
0 & 1 & -1 & 0 & 0 \\
1 & 0 & 0 & 0 & 0
\end{bmatrix}
$$

$$
\mathbf{CM} = \begin{bmatrix}
0 & 0 & 0 & 0 & 0 \\
0 & C_1 & 0 & 0 & 0 \\
0 & 0 & C_2 & 0 & 0 \\
0 & 0 & 0 & -L & 0 \\
0 & 0 & 0 & 0 & 0
\end{bmatrix}
$$

and the right hand side vector is

$$
\mathbf{W}^T = [\, 0\ 0\ 0\ 0\ E\,].
$$

PROBLEMS CHAPTER 17

Networks with any elements can be set up in modified nodal formulation. Here we give a few simple networks to test the skills.

P.17.1–P.17.11.   Write the modified nodal matrix equations.

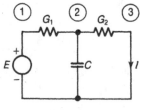

P.17.1

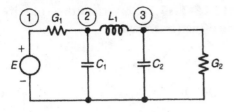

P.17.2

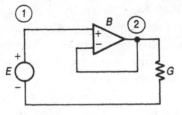

P.17.7

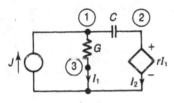

P.17.3

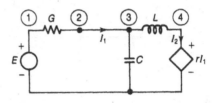

P.17.8

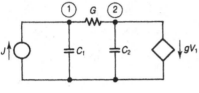

P.17.4

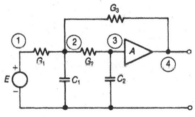

P.17.9

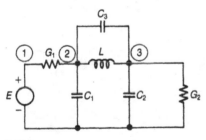

P.17.5

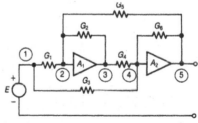

P.17.10

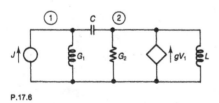

P.17.6

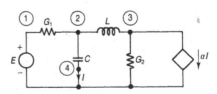

P.17.11

P.17.12–P.17.14. Normally we convert all resistors into conductances, to keep the size of the matrix small. In these examples take the resistors as resistors and write the modified nodal matrix equation. The matrix grows by one row and one column for each such resistor. Write also the matrix equations by converting the resistors into conductances.

P.17.15–P.17.18. Write the modified nodal matrix equations.

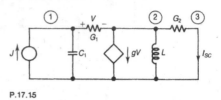

P.17.15

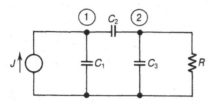

P.17.12

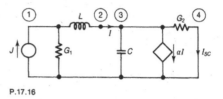

P.17.16

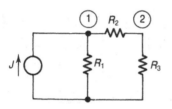

P.17.13

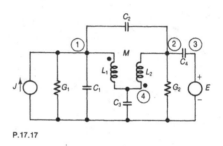

P.17.17

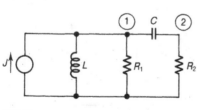

P.17.14

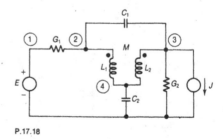

P.17.18

18 FOURIER SERIES AND TRANSFORMATION

INTRODUCTION

Books on basic network theory usually include the Fourier series and Fourier integral theories, but the subject is not strongly related to network analysis. Both Fourier methods are always taught in courses on mathematics and we assume that the student either has been, or will be, exposed to them elsewhere. We cover only those points which have some importance for the material of this book.

In the previous chapters we encountered two types of signals. The first were purely sinusoidal and we discussed them in Chapter 8. We have seen that linear networks influence the amplitude and phase of sinusoidal signals. The other type were signals suddenly starting at $t = 0$ and we considered them in time domain, Chapter 10. In every example the output signal was different from the input signal and the networks had different responses to the same signals. We would like to know why this is so, and perhaps what can be done to either avoid such distortions, or to create distortions according to our wishes. To be able to do so we must have more information about the signals; Fourier series and Fourier integral are the tools which help in answering these questions. Fourier series deals with periodic but nonsinusoidal signals. Fourier integral gives answers for signals of arbitrary shapes, occurring only once. Both methods are the subject of this brief chapter.

18.1. FOURIER SERIES

The Fourier series method tells us what types of components are present in a periodic but nonsinusoidal signal. As we know from Chapter 8, a periodic function satisfies the condition

$$f(t) = f(t + nT) \tag{18.1.1}$$

where T is the time of the period and n is an arbitrary integer. We also defined the frequency,

$$f = \frac{1}{T} \tag{18.1.2}$$

and the angular frequency

$$\omega = 2\pi f = \frac{2\pi}{T} \tag{18.1.3}$$

The theory of Fourier series states that any periodic function is composed of components having various amplitudes and phases, but *all* of them having frequencies which are integer multiples of the fundamental frequency, given by (18.1.2). The most frequently used form of the Fourier series is defined by the equations

$$f(t) = \frac{a_0}{2} + \sum_{n=1}^{\infty}(a_n \cos n\omega t + b_n \sin n\omega t) \tag{18.1.4}$$

where the coefficients are obtained by solving the formulas

$$a_0 = \frac{2}{T}\int_{t_0}^{t_0+T} f(t)dt \tag{18.1.5}$$

$$a_n = \frac{2}{T}\int_{t_0}^{t_0+T} f(t)\cos n\omega t dt \quad n = 1,2,.. \tag{18.1.6}$$

$$b_n = \frac{2}{T}\int_{t_0}^{t_0+T} f(t)\sin n\omega t dt \quad n = 1,2,.. \tag{18.1.7}$$

Of special importance in the above equations is the integration interval from t_0 to $t_0 + T$. We can use an arbitrary point on the periodic function and let it be t_0. If we select $t_0 = 0$, then we will integrate from 0 to T. We can equally well set $t = -\frac{T}{2}$ and integrate from $-\frac{T}{2}$ to $+\frac{T}{2}$ The name *fundamental harmonic* was given to the component with $n = 1$. The other components are called second, third harmonic or just *higher harmonics*.

Fourier series has very few practical restrictions and we quote them for completeness:

(a) $f(t)$ is a single valued function.

(b) The integral $\int_{t_0}^{t_0+T} |f(t)|dt$ is finite.

(c) $f(t)$ has a finite number of discontinuities.

None of these conditions will be violated by any practical function.

Figure 18.1.1a–c shows an impulse of duration δ, repeated with period T. In (a) we selected the interval from the middle of the lower portion. In (b) we placed the origin in the middle of the impulse, while in (c) we placed it at the start of the impulse. This freedom of choice leads to considerable simplifications if the functions are even or odd. As is known from mathematics, an even function satisfies the equation

$$f(t) = f(-t)$$

and choice (a) or (b) makes the function in Fig. 18.1.1 an even function. An odd function satisfies

$$f(t) = f(-t)$$

Consider next the two cases in Fig. 18.1.2. The choice of the origin in (a) makes the function an odd function. The choice in (b) makes it an even function.

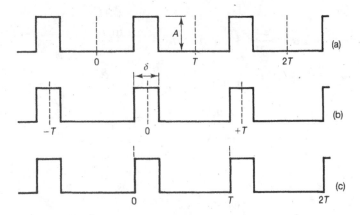

FIGURE 18.1.1 Three choices of the origin for a periodic pulse.

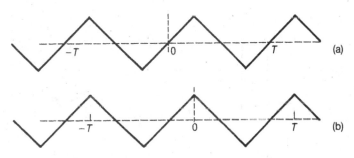

FIGURE 18.1.2. Periodic triangle: (a) As and odd function (b) As an even function.

We give here so much importance to these choices for one very practical reason. An even function will have all b_n coefficients equal to zero and we need not calculate them. Similarly, an odd function will have all a_n terms equal to zero and we can skip the calculations again. We also note that the coefficient $\dfrac{a_0}{2}$ is the average of the function over the period T. Thus, the functions in 18.1.2 will have $a_0 = 0$, because the area above the axis is the same as the area below. The integration to obtain the coefficients of even or odd functions simplifies as well: it is sufficient to calculate the integral over one half of the period and take the result twice.

As an example of a Fourier expansion consider Fig. 18.1.1b. The function has height A and is nonzero from $-\dfrac{\delta}{2}$ to $+\dfrac{\delta}{2}$. The coefficients will be

$$a_0 = \frac{2A}{T} \int\limits_{-\frac{\delta}{2}}^{+\frac{\delta}{2}} dt = \frac{2A\delta}{T} \qquad (18.1.8)$$

$$a_n = \frac{2A}{T} \int_{-\delta/2}^{+\delta/2} \cos n\omega t \, dt = \frac{2A}{T} \left[\frac{\sin n\omega t}{n\omega} \right]_{-\delta/2}^{\delta/2}$$

$$= \frac{4A}{T} \frac{\sin n\omega \frac{\delta}{2}}{n\omega} = \frac{2A\delta}{T} \frac{\sin n\omega \frac{\delta}{2}}{n\omega \frac{\delta}{2}} \tag{18.1.9}$$

The expansion is

$$f(t) = \frac{A\delta}{T} + \sum_{1}^{\infty} \frac{2A\delta}{T} \frac{\sin n\omega \frac{\delta}{2}}{n\omega \frac{\delta}{2}} \cos n\omega t \tag{18.1.10}$$

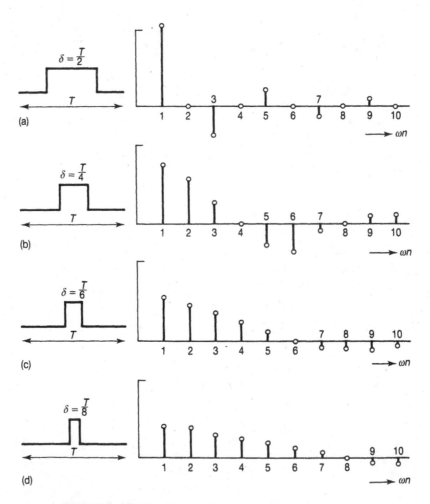

FIGURE 18.1.3   Spectra for the function in Fig 10.1.1.

We will consider four cases with $\delta = \dfrac{T}{2}, \dfrac{T}{4}, \dfrac{T}{6}$ and $\dfrac{T}{8}$ For these cases the dc term, $\dfrac{a_0}{2}$

is equal to $\dfrac{1}{2}, \dfrac{1}{4}, \dfrac{1}{6}$, and $\dfrac{1}{8}$, respectively. The coefficients are given by $a_n = \dfrac{2A}{\pi} \dfrac{\sin \dfrac{\pi}{T} \delta}{n}$

and all four cases are plotted in Fig. 18.1.3a–d for $A = 1$. The coefficients of the harmonic components are indicated as lines with circles at their ends. An envelope can be drawn by connecting the circles: in all cases it has the form $\dfrac{\sin x}{x}$, compare with equation (18.1.10). Fig. 18.1.4 shows approximation of the square wave with $\delta = \dfrac{T}{2}$ by (a) one, (b) two, and (c) three harmonics.

Example 1

Find the Fourier coefficients and the expansion for the wave form in Fig. 18.1.2a. The areas above the horizontal axis are the same as those below and thus the coefficient $a_0 = 0$. In addition, the choice of the origin makes the function and odd one and only b_n coefficients need be considered. We will integrate only from 0 to $\dfrac{T}{4}$ and take the result twice. For the interval from 0 to $\dfrac{T}{4}$ we have the function

$$y = kt \tag{18.1.11}$$

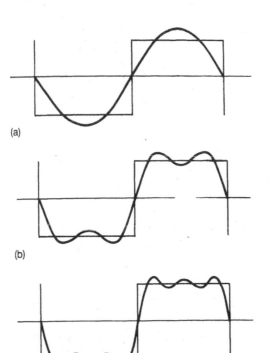

(a)

(b)

(c)

FIGURE 18.1.4   One period of a square wave and its: (s) First harmonic, (b) First two harmonics and (c) First three harmonics.

with the slope

$$k = \frac{4A}{T} \qquad (18.1.12)$$

For the interval from $\dfrac{T}{4}$ to $\dfrac{T}{2}$ we have the function

$$y = -kt + 2A \qquad (18.1.13)$$

with k given by (18.1.12). Altogether we have to solve

$$b_n = 2 \times \frac{2}{T}[\int_0^{\frac{T}{4}} t \sin n\omega t + \int_{\frac{T}{4}}^{\frac{T}{2}} (-kt + 2A)\sin n\omega t] \qquad (18.1.14)$$

where the first coefficient, 2, comes from the fact that we treat the function as an odd function and integrate only over half of the period. In any book on mathematical formulas we find the solution for the indefinite integrals

$$\int t \sin n\omega t = \frac{\sin n\omega t}{n^2\omega^2} - \frac{t \cos n\omega t}{n\omega}$$

and

$$\int \sin n\omega t = -\frac{\cos n\omega t}{n\omega}$$

Somewhat lengthy but simple evaluations for the bounds in (18.1.14) lead to the result

$$b_n = \frac{8A}{n^2\pi^2} \sin n\frac{\pi}{2}$$

All even coefficients are zero. The odd coefficients are

$$b_n = \frac{8A}{n^2\pi^2} (-1)^{(n-1)/2} \qquad n = 1, 3, 5, 7, ... \qquad (18.1.15)$$

and the Fourier expansion is

$$f(t) = \frac{8A}{\pi^2}[\frac{1}{1}\sin \omega t - \frac{1}{9}\sin 3\omega t + \frac{1}{25}\sin 5\omega t - \frac{1}{49}\sin 7\omega t + ...] \qquad (18.1.16)$$

Fourier expansions of periodic functions lead to evaluations of integrals with trigonometric functions and such evaluations are always time consuming. It is not the purpose of this book to practice such integrations. All we wish to achieve is to give some understanding without placing too much emphasis on the mathematical details.

Decomposition of any nonsinusoidal periodic signal into its components is not only a mathematical abstraction; the harmonic components do exist and can be measured, or heard. If different musical instruments play the same steady tone, we hear somewhat different sounds. The difference is caused by different amplitudes of the higher harmonic components generated by the various instruments.

Another important point is the speed with which the coefficients decrease. Our first case, Fig. 18.1.1, has the coefficients given in (18.1.10). The signal has abrupt changes (is not continuous) and amplitudes of the coefficients decrease slowly. Our second example, Fig. 18.1.2, is a continuous signal and its coefficients decrease much more rapidly, see equation (18.1.16). As a rule of thumb we will remember that abrupt changes of the signal result in slowly decreasing coefficients. We will have to take more terms of the expansion if we wish to approach the shape of the original function.

Another point worth noticing is the following. Suppose that we have a signal of constant shape, for instance like the impulses in Fig. 18.1.1. The width of the impulse will be constant but we will start increasing the time of the period, T. The amplitudes will be influenced, but more importantly the spectral lines will start getting closer if the period is increased. If the period becomes very long, the spectral lines will get very close. We can expect that if the period T stretches to infinity, the spectral lines will merge into one continuous spectrum. This is, indeed, the case and the appropriate mathematical theory is the subject of the next section.

18.2.   FOURIER INTEGRAL

Fourier integral studies cases of individual signals which do not repeat. It can be considered as a coupling link between the Fourier series and the Laplace transform theory. Let us first discuss informally what is involved. Similarly to the Laplace transform, the Fourier integral is defined by a pair of integrals. The time function is denoted by $f(t)$ and the Fourier transform, also called the *spectrum*, by $S(\omega)$. The two functions are related by the expressions

$$S(\omega) = \int_{-\infty}^{+\infty} f(t)e^{-j\omega t}\,dt \qquad (18.2.1)$$

$$f(t) = \frac{1}{2\pi}\int_{-\infty}^{+\infty} S(\omega)e^{+j\omega t}\,d\omega \qquad (18.2.2)$$

In all applications relating to this book the signal starts at $t = 0$ and thus the lower limit of the integral in (18.2.1) can be changed to 0. If we accept this restriction, then the definition of the Laplace transform, $F(s)$, equation (9.1.3), and the Fourier transform, equation (18.2.1), differ only in the exponent: the first one uses s while the second uses $j\omega$. There are some subtle differences but we will not go into such details. Other Laplace transform formulas also remain valid. For instance, time shift of a function by an amount of T is expressed by

$$S[f(t-T)] = e^{-j\omega T}S(\omega)$$

like in formula (9.5.15).

Considerable theoretical differences exist between the inverse transformations (9.1.4) and (18.2.2), but we have never used the inverse Laplace transformation formula and we will not need the inverse of the Fourier transformation either. In this book we give them only for the sake of completeness. Because of the similarities of the first formulas, we already have spectra for most practical signals: All we have to do is use our Laplace transform derivations and substitute $j\omega$ for s.

Take as the first example the Dirac impulse. Its Laplace transform was $F_\delta(s) = 1$, a constant. The spectrum, shown in Fig. 18.2.1, will also be a constant and thus *all* frequencies, up to infinity, are present with the same amplitude.

For the unit step, $u(t)$, we derived the Laplace transform $F_u(s) = \dfrac{1}{s}$; substituting $s = j\omega$ we get the spectrum

$$S_u(\omega) = \frac{1}{j\omega} = -j\frac{1}{\omega}$$

All components are shifted by $-90°$ (due to the $-j$). The absolute value

$$|S_u(\omega)| = \frac{1}{\omega} \tag{18.2.3}$$

gives amplitudes of the components at the various frequencies and is plotted in Fig. 18.2.2. It is a continuous curve indicating that all frequencies exist, theoretically up to infinity. Low frequencies are prominently present and the amplitudes get smaller as we move to higher frequencies.

Consider next the pulse, $p(t)$, for which have derived in (9.2.3) $F_p(s) = \dfrac{1 - e^{-sT}}{sT}$ Substituting $s = j\omega$

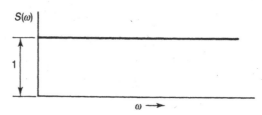

FIGURE 18.2.1   Spectrum of the Dirac impulse.

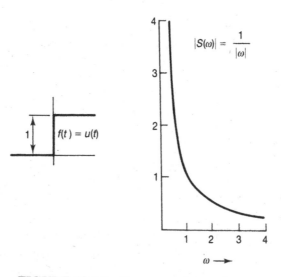

FIGURE 18.2.2   Spectrum of the unit step.

$$S_p(\omega) = \frac{1 - e^{-j\omega T}}{j\omega T} = \frac{\sin \omega T}{\omega T} + j\frac{\cos \omega t - 1}{\omega T}$$

The absolute value is

$$\left| S_p(\omega) \right| = \left| \frac{1 - \cos \omega T + j\sin \omega T}{j\omega T} \right| = \left| \frac{[2(1 - \cos \omega T)]^{1/2}}{\omega T} \right| = \left| \frac{\sin \omega \dfrac{T}{2}}{\omega \dfrac{T}{2}} \right|$$

The plot of the absolute value is in Fig. 18.2.3. Note that frequencies for which $\sin \dfrac{\omega T}{2} = 0$ will not appear in the spectrum. For $\omega = 0$ we find the value of 1 by applying the L'Hospital rule.

The Laplace transform of the ramp function was found to be $F_r(s) = \dfrac{1}{s^2}$ and thus the absolute value of the spectrum is

$$\left| S_r(\omega) \right| = \frac{1}{\omega^2} \tag{18.2.5}$$

It is shown in Fig. 18.2.4; the amplitudes of its components decrease faster than those of for the unit step but otherwise the curve is similar.

Suppose that we take the difference of two ramp functions, one starting at $t = 0$ and one delayed by the amount of T. For $r(t) = t/T$, this combined function will have a linear edge, followed by a constant value equal to 1, see Fig. 18.2.5. Its Laplace transform is $F(\omega) = \dfrac{1 - e^{-sT}}{Ts^2}$. Substituting $s = j\omega$ we get the spectrum

$$\left| S(\omega) \right| = \left| \frac{1}{\omega} \frac{\sin \omega \dfrac{T}{2}}{\omega \dfrac{T}{2}} \right| \tag{18.2.6}$$

The curve decreases faster than in the case of a unit step, because the time function is now continuous. This observation can be extended to a general rule, similarly as in the case of

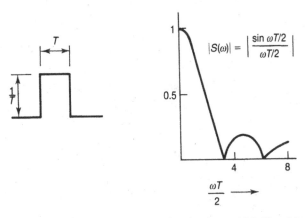

FIGURE 18.2.3   Spectrum of an impulse having width T and height $1/T$.

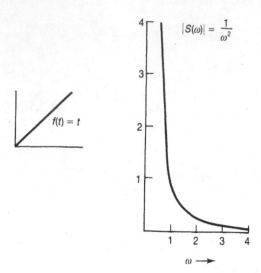

FIGURE 18.2.4   Spectrum of the ramp function.

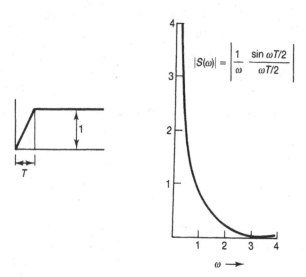

FIGURE 18.2.5   Spectrum of a step with linear transition.

Fourier series: the smoother the curve of the signal, the faster the decrease of its spectral function at higher frequencies.

We will consider two more simple signals, first the function $\sin dt$, suddenly applied at $t = 0$. The Laplace transform was derived in (9.2.10). Substituting $s = j\omega$ wand taking the absolute value,

$$|S(\omega)| = \left| \frac{d}{d^2 - \omega^2} \right| \tag{18.2.7}$$

The spectrum is sketched in Fig. 18.2.6 for $d = 4$.

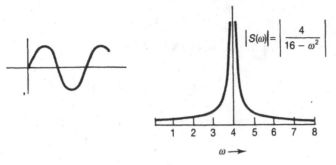

$$|S(\omega)| = \left| \frac{4}{16 - \omega^2} \right|$$

FIGURE 18.2.6   Spectrum of the sine function, switched on at $t = 0$.

For cos dt we derived (9.2.11) and the spectrum is

$$|S(\omega)| = \left| \frac{\omega}{d^2 - \omega^2} \right| \tag{18.2.8}$$

It is similar to Fig. 18.2.5, but is zero at $\omega = 0$.

Let us now discuss what conclusions we can draw from the above results. First of all we see that all suddenly applied nonperiodic signals have a continuous spectrum of frequencies. This is particularly striking in the case of a sine or cosine signal. If these functions are not switched, that is if they started at $t \to -\infty$ and continued for all t unperturbed, their spectrum is just one single line. Once they are suddenly switched, there is a complete spectrum of frequencies, as seen from (18.2.7) or (18.2.8). Compare similarly the spectra of the unit step and of the signal composed of the two ramp functions. The smoother change of the the signal, see Fig. 18.2.5, resulted in smaller spectral components at higher frequencies.

As was the case in periodic functions, the frequencies in the spectrum actually exist. An imperfect but useful explanation can be found in nature by considering a lightning. It is a short pulse of electromagnetic energy. If we have tuned our radio to an AM station, we will hear those spectral components, no matter what station are we tuned to. If we tune to the FM channel, we may not hear them. The FM frequency band is in high frequency ranges where the spectral components may already be so small that they are lost below the noise level.

APPENDIX: SCALING OF LINEAR NETWORKS

Scaling is one of the most useful concepts in linear circuit theory. It gives us the possibility to reduce the frequency band and the impedance level to unit values, without losing any information. It also makes it possible to unscale any scaled network to arbitrary frequency band and impedance. Let us stress here that similar scaling is impossible if any nonlinearities are involved.

If we increase (decrease) the impedance of every element of a linear network, the input–output relationship does not change. It is thus useful to apply scaling which reduces one (any) resistor to the value of 1. We can also scale the frequency so that the cutoff frequency of a lowpass filter, or the center frequency of a bandpass filter, are reduced to 1 rad/s.

To derive the necessary formulas we will denote the scaled values by the subscript s and leave the values to be used for realization without subscript. Impedance scaling means that the impedance of every element must be scaled by the same constant, k. This immediately leads to

$$R_s = R/k$$

For frequency normalization we introduce the formula

$$\omega_s = \frac{\omega}{\omega_0}$$

where ω_0 is a constant by which we wish to scale the frequency. Now consider the impedance of a scaled inductor by writing

$$Z_{L,s} = j\omega\frac{L}{k} = j\frac{\omega}{\omega_0}\frac{\omega_0 L}{k} = j\omega_s L_s$$

Similarly, for a capacitor we obtain

$$Z_{C,s} = \frac{1}{j\omega Ck} = \frac{1}{j\dfrac{\omega}{\omega_0}(\omega_0 Ck)} = \frac{1}{j\omega_s C_s}$$

The results can be collected in the following formulas

$$R_s = \frac{R}{k} \qquad G_s = Gk$$

$$L_s = \frac{L\omega_0}{k}$$

$$C_s = C\omega_0 k$$

Dependent sources must also be appropriately scaled:

- VV and CC remain unchanged
- VC transconductance g is multiplied by k
- CV transresistance r is divided by k

The scaling makes it possible to provide numerous tables of various filters, all normalized so that one resistor is equal to 1 ohm and some frequency is normalized to $\omega_s = 1$.

Because of the scaling possibility, there is no need to bother the students with various practically useful element values, like picofarads, milihenries or kiloohms. All examples of this book use unit element values to make the study easier.

INDEX